Environmental Science in Building

7th Edition

RANDALL McMULLAN

Construction physicist, educational consultant and writer

palgrave
macmillan

First edition 1983
Second edition 1989
Third edition 1992
Fourth edition 1998
Fifth edition 2002
Sixth edition 2007
This seventh edition first published 2012 by
PALGRAVE MACMILLAN

Palgrave Macmillan in the UK is an imprint of Macmillan Publishers Limited, registered in England, company number 785998, of Houndmills, Basingstoke, Hampshire RG21 6XS.

Palgrave Macmillan in the US is a division of St Martin's Press LLC, 175 Fifth Avenue, New York, NY 10010.

Palgrave Macmillan is the global academic imprint of the above companies and has companies and representatives throughout the world.

Palgrave® and Macmillan® are registered trademarks in the United States, the United Kingdom, Europe and other countries

ISBN-13: 978–0–230–29080–8   paperback

This book is printed on paper suitable for recycling and made from fully managed and sustained forest sources. Logging, pulping and manufacturing processes are expected to conform to the environmental regulations of the country of origin.

A catalogue record for this book is available from the British Library.

A catalog record for this book is available from the Library of Congress.

10 9 8 7 6 5 4 3 2 1
21 20 19 18 17 16 15 14 13 12

Printed in China

Contents

Preface

This book gives you information about the science and technologies relating to the comfort of humans in the built environment and to the performance of buildings. Major subject areas include heating, air control, lighting, noise control and water supplies. Additionally, the book connects the broader topics of climate change, sustainability, carbon emissions and energy use. These topics interact with all aspects of the built environment and with the social and policy issues related to our wider environment.

Janet Frame, author and poet, reflected how when she read the science curriculum she felt the mystery and excitement of the subjects listed as 'Heat, Light and Sound'. She thought 'Surely this is the province of poets, painters, composers of music, as well as of scientists...?' I think we can agree that the environmental topics of this book do belong to everyone and do affect us all.

Purpose of the book

The book is intended to be especially useful for people studying architecture, building, engineering, environmental science and surveying at a variety of levels. The contents of the book satisfy the principal requirements of courses for degrees, national diplomas and certificates, and for the examinations of professional institutes. Over three decades of publication have confirmed the usefulness of the text at a range of levels, from introductory to professional reference. The book should also be useful if you live and work in a building, or even in a tent!

There has been continuing support for keeping the term Environmental Science in the title so as to mirror the long-standing use of the term in built environment courses. There are many courses concerned with wider aspects of environmental studies and this book can also support those studies, and provide a focus on the built environment.

Nature of the book

This edition continues the design of the sixth edition which allows narratives on some primary topics to be read and understood in general, while their related science, technology and calculations are located separately for later study as required. In particular, the first two chapters on *the environment* and *energy use in buildings* are intended to be accessible to a wide range of readers. Similarly the final chapter on *green building* can be read as a summation of principles that have been covered in detail in other chapters of the book.

The *Resource* sections at the end of book act as a reference source and 'catch-up' tutorial about terminology and concepts of basic science, especially the introductory principles of heat, light, sound, electricity and fluids. No one has yet complained of being able to remind themselves, or learn for the first time, what exactly is meant by giga, latent heat, decibel, induction, venturi, alkali or other terms that might wrongly be assumed known and remembered from some perfect education in school science.

There are additional resources available on the companion website for the book at:

www.palgrave.com/engineering/builtenvironment/mcmullan

These resources include PowerPoint lecture slides, investigative questions, useful web-links, an online glossary, plus supplementary exercises to complement Part II of the book.

Evolution of the book

Government policies and regulations relating to the topics of this book, such as sustainable buildings, continue to evolve. The policies are applied with variations in techniques and details across the different countries of the European Union, including the administrations within the United Kingdom. These differences in detail of application highlight the importance of learning and understanding the underlying principles and factors that drive design decisions, building codes and any associated calculations. Therefore the text continues to use some 'traditional' simplified calculations involving, for example, R-values, U-values and heat losses as a means of clearly understanding the principles. Also, the methods shown are actually useful.

The book continues its original intention of maintaining readability and clarity without the interruption or intimidation of inline citations, and this approach has been important to the appeal of the book. Most of the text is a selective recasting of material that is in the public domain and ranges from early science principles to current government policies. The origins and details of essential material are readily confirmed via the website links at the end of the book and by an internet search engine. However, I work with students who are taught to cite their sources correctly and who are penalised for plagiarism so, although the content of this book is intended as a resource for student essays, this text is not an exemplar of citation.

The preparation of a new edition of this book involves wide-ranging reviews of information and sources, such as building regulations, for the various countries of the United Kingdom, the Republic of Ireland, the European Union, North America, Australia and New Zealand. I am thankful for the easy access to sources via the internet but that easy access brings challenges for students: how can they assess huge volumes of information and select appropriate material? If the pertinent information was easy to select, to refine and to explain then books like this, and teachers, would be redundant. It isn't easy, and good teachers are not redundant!

Thank you

I am grateful to all those people who have provided feedback about the book and, in particular, to Roger Birchmore of Unitec New Zealand. Many thanks also to the

people in the team at Palgrave Macmillan, led by Helen Bugler, whose wisdom
and skill produce this final book in your hand.

Randall McMullan

How to Use This Book

You can use this book in different ways, depending upon your needs. For example, the topics don't have to be studied in a particular order as their use depends upon your course or professional purpose. All of the subjects in the book are worthy of further study and it is hoped that this book will be a starting point for further investigations. To help you get the most from the book its features are summarised below.

Text

The chapters assume very little previous knowledge of a topic and lead you to a good working understanding. If you have forgotten some school science you will find it useful to read the appropriate Resource sections towards the back of the book, which explain basic science and principles that support the technology of the chapters.

The content of the text itself gives an indication of the depth of knowledge normally expected at this level of study. The style of writing has been kept simple but it uses correct terminology and units. It can therefore act as an example of the type of response expected when you need to display knowledge of a topic, in a project for instance. Make sure that your document includes references to the sources of information you have used or quoted; otherwise you might be accused of plagiarism. Educational institutions provide good guidance on preferred systems of referencing.

Diagrams

The diagrams are intended to help explain the principles behind topics in the book. The drawings have been kept relatively simple and may be the basis for sketches of your own; remember that accurate labels are as important as the drawings. Meanwhile the publications and websites of firms and organisations, including those listed in the reference section, will show you examples – in colour! – of practical equipment and installations.

Calculations

Calculations can give greater understanding of some topics, and you may also need them in your course. The text emphasises those formulas that are especially useful and sometimes worth remembering. Important types of calculation are explained by carefully worked examples, using relatively simple calculations. Where further practice is relevant there are exercises at the end of the chapter.

Margins

Look in the margin for rapid reminders of key terms and ideas and for references to other places in the book where there is related information.

Part II: Resource sections

At the end of the book there is extra information presented in forms that make it easy to look up information such as scientific terms, units and symbols. The basic principles of heat, light, sound, electricity and fluids are also presented in Resource sections to underpin and complement the applied topics in the main text. The section on references also has signposts to many external organisations whose publications and internet sites can give more information on a topic.

Companion website

On the companion website for this book at:

www.palgrave.com/engineering/builtenvironment/mcmullan

you will find investigative questions, useful weblinks and an online glossary, plus supplementary exercises to complement Part II of the book.

Local conditions

The principles behind the applied technologies of environmental science in building are, fortunately for our sanity, the same all over the world. However, different countries and regions express their requirements for the performance of buildings, such as in their use of energy, in a variety of ways. If you understand the principles and examples in this book you will find it easy to interpret particular requirements. You just need to be alert to what is current for the location where you are studying or designing. The reference section lists Internet portals to the building regulations and codes for a variety of countries.

Big picture

In addition to the specialist information about each subject, you should aim to place your knowledge in the context of wider issues concerning construction and the environment. Use the first chapter and the last chapter of the book to help you see the bigger picture.

Acknowledgements

Grateful acknowledgement is made to the following for the photographs reproduced in the book (in most cases, in cropped and adapted form).

Chapter 1, p. 3, the Earth from space, NASA, courtesy of nasaimages.org.

Chapter 2, p. 14, Bahrain World Train Center, photo Omar Chatriwala.

Chapter 3, p. 42, Solar panels in Spain, photo Fernando Tomás.

Chapter 4, p. 84, air conditioning equipment, photo Paul Goyette.

Chapter 5, p. 120, lighting in Regent Street, London, photo Donna Rutherford.

Chapter 6, p. 138, Wikimedia Commons (no photographer credited).

Chapter 7, p. 161, foyer of the Sydney Opera House, photo John O'Neill.

Chapter 8, p. 172, road junction in Chicago, photo Sheila Allen.

Chapter 9, p. 191, building site, photo Paul McIlroy.

Chapter 10, p. 227, Scottish Parliament Building, photo Martin Pettitt.

Chapter 11, p. 246, hydro dam, photo supplied by the author.

Chapter 12, p. 268, reservoir, photo supplied by the author.

Chapter 13, p. 286, clogged drainpipes in El Paso, Texas, photo Robert J. Alvey/ FEMA.

Chapter 14, p. 300, Cabot Circus, Bristol, photo Arpingstone.

Part I

Core Topics

The Environment

It is remarkable to see a 'big picture' view of our environment, such as the entire planet Earth photographed from outer space. Against the darkness of space the Earth's surface is lit by the Sun, that crucial star that governs our planet and supplies it with energy. It is also challenging for us to consider that the Sun and the Earth, like other planetary systems, can happily exist without the current range of life on Earth, including us humans. Compared to such matters, this book has a relatively limited scope!

Used in a broad sense, the term *environment* means the global surroundings that affect our lives. This is obviously a large and complex topic involving factors that range from big events on the Sun to small events within the molecules of living organisms. Many environmental factors also interact with one another in ways that are important, or even vital to life. For example, the oxygen content of the atmosphere is regulated by the plants of the Earth which take up carbon dioxide from the air and then give back oxygen, which we breathe.

Maintaining and improving the quality of our environment is important to life and to the quality of life. Environmental topics can also have important social and political dimensions with difficult choices to be made and issues to be debated. Added to this, the science and technology of the different topics have many interactions and links which can be difficult to understand.

This book focuses on the science, technology and services relating to the comfort of humans in buildings and the environmental performance of those buildings. These aspects of the built environment have many significant interactions with the wider environment and a secure knowledge of facts, terms and principles is a good basis for understanding the environment.

Overarching environmental matters, such as climate and comfort, are considered in this chapter while later chapters examine topics in more detail. The Resource sections of the book contain supporting information that can be used to review the science behind some of the technologies studied and for the further investigation of topics.

ENVIRONMENTS

Basic terms

The components that make up the wider environment can be subdivided according to various systems, but a major distinction for this book is the difference between the natural environment and the built environment.

- **Natural environment** is the entire environment, without human presence or interference.

Notable features of the natural environment include climate, mountains and hills, rivers and lakes, rocks and soil, trees and plants.

- **Built environment** is formed by the buildings and other structures that humans construct in the natural environment.

Notable features of the built environment include buildings, water and drainage systems, transport systems, power systems and communication systems.

> *Related ideas:*
> Green architecture
> Sustainable architecture
> Natural building
> Ecological building
> Environmental building

- **Sustainability** is the general idea of meeting the needs of the present without compromising the needs of the future. It is about enduring.
- A **green or sustainable building** is deliberately designed to minimise impact on the natural environment and to maximise efficiency in the use of resources such as materials, water and energy over the lifecycle of the building.

Although this book focuses on the environment in and around buildings, there is considerable interaction between different environmental factors, as shown in the simple model of Figure 1.1.

Connections between environments

From the earliest times, people have adapted their habitat to provide shelter from the weather and other threats to life. Early humans made use of natural shelters, such as caves, and then they built shelters using available materials like animal hides, stones, straw or wood. Modern buildings involve so many features in design, materials and construction that it is easy to forget that the fundamental aim is to provide an internal environment that is different from the external environment.

> *Vernacular* means the local style

A built environment responds to the local natural environment, and different types of buildings are therefore found in different parts of the world. Climate is a major factor in determining the features of building, together with the availability of building materials and skills. Some of these interactions are summarised in

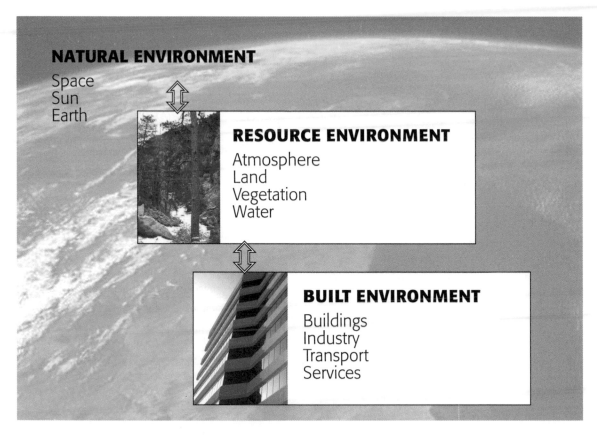

Figure 1.1 Parts of the environment

Table 1.1. Other influences on types of building are local traditions and international architectural styles.

Interactions and issues

Some of the interactions between the built and natural environments have effects that cause concern:

- consumption of non-replenishable resources such as fossil fuel
- consumption of resources without replacement, such as hardwood forests
- harmful changes to local habitat, such as deforestation
- harmful changes to global habitat, such as climate change.

These concerns are connected to various topics in this book and are discussed, where relevant, in other chapters.

Remember also that the natural environment can undergo great changes without the presence of human beings. Britain, for example, has been subjected to ice ages and to warm periods, and that is ignoring an earlier geological time

Table 1.1 Examples of environmental connections

Natural environment features	Built environment features
Hot, dry climates	Light-coloured surfaces Roof overhang to provide shade Openings for breezes Courtyards to trap cooler air
Warm, humid climates	Lightweight materials Buildings on stilts for ventilation
Cold climates	Naturally sheltered sites High insulation Tightly-sealed construction
Snowfalls	Strong roofs for load Sloping roofs to discard snow
High winds	Naturally sheltered sites Low sunken buildings
Forests	Timber as construction material
Loose stone or quarries	Stone as construction material
Clay soil	Mud brick or adobe construction Fired brick as construction material
Earthquake zones	Low-rise flexible construction Reinforced concrete structures Avoidance of unsecured masonry

when Britain was connected to Canada and positioned at the equator. The mountains of the Earth have been eroded and washed into the sea to form new layers of rock which, in turn, have formed new mountains; this has happened four times in the lifetime of the planet. And only 20,000 years ago you could walk on dry land between England and France where the Channel Tunnel now crosses beneath a sea.

The built environment

See also:

Section on
*Life cycle
assessment*
in Chapter 14

Most people on Earth live and move around in a human-made built environment that we have inserted into the natural environment. The features of the built environment include buildings, electricity and water supplies, roads, bridges, tunnels, railways, harbours and airports. The larger features of the built environment can leave their mark in the natural environment for thousands of years after they have been abandoned. Examples include pyramids, stone circles, earthwork forts, old tracks and roads and it is interesting to ask yourself which parts of the current built environment might, if abandoned, be detected thousands of years from now.

The focus of this book is on the buildings in the built environment rather than the roads and bridges. Building types are varied and include houses, schools, shops, office blocks, factories, shopping centres. Most buildings are designed to have people living or working inside and we are therefore interested in those aspects of buildings that affect human comfort. These factors include heating, cooling, lighting, acoustic performance, use of services such as electricity and water. Interlocked with these topics is the sustainable use of resources to construct and to run the buildings.

Considered over time, we can consider buildings to have four major stages, as shown in Figure 1.2 and described below.

1. **Design**: the stage when we think about what we want and specify how best to do it. Decisions made at this stage last for the lifetime of the building, and poor design decisions are usually hard to remedy.
2. **Construction**: during this stage the building is made, requiring resources of land, materials, energy and having an impact on the natural environment.
3. **Performance**: the extended stage of the building as it is provides benefits to people but requires arrangements for energy supplies, water supplies, and waste disposal. This performance continues, for better or worse, over the lifetime of the building.
4. **Disposal**: the stage when the building is disassembled and its materials and fittings are recycled or disposed of with minimum impact on the environment.

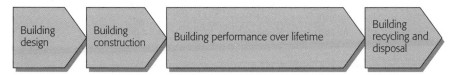

Figure 1.2 Major stages in the life of a building

The technical topics covered in this book are applicable, as appropriate, to all the stages in the total life of a building. The final chapter on 'Green Building' returns to the wider issues of deigning, constructing and using buildings in ways that are environmentally responsible.

CLIMATE

The Earth's atmosphere is a particular environment which produces ever-changing effects of sunshine and clouds, pressure and wind, temperature and humidity, and precipitation in the form of rain, hail and snow. When these short-term variations of *weather* are observed at one place and considered over a period of time they form a climate. The climate varies from place to place on Earth and creates a variety of environments.

The physical mechanisms driving the global climatic system include energy received from the sun, the rotation of the Earth, convection of air masses, heating of land and water masses, and evaporation of water and its subsequent condensation and *precipitation*. These mechanisms are interlinked in complex ways and produce variations that depend on regular cycles of time, such as daily and yearly changes, and variations which are random. Forecasting of local weather on a daily basis is challenging, but there are good data for the probable frequency of events in an area and this information is more important for building design.

Directly or indirectly, climate has an influence on all human activities, as it affects the rocks, soil, vegetation and water resources in a region. Climate is therefore interlinked to traditional social characteristics in a region such as the types of food grown, the clothes worn and the buildings lived in. Even in areas with little tradition the local climate will still affect styles of agriculture, buildings and their services, leisure activities and transport.

The underlying climate of a region can be linked to certain factors which are listed below and described in following sections:

Precipitation: the release of water from the atmosphere. Examples include: rain, snow, hail, dew

- geographical latitude
- season of the year
- altitude and topography
- effects of water
- atmospheric circulation.

Latitude

Higher latitude means farther north or farther south

The geographical latitude of a place on Earth is a measure of its position above or below the equator, and is usually measured by angles in degrees.

- Intensity of solar radiation decreases as latitude increases.

See also:
section on
Cosine law of illumination in Chapter 5

The solar radiation and heating effect received from the Sun is strongest when it strikes the Earth's surface 'straight on', at an angle of 90° to the surface. But most parts of the Earth receive sunlight at an angle of less than 90° to the surface, especially towards the poles of the Earth. The radiation is then less intense by a non-linear factor which varies with the cosine of the angle. The heat received on Earth from the Sun is therefore significantly less in the higher latitudes near the polar regions.

Season of year

The intensity of radiation from the Sun also varies with the season of the year. The angle at which radiation from the Sun falls on a surface changes as the tilt and orbit of the Earth around the Sun changes, as shown in Figure 1.2. In tropical latitudes, near the equator, there is little difference in solar heating between summer and winter. In high latitudes, in the Arctic or Antarctic regions, the Sun never rises for long periods of the year.

The orbit of the earth around the sun is slightly elliptical in shape and the axis of the Earth is tilted by 23.5° with respect to the plane that passes through the Sun and the equator. This tilt causes the change in radiation, length of day and climate between summer and winter. If there was no tilt then there would be relatively uniform climatic conditions throughout the year.

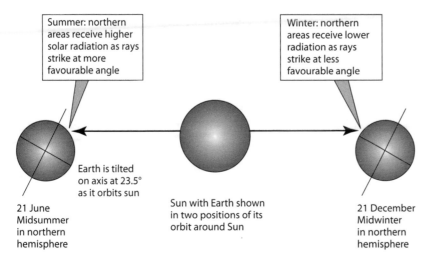

Figure 1.3 Motion of Earth around the Sun

Each day the sun traces an apparent path in the sky. In the northern hemisphere the winter solstice on 21 December marks the lowest path and the shortest day. The summer solstice on 21 June marks the highest path and the longest day. The equinoxes are on 21 March for Spring (Vernal) and 21 September for Autumnal in the northern hemisphere. On these two dates all places on the earth have a 12-hour day and a 12-hour night, with sunrise being exactly due east and sunset exactly due west.

Altitude and topography

The height of a place above sea level affects its climate because the temperature of the air decreases with altitude. Even in the tropics it is possible for tops of high mountains to be snow-capped.

- **Air temperature** drops by 6.5°C for each 1000 metres increase in altitude.

The surface features of the earth, or topography, also influence the local climate by affecting the formation of wind, cloud and rain. For example, as humid air from an ocean sweeps up the slope of a mountain range the air cools, forms clouds and causes rain and snow to fall. As the wind blows down the leeward slopes on the other side of the mountains the air usually warms and clouds tend to disappear.

> *Lapse rate* is the decrease of temperature with altitude

Effects of water

Large masses of water cover the surface of the Earth and have a considerable effect on our climate, both locally and globally. Oceans and large lakes affect climates by reducing the extremes of air temperature at places nearby and downwind of them. The mass of water in an ocean or lake absorbs heat and so takes longer to warm and longer to cool than a landmass. Therefore, the air temperatures over the ocean and in places near oceans have smaller variations in air temperature than at places at the same latitude but well inland. For example, Glasgow and Moscow have similar latitudes, but they have very different climates.

The oceans of the world store great quantities of thermal energy and this heat store stabilises the variations in temperatures over the planet and keeps them within a relatively narrow range of values compared with other planets in the solar system. Within the oceans there are large currents, such as the *Gulf Stream* and *El Niño*, which distribute this energy around the world and interact with weather systems.

Some general climate descriptions:
Tropical
Desert
Arctic
Oceanic/Maritime
Continental
Alpine

Atmospheric circulation

The movements of large masses of air in the atmosphere influence climate by producing winds that distribute heat and moisture. Global belts of wind, such as the *trade winds*, circle the earth and shift north and south as the seasons of the year change. In the spring they move towards the poles and in the autumn they shift towards the equator. These shifts of wind help explain why some areas have distinct rainy and dry seasons.

Climate types

The different climates encountered in the world can be described by various systems of classification which take account of the characteristics of a region, such as vegetation, average temperature and average precipitation. For the purposes of studying the effect of climate upon buildings and human comfort, the four general climate types described in Table 1.2 are useful.

Table 1.2 Climate types

Type of climate	Typical characteristics
Cold	Excessive heat loss for most of year Minimum temperatures: below -15°C
Temperate	Excessive heat loss for part of year Inadequate heat loss for part of year Temperate ranges: -30°C to 30°C Precipitation possible in all seasons
Hot/dry	Overheating for most of year Dry air allows evaporation Temperatures ranges: -10°C to 45°C High radiation Strong winds
Warm/humid	Overheating for most of year Humid air inhibits cooling Temperature often above 20°C Mean relative humidity around 80% High rainfall in certain months

Environment around buildings

From the perspective of a single building, the climate is the set of environmental conditions that surround the building and link into it. The study of climate for the local built environment generally involves the study of smaller systems of climate than those described above, and Chapter 14 considers details of climate measurement and of *microclimates*.

The following general features of the local natural environment are important to our choice of site for buildings and towns:

- availability of drinking water
- drainage of ground
- safety from flooding
- shelter from prevailing weather
- orientation to the Sun, as appropriate.

See also:

section on *Climate around buildings* in Chapter 14

Climate change

Climate change can be a controversial subject both scientifically and politically. The records of global temperature show that the Earth has warmed by at least 0.5°C during the last century. There have been changes in global temperature in past epochs, such as tropical ages and ice ages, which can be linked to natural effects of solar variability, volcanoes and even meteor impacts. Despite the difficulties of being certain about some data, there is considerable scientific evidence to suggest that the activities of humans are causing the current increase in global temperatures.

Whatever the exact causes of global warming are, the effects will change the ecology of many parts of the Earth and bring difficulties for people living there. Possible effects of global warming include:

- melting of polar ice causing a rise in sea levels and disappearance of land
- increase in severity of storms and flooding
- change in rainfall patterns, forming new deserts
- changes in ocean currents, causing changes in local climates
- changes in patterns of snowfall and ice sheets.

Greenhouse gases

The atmosphere surrounding the Earth behaves as a large 'greenhouse' around our world and retains a certain proportion of the heat received from the Sun. There is a balance between the heat absorbed and given off by the planet Earth. Increasing certain 'greenhouse' gases in the atmosphere, such as carbon dioxide from the burning of fossil fuels, increases the greenhouse effect and reduces the quantity of heat that the planet Earth would otherwise radiate back into space. This particular greenhouse effect therefore causes global warming.

Greenhouse gases are those gases that have a large influence on the greenhouse

See also:

section on *The greenhouse effect* in Resource 2

effect. Described below are significant greenhouse gases, including some whose quantities are influenced by human behaviour.

- **Water vapour H_2O**: occurs naturally from the waters of the world, not including clouds, and accounts for most of the greenhouse effect on Earth.
- **Carbon dioxide CO_2**: produced by the burning of fossil fuels and forests and by all organic decay. Chimneys, motor vehicle exhausts and forest fires are major sources.
- **Methane CH_4**: the main component of natural gas supplies. Methane is produced by decay of organic matter and also by the digestion of sheep and cattle.
- **Nitrogen oxides NO_x**: the various oxides of nitrogen, which are mainly produced by motor vehicle emissions.
- **Chlorofluorocarbons CFCs**: families of chemical compounds manufactured for use in refrigerators and spray cans, and for insulation. Although inert at point of use, CFCs escape to the upper atmosphere, where they chemically react and deplete the *ozone layer* which we need for protection from excess ultraviolet radiation from the Sun.

CFCs, described above, are categorised as being among the minority greenhouse gases, but they play a unique role by chemically reacting with and depleting the ozone gas in the upper atmosphere. This *ozone layer* filters out the shorter wavelengths of ultraviolet light that, in large amounts, are harmful to life forms and cause effects in humans such as skin cancer. We do not want to damage the ozone layer, and since agreements in the 1990s the use of CFCs has been dramatically reduced, as described in the section on *Refrigerants* in Chapter 4.

Greenhouse gas emission agreements

The natural greenhouse effect of the Earth's atmosphere is one of the mechanisms that make life possible on this planet and it is also a mechanism that can be changed, for good or ill, by our human activities. There is a generally-agreed need for governments of the world to agree upon targets and mechanisms for limiting greenhouse gas emissions.

The *Kyoto Protocol* is an agreement made between countries to reduce their emission of carbon dioxide and five other greenhouse gases. Most countries in the world have joined the Kyoto agreement and similar agreements such as the *Washington Declaration*. Despite some differences over details in the agreements it is hoped that a unified system of controls will emerge in future years. In general, the international agreements set limits on greenhouse gas emissions for each country. These 'quotas' of emissions can be expressed in terms of 'carbon credits' which may also be traded between countries or companies. For example, a country with large areas of forest, which absorb carbon dioxide, gets carbon credits which other countries can purchase to offset the effects of their emissions.

Major contributors to greenhouse effect:

H_2O = 35–70%
CO_2 = 10–25%
CH_4 = 4–9%

See also:

section on *Carbon management* in Chapter 14

Kyoto agreement objective: to stabilise greenhouse gases in the atmosphere at a level that will avoid dangerous climate change

ENVIRONMENTAL COMFORT

Despite the challenges of using our natural environment with care we do have a need to construct a built environment with comfortable places where we can live and work. The physical comfort of humans greatly depends upon the following physical factors:

- temperature
- quality of air
- lighting environment
- acoustic environment.

See also:

section on *Thermal comfort* in Chapter 2

The technical measurement and control of these factors within buildings is discussed in the various chapters of this book. Figure 1.4 summarises important factors for human comfort within buildings and gives a preview of typical units and comfort ranges.

Outside the focus of our human physical comfort, the features of the built environment interact with many wider themes, as is indicated in the previous sections about climate and global warming. Overarching topics such as health and safety, use of materials, use of land, energy use and carbon management, water use, reduction of waste and pollution occur throughout the various chapters of the book. The final chapter on 'Green buildings' returns to key topics and issues.

See also:

section on *Humidity* in Chapter 4

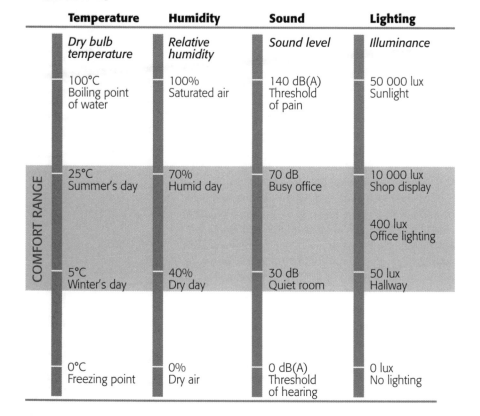

Temperature	Humidity	Sound	Lighting
Dry bulb temperature	*Relative humidity*	*Sound level*	*Illuminance*
100°C Boiling point of water	100% Saturated air	140 dB(A) Threshold of pain	50 000 lux Sunlight
25°C Summer's day	70% Humid day	70 dB Busy office	10 000 lux Shop display
			400 lux Office lighting
5°C Winter's day	40% Dry day	30 dB Quiet room	50 lux Hallway
0°C Freezing point	0% Dry air	0 dB(A) Threshold of hearing	0 lux No lighting

COMFORT RANGE

See also:

section on *Lighting design* in Chapter 5

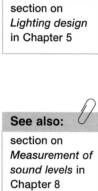

See also:

section on *Measurement of sound levels* in Chapter 8

Figure 1.4 Summary of comfort conditions

2

Energy Use in Buildings

CHAPTER OUTLINE

All buildings use energy for lighting, heating or cooling. Buildings in the British Isles, and in countries with similar climates, suffer an overall loss of heat during the year, and the energy required to replace these heat losses represents a major portion of national energy consumption. Building services, such as heating and cooling, make up 40 to 50 per cent of the national consumption of primary energy, and about half of this is used in domestic buildings.

This chapter begins with an overview about the nature of energy, the different forms of energy and how we measure energy content and usage. One aim is to become comfortable, or at least less confused, when reading and using the various units for energy and power that are found in professional practice.

A major aim of energy use in buildings is to keep its human occupants thermally comfortable in terms of factors like temperature, humidity and ventilation. People vary in their needs and opinions about what is comfortable. Nevertheless, we need to agree on numerical values for human comfort conditions so that we can, for example, design and assess heating and ventilating systems.

The energy conservation and carbon emission strategies of most countries include controlling the energy use of buildings by guiding and regulating their design and use.

The material in the chapter is arranged so that you can:

▶ use the terms associated with the use of fuels and production of energy
▶ appreciate the factors governing thermal comfort
▶ use the design parameters for comfort temperatures, air movement, humidity and ventilation

$\longrightarrow$

▶ understand the causes of heat loss in buildings
▶ understand the causes of heat gain in buildings
▶ understand the overall energy balance of buildings
▶ appreciate the role of building codes relating to energy conservation in buildings
▶ assess individual buildings and building designs for compliance with local building regulations.

The connections between buildings and energy use, carbon emissions and sustainability are also discussed in the final chapter of the book. The Resource sections of the book contain supporting information that can be used for both the revision of principles and for extended investigation of topics.

This chapter focuses on the role of energy in buildings, but you also need to think about the wider issues of using energy. This might include, for example, the wisdom of how we currently use energy sources for the foods we eat, for the vehicles we drive and for the planes we fly in.

USING ENERGY

The concept of energy plays an essential role in the working of our human bodies, in living our daily lives and in running our buildings. We therefore need to consider how to obtain energy, how to use energy, how to convert energy, how to 'count' energy, and how to conserve energy. Energy is not actually a 'thing'. It is a condition or property of things, and can be given the following formal definition:

• **Energy** is the capability of an object, or a system, to do work.

The technical idea of 'work' is connected with doing things and making changes to things. For example, energy can exist as heat, light and sound; it produces motion and electricity; and it produces growth in plants and animals. The Earth constantly receives energy from the Sun, and this energy is used, for example, to warm the oceans and to drive the weather systems of the world that produce effects such as wind and rain.

See also: 📎

Nature of energy and heat, in Resource 2

Energy sources

The Sun is the source of almost all energy in common use on Earth, even if the *solar energy* has been stored for many years in other forms such as oil deposits. Plants and trees use solar energy to grow material which can be burned directly, or decomposed into gases, or preserved underground as coal and oil. The energy in waves, wind and ocean heat is also provided by the Sun.

Nuclear energy is released from small changes in the mass of subatomic particles in the nucleus of an atom during the processes of fission or fusion. Uranium and other materials can be used as a fuel for generating such nuclear reactions. Solar energy also has its origin as nuclear energy, because the sun generates its heat by a fusion reaction that depends on supplies of hydrogen atoms in the sun.

At some distant time all the hydrogen in our Sun will be converted to helium and it will become a dead star.

Energy conversion

We can convert one form of energy to another form of energy. Burning wood is a simple example where the chemical energy of the wood is converted to heat energy. Sometimes we need the help of a machine, such as when petrol (chemical energy) is converted via a car engine into motion on the road (kinetic energy). The total energy involved in a particular system is conserved and remains constant even though changes take place. When we brake a car we 'waste' energy from our point of view, but the kinetic energy of motion is converted to heat by the brakes and the tyres on the road.

Total energy is conserved when we convert one form of energy to another, but *something* changes. That changed thing is termed *entropy*, which is the tendency for systems to go from a state of order to a state of disorder. For example, after burning wood it is not possible to convert the ash, smoke and heat back into wood. From a simple point of view entropy is a measure of 'energy usefulness'. Thermal energy supplied at high temperature (lower entropy) is more useful than the same amount of thermal energy supplied at lower temperature (higher entropy or disorder). For example, we find it easier to make use of heat stored as hot water than to use the same quantity of heat stored in the form of ice.

> *Conservation of energy*: energy can change from one form to another, but the total energy remains constant

Energy terms

The sections on energy and heat in Resource 2 of this book give the scientific and technical definitions and units for heat, energy and power. However, as these have evolved over the years some industries and countries have continued to use older units, or they tend to use units that are considered more convenient.

The following list explains terms used to describe the practical use of energy in national and industrial contexts.

- A **fuel** is a substance that is a source of energy.
- Fossil **fuels** such as coal, crude oil and natural gas were formed by the decomposition of prehistoric organisms, such as plants, during long-term geological changes in the Earth.
- **Non-renewable energy** is from sources which can only be used once. Fossil fuels are examples of non-renewable sources because the decomposition process has taken millions of years and cannot be easily repeated.
- **Renewable energy** is from sources that are replenishable. Wind power, wave power and similar processes are examples of renewal energy sources.
- **Primary energy** is the total energy contained in natural reserves or in flows of primary fuels such as coal, oil or natural gas. The primary energy of a fuel is measured in the 'raw' state, before any energy is used in conversion or distribution.

- **Energy transformation** is an activity that coverts primary energy into another form. Examples of transformation occur when fuels are converted into electricity, or crude oil converted into petroleum.
- **Secondary energy** is the energy contained in a fuel which results from a transformation process. Examples of secondary energy sources include most electricity, manufactured gas, and surplus hot water. The secondary energy of electricity is only 30 to 40 per cent of the primary energy contained in the original fuels; the rest is lost in the conversion and distribution processes.
- **Delivered energy** is energy content as it is received by the consumer. Delivered energy is the delivered energy or 'site energy' for which the consumer generally pays money, although some of this energy will be lost by conversion processes within the building.
- **Useful energy** is the energy required to perform a given task. It is needed to balance the heat losses and heat gains in a typical calculation.
- **Embodied (or embedded) energy** is the total energy needed for making and using a product. Some methods of calculation take account of raw material extraction, transport, manufacture, installation, and eventual deconstruction.

Primary fuels:
coal
crude oil
natural gas
natural gas liquids
nuclear
  electricity
hydro-electricity
renewable energy
sources

Secondary fuels:
coke and other
  manufactured
  fuels
petrol and other
  petroleum
  products
secondary
  electricity

Energy measurements

The scientific unit of energy is the *joule*, as explained in Resource 2. The joule is an inconveniently small unit and therefore larger multiples, such as megajoules ($1 \text{ MJ} = 10^6 \text{ J}$) and gigajoules ($1 \text{ GJ} = 10^9 \text{ J}$), often need to be used. In practical use we also make use of other units, such as those based upon the idea of power or carbon emissions.

Power is the rate at which energy is used or produced. The energy is divided by the time taken to get the rate or speed at which it is used. Higher power performance means that work is done faster and that energy is used more quickly.

The unit of power is the *watt* (W), where 1 watt = 1 joule/second. The watt is another relatively small unit when we need to describe high-power situations such as energy for buildings, so you will more commonly encounter kilowatts (kW) and megawatts (MW).

The *power rating* of a device, such as an electric lamp labelled *20 W*, tells us the lamp's potential for using energy, *if* we switch it on. But to know how much energy the lamp uses we must know the length of time it is switched on.

The formula for power is *Power = energy/time,* and changing the expression around gives a formula for energy:

Energy = power × time

So we calculate energy use by multiplying the power rating by the time that it is used. There is a choice of units for power as shown here.

Prefixes:
kilo, k = 10^3
mega, M = 10^6
giga, G = 10^9
tera, T = 10^{12}
peta, P = 10^{15}
exa, E = 10^{18}

See also:
definitions of
energy and power
in Resource 2

$$\text{Power} = \frac{\text{energy}}{\text{time}}$$

- joules (J) = watts × seconds
  a small unit of energy, so megajoules (MJ) are commonly used.

- kilowatt-hours (kWh) = kilowatts × hours
  a convenient unit of energy, that is often used to pay for electrical energy.

The kWh is equal to 3.6 million J or 3.6 MJ.

Worked Example 2.1

An electric lamp with a power rating of 12 watts (W) runs for 8 hours each day. Calculate:

1. How many kilowatts hours (kWh) of electrical energy does it use each day?
2. How many kWh does it use per year?
3. What is the total cost per year if electricity costs 20 pence per kWh?

Known:
The prefix 'kilo' means 10^3 or 1000 (10^3 is shorthand for 10 x 10 x 10 = 1 000)
    So 1 000 W = 1 kW and 1 W = 1/1000 = 0.0001 kW

Kilowatt hours are calculated using the formula:

    kilowatts × hours = kilowatt-hours (kWh)

Step 1
Convert the 12 W to kW

    12W = 12/1 000 kW = 0.012 kW

Use the formula for kilowatt hours with the values given in the question

    kilowatts × hours = kilowatt-hours (kWh)
    0.012 kW × 8 hours = 0.096 kWh

So the bulb uses 0.096 kWh of electrical energy per day.

Step 2
There are 365 days in a year
So

    total kWh per year = 0.0.96 kWh per day × 365 days = 35.04 kWh

Step 3

The cost of each kWh is 20 pence
So

total cost per year $= 35.04$ kWh $\times$ 20 pence $= 700.8$ pence or about £7

Note: Your own knowledge, or the section on Lamps in Chapter 6, will tell you that 12 W is typical of a modern energy-saving lamp. But many people are still running 60 W tungsten lamps that use electrical energy at five times this rate. The average cost of electricity per kWh depends upon your tariff, but alas, its upward trend does not vary.

Conversion of energy units

There are other units of energy that exist for historical reasons or are used for the large-scale study and statistics of fuel and energy use, such as those involving quantities of oil. Table 2.1 lists some of the units of energy and gives conversion factors.

Table 2.1 Units of energy and conversion factors

Unit of energy	Conversion
Megajoules, MJ 1,000,000 joules	1 MJ = 0.2778 kWh 1 kJ = 0.947 BTU
Kilowatt-hours, kWh 1 kW * 1 hour	1 kWh = 3.6 MJ 1 kWh = 3409 BTU
BTU (British Thermal Unit)	1 BTU = 1.055 MJ 1000 BTU per hour = 0.3 kW
Therm	1 therm = 29.31 kWh
Tonne of oil equivalent, toe	1 toe = 41.87 GJ = 1.163×10^4 kWh
Kilocalorie, kcal	2500 kcal = 3 kWh = 10 MJ

Note: 1000 food 'calories' = 1 kcal.

Calorific values

Although it is not possible to convert or use all the energy contained in a fuel, the theoretical energy content of a fuel is a starting point for understanding and analysing the use of a fuel.

- **Calorific value** is a measure of the primary heat energy content of a fuel expressed in terms of unit mass or volume.

Some typical calorific values are quoted in Table 2.2. These values may be used to compare fuels, to predict the quantity of fuel required for a particular output, and to calculate the cost.

Table 2.2 Calorific values of fuels

Fuel	Calorific value
Coal, average	29 MJ per kg
Crude oil, average	45 MJ per kg
Fuel oil	43 MJ per kg
Natural gas, average	39 MJ per m³ (approx. 22 MJ per kg)
Petrol	47 MJ per kg
Wood	11 MJ per kg

Notes:
The exact calorific value depends on the source of the fuel and its density.
1 MJ per kg = 1 GJ per tonne.

Carbon emissions

Carbon dioxide (CO_2) is one the *greenhouse gases* that interrupt the natural radiation of heat out of the Earth and therefore contribute to global warming. Carbon dioxide itself is not harmful to humans, and indeed we give it out as part of our breathing system. However, we release large quantities of carbon dioxide when we convert energy by burning carbon-based fuels such as timber, oil and gas.

The quantities of carbon dioxide gas (or carbon the element) generated and emitted when fossil-based fuels are burned give us another measurement for energy use of a certain kind:

See also:

section on *Greenhouse gases* in Chapter 1

- *Carbon emission* can be used as an alternative indicator of the use of energy derived from fossil-based fuels.

  Units: kilograms of carbon dioxide, kg CO_2
  or
  kilograms of carbon, kg C

To convert from kg C to kg CO_2 – multiply by 44/12 (the ratio of their atomic weights)
To convert from kg CO_2 to kgC – multiply by 12/44.

Emissions of other greenhouse gases, such as methane (CH_4) and nitrous oxide (N_2O), can be expressed as *equivalent* amounts of carbon dioxide emission. In this case equivalent is defined as the amount that produces the same global warming effect over a period of 100 years. This long period allows for the different behaviours over time of different greenhouse gases.

Carbon emission figures are not really units of energy, and they are not relevant to renewable sources of energy such as wind power. They are however useful for measuring and setting targets connected with energy in buildings, and some current Building Regulations use carbon emission targets, as shown in later sections of this chapter.

National energy use

Table 2.3 shows typical use of energy types on the large scale of national energy use over a year. These patterns of national energy use change with time. For example, in the United Kingdom over a period of 50 years to 2010, the use of solid fuels, mainly coal, declined from around 60 per cent to a few per cent of the total, while the use of gas rose from 5 per cent to over 30 per cent. These proportions are changing again as natural gas supplies from the North Sea decline. Meanwhile, the use of electricity has increased from 7 per cent to near 20 per cent of the total.

Table 2.3 National energy consumption for the United Kingdom

By final user		By fuel as consumed	
Transport	37.5%	Oil	42.5%
Domestic	32.5%	Gas	34.5%
Industry	18%	Electricity	19%
Services (incl agriculture)	12%	Renewables and heat	2.5%
		Coal and manufactured fuels	1.5%

Note: Figures are typical of the United Kingdom 2010, published in the *Digest of UK Energy Statistics* – available online from various government websites.

THERMAL COMFORT

The thermal comfort of human beings is governed by many physiological mechanisms of the body, which vary from person to person. In any particular thermal environment it is difficult to get more than 50 per cent of the people affected to agree that the conditions are comfortable!

The body constantly produces heat energy from the food energy it consumes. This heat needs to be dissipated at an appropriate rate to keep the body at constant temperature. The transfer of the heat from the body is mainly by the processes of convection, radiation and evaporation. Evaporation transfers heat from our bodies by using the latent heat content of the water vapour given out on the skin (perspiration) and in the breath (respiration).

See also:

section on *Nature of heat* in Resource 2

The total quantity of heat produced by a person depends upon the size, age, sex, activity and clothing of the person. A further complication is the ability of the body to become accustomed to the surrounding conditions and to adapt to them. For example, everyone can tolerate slightly lower temperatures during winter. This adaptation can be influenced by the type of climate and the social habits of a country.

See also:

section on *Environmental comfort* in Chapter 1

Factors affecting thermal comfort

The principal factors affecting thermal comfort can be considered conveniently under the headings below, and are discussed further in the following sections.

Personal variables

- Activity
- Clothing
- Age
- Gender

Physical variables

- Air temperature
- Surface temperatures
- Air movement
- Humidity

Activity

The greater the activity of the body the more heat it gives off. The rate of heat emission depends upon the individual metabolic rate and upon the surface area of each person. People who seem similar in all other respects can vary by 10 to 20 per cent in their heat output.

The average rate of heat emission decreases with age. Table 2.4 lists typical heat outputs from an adult male for a number of different activities. The output from adult females is about 85 per cent that of males. Although there are differences in heat emissions between different groups of people such as female and

Table 2.4 Typical heat output of the human body

Activity	Example	Typical heat emission of adult male
Immobile	Sleeping	70 W
Seated	Watching television	115 W
Light work	Office	140 W
Medium work	Factory, dancing	265 W
Heavy work	Lifting	440 W

male, surveys show there is often little difference between groups in what they perceive to be a comfortable temperature.

Clothing

Clothes act as a thermal insulator for the body and help to maintain the skin at a comfortable temperature. Variations in clothing have a significant effect on the surrounding temperatures that are required for comfort.

To enable heating needs to be predicted, a scale of clothing has been developed: the clo-value. On this scale, 1 clo represents 0.155 m^2 K/W of insulation and values range from 0 clo to 4 clo. Table 2.5 shows the value of different types of clothing and indicates how the room temperature required for comfort varies with clothing. On average, women prefer slightly higher temperatures.

> Thermal comfort: 'that condition of mind which expresses satisfaction with the thermal environment' (BS EU ISO 7730)

Table 2.5 Clothing values

Clo value	Clothing example	Typical comfort temperature when sitting (°C)
0 clo	Naked; swimwear	29
0.5 clo	Light trousers; shirt; light dress, blouse	25
1.0 clo	Business suit; dress, jumper	22
2.0 clo	Heavy suit, overcoat, gloves, hat	14

Room temperature

The temperature of the surrounding surfaces can affect the thermal comfort of people as much as the temperature of the surrounding air. This is because the rate at which heat is radiated from a person is affected by the radiant properties of the surroundings. For example, when someone is sitting near the cold surface of a window, the heat radiated from the body increases and can cause discomfort.

A satisfactory design temperature for achieving thermal comfort needs to take account of both air temperatures and radiant effects. Different types of temperatures are described below.

> See also: section on *Temperature* in Resource 2

Inside air temperature t_{ai}

The inside air temperature is the average temperature of the bulk air inside a room. It is usually measured by an ordinary dry-bulb thermometer which is suspended in the centre of the space and shielded from radiation.

Mean radiant temperature t_r

The mean radiant temperature is the average effect of radiation from surrounding surfaces. At the centre of the room this temperature can be taken as being equal to the mean surface temperature as calculated by

$$t_r = \frac{A_1 t_1 + A_2 t_2 + \ldots}{A_1 + A_2 + \ldots}$$

where t_1, t_2 ... are the surface temperatures of the areas A_1, A_2 ... , etc.

The mean radiant temperature should be kept near the air temperature but not more than about 3°C below it, otherwise conditions are sensed as 'stuffy'.

Inside environmental temperature t_{ei}

The environmental temperature is a combination of air temperature and radiant temperature. The exact value depends upon convection and radiation effects. For average conditions t_{ei} can be derived from the following formula.

$$t_{ei} = 2/3\ t_r + 1/3\ t_{ai}$$

Environmental temperature is recommended for the calculation of heat losses and energy requirements.

Dry resultant temperature t_{res}

Dry resultant temperature is a combination of air temperature, radiant temperature and air movement. When the air movement is low, t_{res} can be derived from the following formula.

$$t_{res} = 1/2\ t_r + 1/2\ t_{ai}$$

Table 2.6 Typical design temperatures for comfort

Type of environment	Winter (heating season)	Summer (cooling season)	Air supply
Domestic: living room	22–23	23–25	0.4–1 ach*
Domestic: bedroom	17–19	23–25	0.4–1 ach
Domestic bathroom	26–27	26–27	5 ach minimum
Office	20–24	23–25	8 litres/s/person
Classroom	20–24	23–25	8 litres/s/person
Restaurant	20–24	24–25	8 litres/s/person
Department store	16–22	21–25	

Notes:
* ach = air changes per hour
The figures assume typical activity and clothing levels for the space, no excessive air movements, and no smoking.
Sources: composite from CIBSE, BS, ASHRAE.

Room-centre comfort temperature t_c

A comfort temperature is a measure of temperature which gives an acceptable agreement with thermal comfort. When air movement is low, the dry resultant temperature at the centre of a room is a commonly used comfort temperature. This temperature also approximates to the *operative temperature* used in some design codes.

The *globe thermometer* is a regular thermometer fixed inside a blackened globe of specified diameter (150 mm is one standard). This globe temperature can be used to calculate other temperatures, and when air movement is small it approximates to the comfort temperature.

Air movement

The movement of air in a room helps to increase heat lost from the body by convection, and can cause the sensation of draughts. The back of the neck, the forehead and the ankles are the most sensitive areas for chilling.

- Air movements above 0.1 m/s in speed require higher air temperatures to give the same degree of comfort.

For example, if air at 18°C increases in movement from 0.1 m/s to 0.2 m/s, then the temperature of this air needs to rise to 21°C to avoid discomfort.

The air movement rate is not the same thing as the air change rate, and is not always caused by ventilation. Uncomfortable air movement may be caused by natural convection currents, especially near windows or in rooms with high ceilings.

> Natural convection in a room can cause uncomfortable air movement

A *hot-wire anemometer* and a *Kata thermometer* are devices used to measure air movement. Both devices make use of the cooling effect of moving air upon a thermometer.

Humidity

Humidity is caused by moisture in the air, and is treated more fully in later chapters.

- Relative humidity within the range of 40 to 70 per cent is required for comfortable conditions.

High humidities and high temperatures feel oppressive and decrease natural cooling by perspiration. High humidity and low temperatures cause the air to feel chilly.

Low humidities can cause dryness of throat and skin. Static electricity can accumulate with low humidity, especially in modern offices with synthetic carpet, and cause mild but uncomfortable electric shocks.

Ventilation

See also:
sections on
Ventilation and
Humidity in
Chapter 4

In any occupied space ventilation is necessary to provide oxygen and to remove contaminated air. Fresh air contains about 21 per cent oxygen and 0.04 per cent carbon dioxide, while expired air contains about 16 per cent oxygen and 4 per cent carbon dioxide. The body requires a constant supply of oxygen but in unventilated conditions the air would become unacceptable well before there was a danger to life. As well as being a comfort consideration, the rate of ventilation has a great effect on the heat loss from buildings and on condensation in buildings.

The normal process of breathing gives significant quantities of latent heat and water vapour to the air. Body odours, bacteria, and the products of smoking, cooking and washing also contaminate household air. In places of work, contamination may be increased by a variety of gases and dusts.

A number of statutory regulations specify minimum rates of air-supply in occupied spaces. Recommended rates of ventilation depend upon the volume of a room, the number of occupants, the type of activity and whether smoking is expected. It is difficult therefore to summarise figures for air-supply but Table 2.7 quotes some typical values.

Table 2.7 Typical fresh air supply rates

Type of space	Recommended air supply
Residences, offices, shops	8 litres/second per person
Restaurants, bars	18 litres/second per person
Kitchens, domestic toilets	10 litres/second per m² floor

Predicting thermal comfort, PMV index

Thermal comfort factors:
 air temperature
 mean radiant
 temperature
 humidity level
 air movement
 activity
 clothing

A standard method of predicting the comfort of people working in a given environment uses surveys to obtain a *predicted mean vote* (PMV) index, or *percentage people dissatisfied* (PPD) index. The index takes account of six factors affecting themal comfort and uses a seven-point scale ranging from +3 for hot through 0 for neutral and –3 for cold.

ENERGY TRANSFER IN BUILDINGS

Heat losses for buildings

Heat energy flows out of a building when the temperature inside is higher than the temperature outside, which is most of the time in temperate climates. Heat is transferred through an element of a building, such as a wall, by a number of mechanisms within that element. For example, layers of different materials in a wall conduct heat at different rates, and within any cavity there is heat transfer by conduction, convection and radiation. At the inside and outside surfaces of the wall the heat transfer by radiation and convection is affected by factors such as surface colour and exposure to climate.

These various processes of heat transfer through a building envelope can be represented by combinations of the various physics formulas for thermal transfer. In addition to this the variations in conditions, such as inside and outside temperature changes, need to be applied. We can also measure the actual performance of a 'live' building in terms of heat lost or gained under various conditions over time, and from this performance we can deduce insulation values for the various types of wall, roof and window used in modern construction. Building regulations and codes also use these values to specify targets and limits for thermal insulation and energy use.

Measuring insulation and heat loss

It is convenient in practice to combine all the heat transfer factors into an ambitious single measurement which describes the air-to-air behaviour of a particular construction. This measurement is called the *overall thermal transmittance coefficient*, or *U-value*.

See also:
sections on *Heat transfer in* Resource 2

- A *U-value* is a measure of the overall rate of heat transfer, by all mechanisms under standard conditions, through a particular section of construction.
  Unit: W/m² K

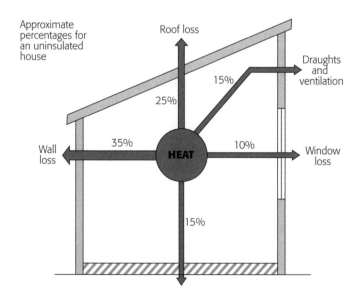

Figure 2.1 Typical heat losses from a building

The coefficient, or U-value, is measured as the rate of heat flow in watts through 1 m² of a structure when there is a temperature difference across the structure of 1 degree (K or C).

For example, a wall with a U-value of 0.3 W/m² K loses heat at half the rate of a wall with a U-value of 0.6 W/m² K. Therefore, the cost of replacing heat lost through the first wall (0.3) will be half the cost of replacing heat lost through the second wall (0.6).

U-values are calculated from the thermal resistances of the different materials and parts of a structure through which heat flows. The different materials, cavities and layers of a building element, such as a wall, *oppose* the transmission of heat by varying amounts. These differences are described by the thermal resistance, or R-value.

Lower *U*-values mean better thermal insulation

See also:
section on *Calculating insulation values* in Chapter 3

- An R-value is a measure of the opposition to heat transfer offered by a particular building element, such as a wall, or by parts of that element.

Unit: m² K/W

Higher *R*-values mean better thermal insulation

Factors affecting heat loss

The principal factors that affect the rate at which heat is lost from a building are listed here and summarised in in the following sections:

- insulation of the building shell
- exposed area of the building shell
- temperature difference between inside and outside
- air change rate
- exposure to external climate
- efficiency of services in the building
- patterns of use for the building.

These various forms of heat loss from buildings can be grouped into the following two broad types:

- **Fabric heat loss** – by transfer through the external shell of the building
- **Ventilation heat loss** – by both purposeful and unintentional changes of air in the building.

Insulation of the building shell

Building regulations and codes use U-values or R-values as one mechanism to specify targets and limits for thermal insulation and energy use. R-values and U-values are analysed and calculated in Chapter 3, but we can make immediate use of typical figures to compare insulation values and heat losses for different buildings.

Table 2.8 compares the insulation values for older housing, such as pre-1960s houses, with modern housing using current standards of insulation. The most dramatic change is that the insulation of walls and roofs has *improved* by a factor of least 10. The *U*-values have decreased and the *R*-values have increased. This means that the heat energy flowing per second out through that particular wall or roof has reduced to at least one tenth of what it was in an older construction. Therefore only one tenth of the heating is needed in order to replace the heat flowing out through that wall or roof. This gives great reductions in energy use and the related costs of that energy.

Windows do not, and cannot, gain such impressive improvements in insulation, even when they have high-performance features. Table 2.8 shows how modern double glazing has an insulation value over ten times worse than a well-insulated wall (*U*-value of 2.0 compared with 0.15). Even high-performance triple glazing which includes insulated frames will still be the 'weakest link' in the fabric of the building. Obviously we need and enjoy the benefits of windows in homes, but modern building regulations and codes usually restrict the total area of windows allowed in a dwelling.

Table 2.8 Typical *U*-values and *R*-values for dwellings

Element of building	Type of construction	U-value	R-value (total)
Walls	Older (min insulation)	2.5	0.4
	Modern common practice	0.25	4
	High performance	0.15	6.67
Roofs	Older (min insulation)	1.9	0.53
	Modern common practice	0.15	6.67
Windows	Older single glazing	4.8	0.21
	Modern double glazing	2.0	0.37
	High-performance triple glazing	0.80	1.25

Double glazing has relatively poor insulation values

Notes:
Values are indicative of ranges of values.
To convert: U = 1/R or R = 1/U.

Area of the shell

The greater the area of external surfaces, the greater is the rate of heat loss from the building. A terraced house, for example, loses less heat than a detached house of similar size. Table 2.9 compares exposed perimeter areas for different shapes of dwelling.

The basic plan shape of a building is one of the first design decisions to be made, although choices may be restricted by the nature of the site and by local regulations.

Table 2.9 Exposed areas of dwellings

Type of dwelling (each of same floor area)	Exposed perimeter area (per cent)
Detached house	100
Semi-detached house	81
Terraced house	63
Flat on middle storey (2 external walls)	32

> Sharing walls and floors reduces heat losses

Temperature difference

A large difference between the temperatures inside and outside the building increases the rate of heat lost by conduction and ventilation. This loss is affected by the design temperature for the inside air, which in turn depends upon the purpose of the building. Recommended comfort temperatures for different types of buildings are given in Table 2.6 earlier in this chapter.

Air change rate

When warm air leaves a building it carries heat energy which is therefore wasted and must be replaced by heating the incoming air, which is usually cooler. The airflow from a building occurs through windows, doors, gaps in construction, ventilators and flues. This air leakage or air permeability may be by intentional ventilation required for comfort and safety, or it may be by unintentional 'infiltration' such as through gaps and cracks in the fabric of the building, and by exposure to wind.

The actual air change or ventilation rate therefore depends on a number of factors, some of which cannot be accurately predicted from plans. However, it is possible to estimate leakage by using typical figures for a particular form of construction. It is also possible to measure the actual air leakage of a completed building by a pressurisation test, which uses a large fan to measure the airflow into a building at a reference pressure difference of 50 pascals (Pa) between inside and outside the building. A typical design limit for air permeability in a dwelling is a maximum of 10 m³/h/m² at 50 Pa, and a high-performance house can perform 10 times better than this limit.

Exposure to climate

> Higher storeys of a building lose more heat

When a wind blows across a wall or roof surface the rate of heat transfer through that element increases. This factor in effective insulation can be included when calculating a U-value by varying the value of the external surface resistance. A system of standard surface resistances is available for the following three groups of building exposure in the United Kingdom:

- *sheltered*: buildings up to three storeys in city centres.

- *normal*: most suburban and country buildings.
- *severe*: buildings on exposed hills or coastal sites, floors above the fifth in suburban or country sites, and floors above the ninth in city centres.

Ventilation rates are also affected by exposure to climatic effects such as wind, and these effects are reduced when a building is sheltered by other buildings or high hedges. For example, the SAP energy rating described in Chapter 3 takes account of the number of *sheltered sides*, which for normal urban buildings is assumed to be two.

Efficiency of services

There is usually some wastage of heat energy used for water heating and space heating, and the design of the services can minimise or make use of this waste heat. For example, some of the heat from the hot gases passing up a flue can help heat the building if the flue is positioned inside the building rather than on an external wall. More advanced techniques include the recovery of the latent heat contained in the moist flue gases produced by gas burners.

> Burning natural gas gives off water vapour that contains energy

The heat given off by hot water storage cylinders and distribution pipes, even well-insulated ones, should be used inside the building if possible rather than wasted outside. Even the draining away of hot washing water, which is dirty but full of heat energy, is a heat loss from the building and a heat gain to the sewerage system! Methods of heat recovery are discussed in later sections on energy conservation.

Patterns of use

The energy consumption of a building is greatly affected by the number of hours per day and the days per year that the building is occupied and used. Many buildings that are unoccupied at some times, such as at night, need to be preheated before occupancy each morning.

These patterns of building use and occupancy vary greatly, even for similar buildings. When a building has separate areas with different patterns of occupancy, each part needs to be considered as a separate building for heating calculations.

Heat gains for buildings

A building gains heat energy as well as losing it, and both processes usually occur at the same time. In locations with a temperate climate, such as in the British Isles, the overall gains are less than the overall losses, but the heat gains may still provide useful energy savings. The factors affecting heat gains are indicated in Figure 2.2. They can be considered under the following two broad categories.

- **Solar heat gains** – from the sun
- **Casual heat gains** – from occupants and equipment in the building.

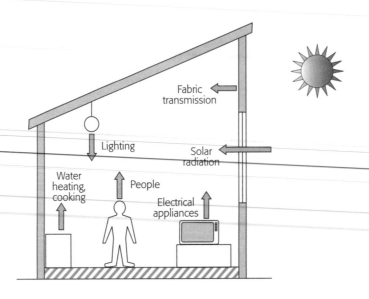

Figure 2.2 Typical heat gains in a building

See also:
section on
Climate in
Chapter 1

Solar heat gains

The heat gained in a building by radiation from the sun depends upon the following factors.

- the geographical latitude of the site, which determines the height of the sun in the sky
- the season of the year, which also affects the height of the sun in the sky
- the orientation of the building on the site, such as whether rooms are facing south or north
- the local cloud conditions, which can block solar radiation
- the angles between the sun and the building surfaces, because maximum gain occurs when surfaces are at right angles to the rays from the sun
- the nature of the window glass and whether it absorbs or reflects any radiation
- the nature of the roof and walls, because heavyweight materials behave differently to lightweight materials.

Windows that collect solar energy for the building are an example of *passive design*

The rate at which heat from the sun falls on a surface varies throughout the day and the year: more information is given in Chapter 3 where heat gains are calculated. Most solar heat gain to buildings in the British Isles, and in countries with similar climates, is by direct radiation through windows. The maximum gains through south-facing windows tend to occur in spring and autumn when the lower angle of the sun causes radiation to fall more directly onto vertical surfaces. This heat gain via windows can, if used correctly, be useful for winter heating.

The fabric solar-heat gains through walls and roofs are considered negligible for masonry (brick and block) buildings during the British winter. Little solar heat reaches the interior of the building because the high thermal capacity of heavy-weight construction tends to delay transmission of the heat until its direction of flow is reversed with the arrival of evening. Lightweight construction, for example with a timber or steel frame, transfers heat gains and losses more quickly.

Casual heat gains

Casual heat gains arise from the heat given off by various activities and equipment in a building that are not primarily designed to give heat. The major sources of such heat are:

- heat from people
- heat from lighting
- heat from cooking and water heating
- heat from machinery, refrigerators, electrical appliances.

In commercial or public buildings this type of heat gain must be allowed for in the design of the heating/cooling system. Where possible this heat should be used rather than wasted. In houses, the casual heat gain in winter is useful, and as dwellings become better insulated, it forms a higher proportion of the total heating needed. Chapter 3 covers the calculations of casual heat gains.

Energy balance of buildings

We normally require the temperature inside a building to be kept at a constant comfort level, with appropriate variations between areas of the building and the times of their use. To keep inside temperatures constant, the heat flowing out of the building needs to be balanced by the same amount of heat energy put into the building. In the case of a building in Europe, for example, the natural heat gains of a building described in the previous sections cannot balance the heat losses. We therefore need to add extra heat energy by using heating services equipment. Even if we need to subtract heat from the building using air-conditioning equipment, that equipment still uses energy.

Figure 2.3 shows the balance needed in a building between the heat energy losses, the heat energy gains, and the energy used by the heating and/or cooling

See also:
section on *Thermal admittance* in Chapter 3

See also:
section on *Lighting design* in Chapter 6

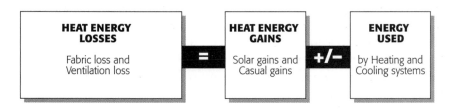

Figure 2.3 Energy balance in buildings

equipment. The balance is broader than just heat energy as the services equipment will have supplied the heating or cooling effects by converting other forms of energy such as electricity or gas. Chapter 3 shows how to calculate typical figures for the energy quantities involved in the energy balance of a building during the course of a year.

See also:
section on
Climate change in
Chapter 1

ENERGY REGULATIONS

Many countries have introduced measures that encourage or enforce the efficient use of energy. One reason for the *conservation* of fuel and energy is that the Earth's total reserves of convenient fossil fuels, such as oil and gas, are limited. Another motive for restricting our use of fuels is an environmental concern about the emission of carbon dioxide and other greenhouse gases produced when fuels are used by transport, by electricity generating power stations, and in buildings.

Energy use in buildings is one of the largest categories of national energy use, and therefore any regulations controlling the design of new buildings are an opportunity to make sure that buildings have features that minimise energy use and carbon emissions. Good standards of thermal insulation in new buildings, as discussed in this chapter, are an important method of controlling heat loss from buildings, and these standards feature in building regulations that promote energy efficiency.

In addition to changing the energy-efficiency requirements for new buildings, the energy efficiency in *all* buildings can be improved by actions in other areas, such as the following.

- *Improving existing building stock.* A variety of incentives and requirements can be used to encourage owners to improve their properties, such as by increasing thermal insulation.
- Requiring that *changes to existing buildings*, such as constructing extensions, must meet modern standards.
- *Planning* for energy efficiency at the earliest stages of a project when site orientation, infrastructure and building form can be considered, and sustainable and energy-efficient designs generated.

See also:
section on
*Carbon and
energy
management* in
Chapter 14

Most countries are evolving methods for encouraging or enforcing national energy efficiency in buildings. The European Union, for example, has issued a Directive for Energy Performance in Buildings (EPBD) which obliges member states to have appropriate regulations and systems in place to ensure that the overall objectives of saving fuel and energy are met. The United Kingdom and the Republic of Ireland have chosen to use their building codes and regulations to help control energy use and carbon emissions. Some EU countries are using other systems and incentives.

In the United Kingdom the detail of building legislation and the promotion of energy efficiency is devolved to the three Administrations of England and Wales, Scotland and Northern Ireland. The three sets of UK building regulations, and also those in the Republic of Ireland, share technical approaches and techniques for defining the performance of buildings with regard to the conservation of fuel and power. This section outlines concepts and techniques typically used in the building regulations of the United Kingdom and of the Republic of Ireland. There are similar approaches in Australia and New Zealand. All such regulations evolve, especially in their performance targets and current dates, so it is good and wise practice to check the latest version of any regulations that apply in the area where you live or work.

Building regulations

Building regulations or building codes are rules that specify minimum levels of performance for the construction and use of buildings. The rules are intended to protect public safety and general welfare, and they typically cover the areas of structural stability, durability, fire safety, accessibility, noise control, ventilation and thermal comfort. Building codes do not generally deal with the visual appearance or other aesthetic values of buildings, although planning regulations usually exist to protect and improve the look and feel of our built environment.

In devising the technical content of the building regulations, the methods for achieving energy efficiency must also be balanced against other needs such as the following:

- other essential performances such as structural stability, resistance to rain penetration, and fire safety
- the need for design details that are practical and within the capabilities of the construction workforce
- building services that are easy for occupiers to manage successfully
- regulations that are simple to interpret and enforce.

The thermal insulation sections of the Building Regulations in the United Kingdom originally restricted themselves to the control of thermal insulation in domestic buildings, and the early aim was to ensure minimum standards of health and comfort, and to reduce the risk of condensation in housing. More recent Building Regulations in the United Kingdom include controls for the 'conservation of fuel and power' in all new and converted buildings, and for the reduction of carbon dioxide emissions. Lowering emissions by using less fuel fortunately harmonises with both conserving energy and saving money on running the building.

Typical building regulations achieve their objectives by controlling the following areas of building design and performance:

- The annual rate of carbon dioxide emission by the completed building is kept to a minimum.

See also:

section on *Sustainable buildings* in Chapter 14

See also:

Links to individual countries in reference section

- Thermal insulation of the fabric of the building is maximised.
- Airtightness of the fabric of the building is maximised.
- Summer overheating is limited by shading and other measures.
- Performance of fixed services such as heating, ventilation and lighting equipment is optimised.
- The actual quality and performance of construction is verified.
- Operating and maintenance instructions are available for users.

See also:

section on *Carbon and energy management* in Chapter 14

There are variations in requirements depending upon whether the building is a dwelling or a non-dwelling, and whether the work is for new buildings or conversions of existing buildings.

The following sections describe some methods used by building codes to calculate and define energy targets for the performance of buildings. Some of the specialised definitions and techniques are centred on the Building Regulations for England and Wales, but Scotland and Northern Ireland have similar regulations using the same methods. The building codes in most countries also share the objectives and the technical methods described here.

Insulation of the building fabric

A simple method of setting performance targets for thermal comfort and energy use in buildings is to specify the insulation of the building fabric. The following specifications can be used for optimum insulation:

- target U-values or total R-values for elements of the construction, such as walls and roofs
- thicknesses of common insulating materials within elements
- maximum areas of glazing, such as by percentage of floor area or wall area.

Lower U-values mean better thermal insulation

To show compliance with insulation targets the following relatively simple methods can be used:

- following an approved type of construction
- calculations to show that target U-values or R-values for each element have been matched or bettered – sometimes termed an *elemental method*
- calculations to show that under-performance in one area is compensated for by better performance in another area – sometimes termed *trading off.*

Higher R-values mean better thermal insulation

Trading off

Trading off is an informal way of describing how a shortfall in performance in one area is balanced by extra performance in another area. This system, if permitted, gives more design flexibility; such as allowing larger areas of glazing to be used, providing that higher standards of insulation are used in the walls and roof.

An acceptable trade-off needs to be proved by calculations of total heat loss for the overall building, using agreed standard methods related to the ones shown earlier in the chapter. In a simple system, the total heat losses and gains calculated for the proposed building must not exceed the values calculated for a notional building that uses standard quoted values of insulation. Systems of energy performance rating (SAP and SBEM), described below, use the principle of trading off between areas that include fabric insulation, glazing, air infiltration, efficiency of services and carbon dioxide emissions.

Limiting U-value standards

Building designers welcome the flexibility allowed by trading off, but in order to conserve national fuel and power, limits are placed on the system by specifying minimum performance standards in certain areas. For construction elements of the building fabric, the 'poorest acceptable', or limiting, U-values are specified for each element. Some examples are shown in Table 2.10. In order to achieve other targets, such as the carbon dioxide emission rate, the values for most of the elements will need to be significantly better than those listed in the table of limiting U-value standards.

> **See also:**
> section on
> *Calculating insulation values* in Chapter 3

Table 2.10 Limiting U-value standards

Element type	Area-weighted average U-value (W/m² K)
Roofs	0.20
Walls	0.30
Floors	0.25
Windows/rooflights/doors	2.00

Notes: The values for new dwellings are summarised from typical building regulations (England and Wales 2010). These values change periodically, and the latest regulations applicable to a particular situation should be consulted.

Target carbon dioxide emission rate, TER

The target carbon dioxide emission rate (TER) is an energy performance target expressed in terms of the rate of emission of carbon dioxide gas (CO_2) in units of kilograms per square metre of floor area per year (kg/m²/yr). The emissions are associated with the heating, hot water, ventilation and internal lighting in the building, and may take place away from the building, such as at the power station generating electricity used by the building.

The Building Regulations for England and Wales use values of TER to specify minimum acceptable energy performance values for various classes of buildings, such as dwellings and commercial buildings. The regulations also specify approved methodologies that can be used to show that the building as designed, and as built, has an actual TER lower than the minimum acceptable rate. For dwellings the approved methodology makes use of the SAP procedure described in the next section.

Energy performance rating, SAP

The overall energy efficiency of a dwelling can be given an *energy rating* by using a *standard assessment procedure* (SAP). SAP calculations generate figures for carbon dioxide emissions and energy costs associated with a building. The SAP 'rating' of a dwelling is found by using a specified standard method to calculate the annual energy costs for space heating, water heating, ventilation and lighting, less any cost savings from energy generation. It is a composite method which includes the calculations of heat loss based on insulation, as shown earlier in the chapter, although this simple calculation can be obscured by the many other factors included in the generation of SAP results. Fortunately these results are usually calculated by computer programs.

The SAP used in England and Wales includes the following design factors that affect energy balance and costs:

- types of materials used for construction
- thermal insulation of the building
- efficiency and control of the heating system
- ventilation characteristics of the building
- solar gain by the building
- fuels used for space and water heating, ventilation and lighting
- renewable energy technologies.

A SAP rating is based on energy costs and expressed on a scale of 0 to 100; the higher the SAP number the better the performance and the lower the running costs. Ratings are calculated so that they are not affected by differences in the number of people in the dwelling, the floor area, the ownership of particular domestic appliances, and the geographical location of the dwelling. The exact methodologies for calculating a SAP have changed at intervals over the years (pre 2005, 2005, 2009 for example) as assumptions about fuel energies and emission have evolved. The different SAP ratings before 2005 cannot be compared with modern ratings, which run on a scale of 1 to 100.

The average energy efficiency, by SAP rating, of new homes in England and Wales over a period in 2009 was in the range 70–80.

See also:

Environmental appraisal of buildings, Chapter 14 for details of schemes like BREEAM, PassivHaus, Green Star

SAP	Energy costs
SAP = 0	Maximum energy costs
SAP = 100	Zero energy costs
SAP = 100 +	Building is a net exporter of energy

Practical details for energy conservation

Controlling the insulation of the building fabric is a major factor for conserving energy in buildings, but other aspects of building and running buildings also have a significant effect on energy use. The following headings indicate areas that affect energy use. These aspects of a building can usually be improved in order to meet energy performance targets, and might also be subject to building regulations.

See also:
section on *Sustainable buildings* in Chapter 14

Thermal bridging around openings

Area of thermal bridging	Measures available
Lintels, jambs and sills	Insulating blockwork or internal insulation on the inside face of the wall and its return or partial cavity fill in the wall or full cavity fill in the wall

Airtightness

Area of air leakage	Measures required
Dry lining on masonry walls	Sealing of gaps with windows and doors; sealing of gaps at junctions with wall, floors and ceilings
Timberframe construction	Complete sealing of vapour control membrane
Windows, doors, rooflights	Fitting draught-stripping in frames of openable elements
Loft hatches	Sealing around the hatch
Service pipes	Sealing around boxing at floor and ceiling levels. Sealing around pipes which project or penetrate into voids

Space heating controls

Feature	Aim
Zone controls, such as room thermostats, thermostatic radiator valves	To control the temperatures independently in areas that require different temperatures, such as living and sleeping areas
Timing controls	To control the periods when heating systems operate
Boiler control interlocks	To switch boiler off when no heat is required. For boiler to fire only when thermostat is calling for heat. To prevent unnecessary boiler cycling

Hot water controls and insulation of storage

Feature	Aim
Heat exchanger of appropriate capacity	To allow effective control
Thermostats	To shut off the supply of heat when the storage temperature is reached To switch off the boiler when no heat is required in the case of a central heating system
Timer, which may be part of central heating system	To shut off the supply of heat when water heating is not required
Insulation of hot water vessels	To reduce heat loss
Additional insulation of unvented hot water systems	To control heat loss through the safety fittings and pipework
Insulation of pipes and ducts, which do not contribute to useful heat requirement	To reduce heat loss
Insulation of hot pipes connected to hot water storage vessels	To reduce heat loss
Spray taps Motion sensor taps	Reduces water consumption and energy used for hot water

Lighting

See also:
Chapter 6 on
Artificial lighting

Feature	Detail
Lamps of high efficiency	Reduces energy use and heat gain Types such as high pressure sodium, metal halide, induction lighting, tubular fluorescent, compact fluorescent, LED
Lighting controls: local switching, time switching, motion sensors, daylight sensors	To switch off lighting when there are no occupants To encourage the maximum use of daylight without endangering the movement of occupants

EXERCISES

1. All forms of energy can be classified as either potential energy or kinetic energy, as shown in Table R2.1 in the *Resource Section* at the back of the book. Consider the forms of energy listed in that table, together with the given examples. Make a note whether you think each example is potential or kinetic energy.

2. Table 2.8 shows how insulation values have improved over the decades. The earliest measures to increase insulation started with buildings containing minimum insulation in building so they produced large gains and percentage changes. But at modern levels of thermal insulation further increases in insulation involve expensive efforts for much smaller gains. Therefore, in a well-insulated building other energy factors, such as ventilation heat loss or efficiency of services, play a relatively larger role.

   Consider the areas outlined on page 39 and compare them to a real building environment such as in your home or office. Select and appraise improvements to areas and features of your chosen building that might improve overall energy conservation and reduction of carbon emissions.

3

Thermal Effects in Buildings

CHAPTER OUTLINE

Sections of the previous chapters looked at our human need for a comfortable environment, and considered some of the external interactions of buildings and the wider environment. Heat energy and other thermal properties are major factors in maintaining our human body comfort and therefore also play major roles in the performance of buildings.

The transfer of thermal energy through the fabric of buildings is a dominant factor in the energy and carbon profile of buildings, and it is the focus of this chapter. The broader subject of energy use in buildings was examined in Chapter 2, and you may find it useful to consider these two chapters together.

Thermal insulation is a major factor in reducing the heat loss from buildings and so minimising energy use and carbon emissions. Adequate insulation should therefore be a feature of good initial design. The relatively small cost of extra insulating materials is quickly paid for by the reduction in the size of the heating plant required and by the annual savings in the amount of fuel needed. These fuel savings, and the related reduction in carbon emissions, continue throughout the life of the building.

This chapter explains the technical basis of thermal insulation and associated effects so that you can:

▶ appreciate the need for insulating buildings
▶ examine the types of materials used to insulate buildings
▶ make use of the terms and units associated with thermal insulation
▶ calculate *R*-values and *U*-values for buildings
▶ compare the insulation in different parts of buildings

▶ assess buildings for insulation qualities
▶ assess individual building elements for relative insulation value
▶ understand the causes and effects of thermal bridging
▶ know about the temperature profiles inside structures, and make calculations that predict possible effects such as condensation
▶ understand why different structures respond to temperature changes at different rates
▶ understand the causes of heat loss in buildings and make practical calculations
▶ understand the causes of heat gains and make practical calculations
▶ understand the overall energy balance of buildings and make practical calculations.

The resource sections in this book, especially Resource 2 about the *Principles of heat*, contain supporting information that can be used for revision of principles and for extended investigation of topics.

THERMAL INSULATION

In order to maintain a constant temperature within a building we need to restrict the rate at which heat energy is exchanged with the surroundings. Keeping heat inside a building for as long as possible conserves energy and reduces heating costs.

See also: 📎

Part II: Resources sections

Good thermal insulation will also reduce the flow of heat into a building when the temperature outside is higher than the temperature inside. In other words a well-insulated structure will, if ventilation and direct solar gain are controlled, stay cooler in the summer than a poorly insulated one. Some people are reluctant to believe that insulation will help keep a building cooler in summer, but it may be helpful to imagine the discomforts of living in a tent or garden shed. This 'building' will be very hot in summer for the same reason that it will be very cold in winter: lack of insulation. In a large building good thermal insulation will give savings in the energy needed to run the cooling plant. Some current office buildings use more energy for summer cooling than for winter heating.

The terms *heat, heat energy, thermal energy* are used here to describe the same concept

Another benefit of good thermal insulation is that the risk of surface condensation is reduced because the internal surfaces of a room are kept at a temperature which is above the dew-point of the air. Surface condensation is unsightly, unhealthy and damages decorations. Well-placed thermal insulation also reduces the time taken for a room to heat up to a comfortable temperature if, for example, the room is unoccupied during the day.

Insulating materials

A thermal insulator is a material which opposes the transfer of heat between areas at different temperatures. In present-day buildings the main method of heat transfer is by conduction, but the mechanisms of convection and radiation are also relevant. Sometimes there is also a contribution from the process of condensation, which releases heat energy when water vapour changes to water liquid.

A vacuum provides perfect insulation against *conduction* but is not practical for everyday purposes. The best practical arrangement is to look for materials that have atoms spaced well apart; such materials will also have low densities. Gases have widely spaced atoms and therefore provide good insulation against conduction. Air, which is a mixture of gases, is commonly used as the 'active ingredient' for insulation in materials such as glass fibre and aerated concrete. Air is not, of course, the active construction material, and the need for insulation must be balanced against other requirements such as strength and rigidity.

For air to act as an insulator it must be held still, because if it is allowed to move it will transfer heat by *convection*. The primary purpose of fibreglass or expanded plastic is therefore to trap and to hold air still. Also, any surface such as a wall or human skin holds a *boundary layer* of stationary air which provides a certain amount of thermal insulation against conduction.

Heat transfer by *radiation* can be restricted by using surfaces that do not readily absorb or emit radiant heat. Such surfaces, which look shiny, reflect the electromagnetic waves of heat radiation, and this insulation against radiant heat depends only on the surface appearance. When we use aluminium foil as an insulator, for example, it is the shiny surface of the material that is important. Aluminium is actually a good conductor of heat, but because the foil is thin the conduction effect is small.

Types of thermal insulator

Thermal insulators used in construction are made from a wide variety of raw materials and marketed under numerous trade names. These insulation products can be grouped by form under the general headings given below.

- **Rigid preformed materials**. Example: aerated concrete blocks.
- **Flexible materials**. Example: fibreglass quilts.
- **Loose fill materials**. Example: expanded polystyrene granules.
- **Materials formed on site**. Example: foamed polyurethane.
- **Reflective materials**. Example: aluminium foil.

The materials considered here are designed for the insulation of heat transfer at the relatively low temperatures of weather conditions and human comfort. There are other specialised materials for insulating against heat transfer under the high-temperature conditions encountered in boilers, furnaces and flues.

Properties of thermal insulators

When choosing materials for the thermal insulation of buildings, the physical properties of the material need to be considered. An aerated concrete block, for example, may need to be capable of carrying a load if it is supporting blocks

---

See also:

section on *Condensation* in Chapter 4

Mechanisms for heat transfer:
Conduction
Convection
Radiation
See also
Resource 2

Typical building insulators:
blocks of aerated concrete
slabs of foamed plastic
granules of foamed plastic
quilts of fibreglass
sheets of aluminium foil

above it. The properties listed below are relevant to many situations, although different balances of these properties may be acceptable for different purposes:

- thermal insulation suitable for the purpose
- strength or rigidity suitable for the purpose
- moisture resistance
- fire resistance
- resistance to pests and fungi
- compatibility with adjacent materials
- being harmless to humans and the environment.

The measurement of thermal insulation is described in the following sections. As well as resisting the passage of moisture, it is important that a material is able to regain its insulating properties after being made wet, perhaps during the construction of a building. The fire resistance of many plastic materials, such as ceiling tiles, is seriously altered by the use of certain types of paints, and therefore manufacturers' instructions must be followed. Some materials are incompatible with one another, and this should be considered when installing the materials. For example, bituminous products may attack plastic-based materials.

Thermal conductivity – lambda or *k* values

In order to calculate heat transfer and to compare different materials, we need a measurement to quantify just how well a material conducts heat.

- **Thermal conductivity** (k or λ) is a measure of the rate at which heat is conducted through a particular material under specified conditions.

  Unit: W/m K

The coefficient of thermal conductivity (k-value or lambda value), is measured as the heat flow in watts across a thickness of 1 m of material for a temperature difference of 1 degree K and a surface area of 1 m², as shown in the Resource 2 section later in the book.

- **Resistivity** (r) is an alternative index of conduction in materials and is the reciprocal of thermal conductivity, so that $r = 1/k$. Similarly, the unit of resistivity is the reciprocal of the unit k-value unit: m K/W.

  Resistivity $r = 1/k$ or r $= 1/\lambda$

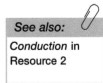

See also:

Conduction in Resource 2

Values of thermal conductivity

Values for the thermal conductivity of some typical building materials are given in Table 3.1. These values are a selection of measured values which can be used for standard calculations. It is important to remember that the exact thermal conductivity of practical building material can vary for the following reasons:

Table 3.1 Thermal conductivity of typical building materials

	Density (kg/m³)	Thermal conductivity λ or k (W/m K)
Aluminium alloy, typical	2700	190
Asphalt roofing	2100	0.70
Bitumen/felt layers	1100	0.23
Brickwork, exposed	1700	0.77
Brickwork, internal	1700	0.56
Concrete, reinforced 1% steel	2300	2.3
Concrete, high density	2400	1.93
Concrete, medium density	1800	1.13
Concrete block, lightweight aggregate	1400	0.57
Concrete block, aerated	600	0.18
Copper, commercial	9840	160
Fibreboard	400	0.1
Glass, window pane	2560	1.022
Mortar, exposed	1750	0.94
Mortar, protected	1750	0.88
Plaster, dense	1300	0.57
Plaster, lightweight	600	0.18
Insulation, expanded polystyrene (EPS) board	15	0.04
Insulation, mineral wool batt	25	0.038
Insulation, mineral wool quilt	12	0.042
Insulation, phenolic foam board	30	0.025
Insulation, polyurethane board	30	0.025
Rendering, external	1300	0.57
Screed, floor	1200	0.41
Steel, mild structural	7900	60
Stone, chippings	2000	2.0
Stone, sandstone	2600	2.6
Stone, soft limestone	1800	1.1
Timber, softwood	500	0.13
Timber, hardwood	700	0.18
Timber, plywood, chipboard	500	0.13
Tiles, ceramic wall	7900	17
Tiles, clay roof	2000	1
Tiles, concrete roof	2100	1.5

Lower k values mean better thermal insulation

Note: use manufacturers' data for precise details.

- manufacturing variations in density and thickness
- changes in moisture content
- effects of time on insulating properties.

Variations in density and production batches have significant effects on the k-values of brickwork, concrete and stone. The presence of water increases conduction, and therefore as moisture content increases the k-value increases and insulation effects decrease. The moisture content of brickwork, concrete and stone depends upon their position in the structure, and after a period of time they will have a moisture content which is adjusted to the surrounding conditions. For example, the brickwork of a sun-facing wall will have lower average moisture content than that of a shaded wall. The moisture content of internal masonry will be even lower, and Table 3.1 gives typical values for external and internal brickwork and mortar.

Some insulating products, such as expanded plastic-based materials, suffer from deterioration in performance as they age. For example, the k-value of expanded polystyrene typically changes from 0.035 to 0.04 with time. The aged lambda value takes this effect into account, and some building codes require the use of aged lambda or k-values. Manufacturers of insulating materials publish the k-values of their particular products, and when supported by recognised test certificates, these values should be used in calculations.

> Lambda value
> =
> k values under
> aged conditions

Emissivity and absorption

The ability of a material to absorb or give off radiant heat is a property of the surface of the material. Rough black surfaces absorb most heat and emit most heat. Conversely, shiny silvered surfaces both absorb and emit least heat.

To specify these properties of a surface, coefficients of emission and absorption are used. They compare the behaviour of a particular surface to a theoretically perfect absorber and emitter called a 'black body', whose coefficient is given a value of one.

> Good emitters of
> radiant heat are
> also good
> absorbers of
> radiant heat

- **Emissivity** is the fraction of energy radiated by a body compared with that radiated by a black body at the same temperature.

Similarly, the absorptivity, or absorption factor, is the fraction of radiant energy absorbed by a body compared with that absorbed by a black body.

Values of emissivity and absorptivity depend upon the wavelength of the radiation, and this is determined by the temperature of the source of the radiation. The sun is a high-temperature source of radiation and building materials are low-temperature sources, so different sets of values may be quoted for the same surface. Table 3.2 gives typical values for common building surfaces.

In general, the colour of most building materials has an important effect on the heat absorbed by the building from the sun (high-temperature radiation) but has little effect on the heat emitted from buildings (low-temperature radiation).

> **See also:**
>
> *Radiation* in
> Resource 2

Table 3.2 Surface coefficients for building materials

Surface	Emissivity (low-temperature radiation)	Absorptivity (solar radiation)
Aluminium	0.05	0.2
Asphalt	0.95	0.9
Brick – dark	0.9	0.6
Paint – black	0.9	0.9
Paint – white	0.9	0.3
Slate	0.9	0.0

Certain types of window glass have special coatings which give the glass *low emissivity* values of less than 0.2. This 'low-E' glass is designed to transmit the maximum possible amount of visible light, but also to reject the maximum amount of solar energy from outside the building, and to reflect the maximum amount of room temperature energy back into the room.

Clear-sky radiation

Clear-sky radiation is a mechanism that causes surface temperatures to fall because a clear dark sky behaves more like a theoretical black body than does a cloudy sky. The clear sky therefore acts as a better absorber of radiant heat and can be informally thought as providing radiant 'suck' upon objects on the Earth's surface. This effect can cause the structural temperature of a roof, for example, to fall significantly below the temperature of the surrounding air at night-time. A roof can therefore suffer from great thermal stress as the temperatures experienced by the structure range from those caused by sunshine to those caused by clear-sky radiation at night.

The same radiant mechanism causes dew or ground frost to occur during a clear starry night. Dew or frost commonly occurs on grass or other vegetation because the surfaces of such plants radiate heat to the clear sky faster than heat can be replaced from their surroundings. This effect does not occur when the sky is clouded over because the colour and surface of clouds do not encourage emission of heat from objects below.

INSULATION CALCULATIONS

U-values and *R*-values are used to quantify how quickly heat energy is transferred through a particular section of construction, such as a wall. This concept was introduced and used in Chapter 2, and now this section explores how these values are built up and calculated.

When heat is being transferred by conduction through a building material, the insulation effect depends upon the following two properties:

- **thermal conductivity** values (λ-value or k-value) of each building material in the construction
- **thickness** used of each building material.

The two properties work in opposite ways, in the sense that for good insulation we need minimum thermal conductivity value and maximum thickness. In practice we usually try to reduce thickness of construction by choosing insulating materials, such as expanded plastics, with low thermal conductivity. But it is possible to achieve the same insulation effect by having a large thickness of a material with relatively poor thermal conductivity, and sometimes this option occurs naturally if a building is built into the ground. The earth or rock of the ground has a high relatively thermal conductivity, comparable to concrete or brick, but that is balanced by the large 'thickness' of ground that the heat will have to flow to reach the air.

Figure 3.1 lists some common construction materials and shows the relative thickness of each material needed to achieve a similar insulation value. This comparison is valid for the *conduction* of heat inside the material. In real construction situations there are also effects of still air layers, air movements (convection) and radiation at surfaces. U-values and R-values take these effects into account, as shown in the following sections.

> Lower *U*-values mean better thermal insulation

> Higher *R*-values mean better thermal insulation

See also:

Chapter 2, Table 2.8 for typical *U*-values and *R*-values

Thermal insulation material	Relative thermal conductivity	Relative thickness to achieve same thermal insulation value (U or R)
Insulation board (EPS)	1	
Timber (softwood)	3.3	
Concrete block (aerated)	4.5	
Brickwork (external)	19	
Glass window pane	25	
Concrete (high density)	48	

Figure 3.1 Thermal insulation and thickness

Thermal resistance, *R*-value

U-values are calculated from the thermal resistances of the parts making up a particular part of a structure. The different layers and surfaces of a building element such as a wall oppose the transmission of heat by varying amounts. These differences are described by thermal resistances, or R-*values*.

- **Thermal resistance** R is a measure of the opposition to heat transfer offered by a particular component in a building element.

    Unit: $m^2\ K/W$

The idea of thermal resistance is comparable to electrical resistance. The term *conductance* (C) is sometimes used to express the reciprocal of thermal resistance, where $C = 1/R$.

There are three general types of thermal resistance which need to be determined, either by calculation or by seeking published standard values.

Material resistances

The thermal resistance of each layer of material in a structure depends on the rate at which the material conducts heat and the thickness of the material. Assuming that a material is homogeneous, this type of resistance can be calculated by the following formula.

$$R = d/k$$

where
R = thermal resistance of that component (m² K/W)
d = thickness of the material (m)
k = thermal conductivity of the material (W/m K)
Alternatively

$$R = r \times d$$

where $r = 1/k$ = *resistivity* of that material (m K/W)

> Higher *R*-values mean better insulation

Surface resistances

The thermal resistance of an open surface depends upon the conduction, convection and radiation at that surface. The air in contact with a surface forms a stationary layer which opposes the flow of heat. Surface resistances are usually found by consulting standard values that have been found by measurement or by advanced calculations. Some useful values are given in Table 3.3.

Factors that affect surface resistances are:

- **direction of heat flow**: upward or downward
- **climatic effects**: sheltered or exposed
- **surface properties**: normal building materials with high emissivity or polished metal with low emissivity.

Total thermal resistance

The thermal resistances of the consecutive layers in a structural element, such as a wall or roof, can be likened to electrical resistances connected in series. Thus the total thermal resistance is the sum of the thermal resistances of all the components in a structural element. In the case of the wall shown in Figure 3.2, the total resistance is:

$$R_T = R_{si} + R_1 + R_2 + R_a + R_{so}$$

Table 3.3 Standard thermal resistances

Type of resistance	Construction element	Heat flow	Surface emissivity[1]	Standard resistances (m^2 k/W)
Inside surfaces	Walls	Horizontal	High	0.12
			Low	0.30
	Roofs (pitched or flat) Ceilings/ floors	Upward	High	0.10
			Low	0.22
	Ceilings/ floors	Downward	High	0.15
			Low	0.56
Outside surfaces	Walls	Horizontal	High	0.06/0.04[2]
			Low	0.07
	Roofs	Upward	High	0.05
			Low	0.05
Airspaces (including boundary surfaces)	Unventilated, 5mm	Horizontal or upward	High	0.11
			Low	0.18
	Unventilated, 20mm or greater	Horizontal or upward	High	0.18
			Low	0.35
	Ventilated loft space with flat ceiling, unsealed tiled pitched roof			0.11

Notes:

1 High emissivity is for all normal building materials, including regular glass. Low emissivity is for untreated metallic surfaces such as aluminium or galvanised stell, and for specially coated glass.

2 Use the higher value for urban and sheltered sites, otherwise use the lower figure. Further standard resistances are available in publications by CIBSE, BRE, and Scottish Building Standards.

The general expression calculating total thermal resistance can be written:

$$R_T = \Sigma R$$

where ΣR is the total of the thermal resistances of all components occurring in a section through the element being calculated.

Figure 3.2
Thermal resistances

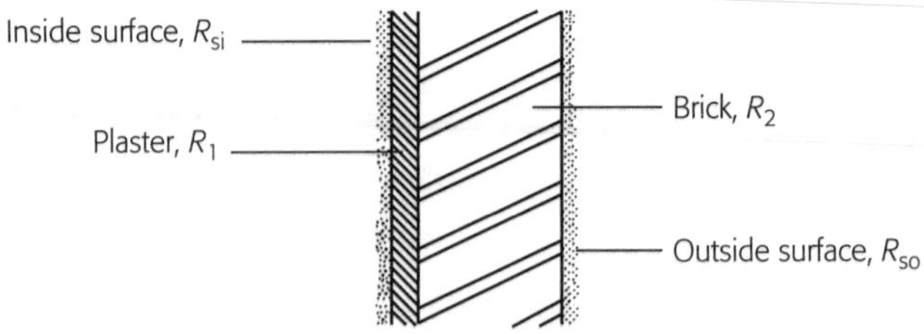

Inside surface, R_{si}

Plaster, R_1

Brick, R_2

Outside surface, R_{so}

Calculation of *U*-values

The thermal transmittance, or *U*-value, is calculated as the reciprocal of the total thermal resistance using the following formula:

$$U = \frac{1}{R_t}$$

Calculation guide

The following rules are useful in the calculation of practical *U*-values.

- Remember that there are always at least *two* surface resistances, the inside surface and the outside surface, even for the simplest element such as a window pane.
- Convert the thickness of all materials from millimetres to metres.
- Use the tables given later in the chapter to determine the *U*-values of ground floors. Only exposed floors, such as an overhang, can be calculated like the worked example.
- Work to a final accuracy of two decimal places. The *k*-values and standard resistances used in the calculations are not precise enough for higher accuracy.
- Ignore the effect of timber joists or frames, wall ties, thin cavity closures, damp-proof membranes and other thin components.
- Lay the calculation out in a table, as shown in the worked example. This method encourages clear accurate working, allows easy checking, and is also suitable for transfer to a computer spreadsheet.
- List the layers of the structure in correct order from high temperature to low temperature. Although the order makes no difference to the simple addition of resistances, it will affect the later use of the calculation for temperature profiles.

U-value calculations have simple mathematical components but can become laborious as the number of construction factors increases. However, the tabular forms of insulation calculations are easily translated to computer spreadsheets, which then make it easy to rapidly trial the effects of different types and thickness of materials.

where

$U = U$-value (W/m² K)

R_T or ΣR = sum of thermal resistances (R-values) of all components in the element.

Standard *U*-values

Previous sections have explained how variations in conditions, such as moisture content, will cause insulation properties to vary with position in a structure. The U-values for an element such as a wall or roof are calculated by using standard values for factors such as moisture content of materials and rates of heat transfer at surfaces and in cavities. Although the standard assumptions represent practical conditions as far as possible, they will not always agree exactly with U-values measured on site.

Standard U-values of elements are also needed as a common basis for comparing the effectiveness of different forms of wall, window, roof and floor, for example. The U-values for some common forms of construction element are given in Table 3.4. The table includes constructions that do not meet modern standards of insulation but are still found in existing buildings in Britain and Ireland.

Building regulations and codes often use U-values as one mechanism to specify targets and limits for thermal insulation and energy use. The various sections of this chapter show how to calculate the U-value of simple multi-layer elements, such as walls and roofs, and then consider how to treat combinations of elements such as the glass and frame components of a window.

See also:

Energy regulations,
Chapter 2

Worked Example 3.1

Calculate the U-value of a cavity wall with a 103 mm thick brick outer leaf, 50 mm of clear cavity, 40 mm of insulation board, then a 115 mm high-performance aerated concrete block inner leaf with a 15 mm layer of lightweight plaster. Values of thermal conductivity in W/m K are: lightweight plaster 0.18, aerated concrete blockwork 0.11, insulation board 0.025, brickwork 0.77. Standard thermal resistances in m² K/W are: internal surface 0.12, external surface 0.06, cavity 0.18.

Step 1

Sketch a diagram indicating all parts of the construction and surface layers. See Figure 3.3.

Step 2

Tabulate all information and, where necessary, calculate thermal resistance using formula $R = d/k$. See the example table overleaf.

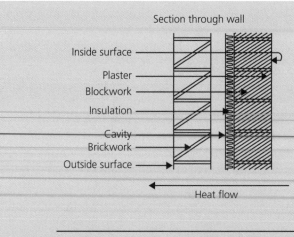

Section through wall

Inside surface

Plaster

Blockwork

Insulation

Cavity

Brickwork

Outside surface

Heat flow

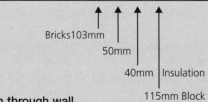

Bricks103mm

50mm

40mm | Insulation

115mm Block

Figure 3.3 Section through wall

Layer	Thickness (m)	Thermal conductivity (W/m K)	Resistance (m² K/W)	
Internal surface	n/a	n/a	Standard	= 0.120
Lightweight plaster	0.015	0.18	0.015/0.18	= 0.083
Aerated concrete block	0.115	0.11	0.115/0.11	= 1.045
Polyurethane insulation board	0.040	0.025	0.040/0.025	= 1.600
Clear cavity	0.050	n/a	Standard	= 0.180
Exposed brickwork	0.103	0.077	0.103/0.077	= 0.134
External surface	n/a	n/a	Standard	= 0.060
Total resistance, R_T				**= 3.222**

Step 3

Use U-value formula:

$$U = \frac{1}{R_t}$$

$$U = \frac{1}{3.222} = 0.310$$

Table 3.4 *U*-values of common forms of construction

Element	Composition	U-value (W/m² K)
Solid wall	brickwork, 215 mm plaster, 15 mm	2.3
Cavity wall	brickwork, 103 mm clear cavity, 50 mm brickwork, 103 mm	1.6
Cavity wall	brickwork, 103 mm clear cavity, 50 mm lightweight concrete block, 100 mm lightweight plaster, 13 mm	0.58
Cavity wall	brickwork, 103 mm insulation in cavity, 50 mm lightweight concrete block, 100 mm lightweight plasterboard, 13 mm	0.48
Cavity wall	brickwork, 103 mm clear cavity, 50 mm aerated concrete block, 115 mm insulation board, 55 mm	less than 0.30
Timber frame wall	brickwork, 103 mm clear cavity, 50 mm OSB or sheathing ply, 9mm timber frame filled with insulation, 120 mm plasterboard, 13 mm	less than 0.30
Pitched roof	tiles on battens and felt ventilated loft airspace mineral wool across joists, 100 mm mineral wool between joists, 100 mm plasterboard, 13 mm	0.15
Window	single glazing, metal frames	5.7
Window	single glazing, wood or PVC frames	4.8
Window	double glazing, wood or PVC frames, low emissivity, 6 mm gap	2.7
Window	triple glazing, wood or PVC frames, low emissivity, 12 mm gap	1.6
Door	solid wooden	3.0
Floors	See Table 3.3	

Notes:

Values are indicative only – consult manufacturers' details for exact values.

'Insulation' materials in this table are taken to have thermal conductivity values of 0.03 W/m K or less.

Non-insulating layers such as thin membranes not listed.

> *Step 4*
>
> State result with units: U-value = **0.31 W/m² K**

U-values for floors

There is no heat loss through a floor if the space below is normally heated to a similar temperature. Therefore heat exchanges through intermediate floors in blocks of apartments, for example, are negligible. For ground floors however, there is heat loss by a flow path that starts downwards but then curves through the ground back to the outside air at the edges of the building. Heat losses through the floor are therefore highest near the exposed edges of the floor where the path is shortest, and heat losses are lowest near the centre of the floor where the extra path through the ground itself adds to the insulation.

Floor insulation factors:
Distance to edge
Wall thickness
Soil type
Height above
  ground
Ventilation

The calculation of the U-values for ground floors can be made by using a standard method defined in BS EN ISO 13370 which takes account of parameters including wall thickness and soil type for solid floors, and height, ventilation and wind shelter for suspended floors. It is also common to use tables or graphs that relate insulation thickness and U-values; manufacturers of insulating materials provide tables or other calculators customised for their products. Table 3.5 gives some generalised examples for a range of insulating materials. To get a feel for the figures remember that glass fibre and expanded polystyrene products, with various trade names, have a thermal resistance of around 0.04 m² K/W, and that a target U-value for floors is 0.25 W/m² K maximum.

In calculating U-values for floors it is first necessary to calculate the floor perimeter to area ratio which is expressed by the expression

$$P/A$$

As floor area *increases* the P/A value *decreases*

where

P = floor perimeter length (m)
A = area of floor (m²)

Adjustments to U-values

It is sometimes necessary to calculate the effect that additional insulating material has on a U-value, or to calculate the thickness of material that is required to produce a specified U-value. Use the following guidelines to make adjustments to U-values.

- U-values *cannot* be added together or subtracted from one another.
- Thermal resistances, however, can be added and subtracted. The resistances making up a particular U-value can then be adjusted to produce the new U-value.

Table 3.5 *U*-values for floors

Solid floor in contact with ground		
U-values in W/m² K		
Thermal resistance of insulation (m² K/W)		
P/A	0	1	2
0.20	0.37	0.25	0.19
0.60	0.78	0.43	0.29
1.00	1.05	0.50	0.33
Solid floor in contact with ground		
Insulation thickness (mm) to achieve *U*-value of 0.25 W/m² K		
Thermal conductivity of insulant (W/m K)		
P/A	0.02	0.03	0.05
1.00	62	108	155
0.60	56	98	139
0.20	24	42	60
Suspended timber floor		
Insulation thickness (mm) to achieve *U*-value of 0.25 W/m² K		
Thermal conductivity of insulant (W/m K)		
P/A	0.02	0.03	0.05
1.00	95	140	184
0.60	85	126	166
0.20	33	52	69

Note: P/A is the ratio of the perimeter (m) to floor area (m²).

A computer spreadsheet makes easy work of adjusting *U*-values. Worked Example 3.2 illustrates the technique.

THERMAL BRIDGING

The insulation of building elements considered in previous sections has been for heat flow through simple sections where the construction is uniform and uninterrupted. However, in practice buildings have structural details such as frames and openings and junctions that interrupt insulating materials. As the layout of the element changes, from insulation to roof beam for example, then heat flow increases in that area and the overall insulation can be considered 'bridged'.

- **A thermal bridge** is a portion of a structure with higher thermal conductivity that increases heat flow and lowers the overall thermal insulation of the structure.

Worked Example 3.2

A certain uninsulated cavity wall has a U-value of 0.91 W/m² K. If insulation board is added to the construction what minimum thickness of this board is needed to reduce the U-value to 0.35 W/m² K? Given that the thermal conductivity of the insulation board is 0.025 W/m K.

Step	Calculation
Target U-value	$U_2 = 0.35$
Calculate target total resistance (using $R = 1/U$)	$R_2 = 1/0.35 = 2.857$
Existing U-value	$U_1 = 0.91$
Calculate existing total resistance (using $R = 1/U$)	$R_1 = 1/0.91 = 1.099$
Extra resistance required is $R_2 - R_1$	$R_e = 2.857 - 1.099$ $= 1.758$
k-value of proposed insulating material is $k = 0.025$	$R = d/k$
Use formula involving thickness of material d	So $d = R \times k$
Calculate thickness of material	$d = 1.758 \times 0.025$ $= 0.044$ metres $= 44$ mm

So the minimum thickness of insulating board needed to give 0.35 U-value is 44 mm. (A suitable manufactured size greater than 44 mm would be chosen.)

Even in well-insulated modern construction, thermal bridging is inevitable, and some details and areas are listed below. Some areas have *repeating thermal bridges*, such as in timber framing which is constructed at regular intervals:

- junctions of walls with roofs
- junctions of walls with floors
- mortar joints around high-performance concrete wall blocks
- timber framing between sections of wall insulation
- timber joists and beams between sections of roof and floor insulation
- steel lintels above windows and doors
- window frames and sills
- external door frames and sills.

As insulation standards have improved over recent decades the relative importance of thermal bridging has greatly increased. Earlier designs concentrated on installing thermal insulation where there had been none at all, so the prevention of thermal bridging may have seemed like an optional extra. But for modern design, with fundamentally high levels of insulation, thermal bridges become a major source of heat loss. The effects of thermal bridging can be minimised by correct design and installation of thermal insulation, and information from insulation manufacturers is a good source of practical advice.

Calculating thermal bridge effects

Manufacturers of insulation material often quote *U*-values for typical constructions which include the effects of thermal bridging. An example calculation is shown below in the section on *Combining U-values*. If the effect of thermal bridging is not easily incorporated into a combined *U*-value it can be calculated separately using the idea of linear thermal transmittance.

- The *linear thermal transmittance* (Ψ) of a thermal bridge is the rate of heat flow per degree per metre of the bridge that is not accounted for in the *U*-values of the building elements around the thermal bridge.

    Unit: W/m K   Symbol: Ψ

The extra heat loss (in watts) caused by the thermal bridge is calculated by multiplying the linear thermal transmittance by the length of the thermal bridge by the difference in temperature between each side of the bridge. This calculation can be used as a correction factor or can be placed within a standard heat loss calculation shown in Chapter 3. Some typical figures for linear thermal transmittance values are given in Table 3.6, taken from limiting values in building regulations.

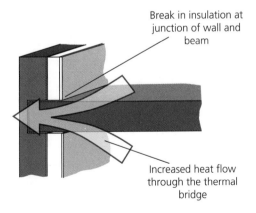

Break in insulation at junction of wall and beam

Increased heat flow through the thermal bridge

Figure 3.4 Thermal bridge

Table 3.6 Linear thermal transmittance values

Maximum values of linear thermal transmittance (Ψ) for selected locations	
Junction or detail in external wall	Maximum value Ψ (W/m K)
Steel lintel with perforated steel base plate	0.50
Other lintels	0.30
Window/door sills and jambs	0.06
Ground or intermediate floor, party wall	0.16
Eaves (ceiling level)	0.06
Gable (ceiling level)	0.24

Cold bridges

See also:

section on
Condensation in
Chapter 4

A distinctive thermal bridge, such as an uninsulated lintel above a window, conducts heat away more quickly than surrounding areas, and the surfaces on the interior side of the bridge therefore have a lower temperature. This type of thermal bridge is informally termed a 'cold bridge'. There is an increased risk of condensation and mould growth on the internal surfaces on and around such cold bridges.

Pattern staining

Pattern staining on a ceiling is the formation of a pattern, in dirt or dust, which outlines the hidden structure of the ceiling. It is a particular result of thermal bridging and also depends upon the frequency of redecoration. The areas of lower insulation transmit heat at a higher rate than the surrounding areas, and the resulting convection activities then cause a greater accumulation of airborne dirt on these areas. There is a risk of pattern staining if the difference between surface temperatures exceeds 1°C. This condition is unlikely in a modern well-insulated roof, but it can be seen in older buildings and it can be a useful indicator that the roof needs upgrading.

Combining *U*-values

Average U-values

If a wall or other element is composed of different constructions with different *U*-values, then the overall insulation of the wall depends upon the relative areas of the different constructions. For example, if a wall contains two-thirds brickwork and one-third windows, then the *U*-values of the brickwork and windows must be combined in the proportions of 2/3 and 1/3 to give the average *U*-value for the wall.

The general formula is as follows:

$$U \text{ (average) } = \frac{A_1 U_1 + A_2 U_2 + \ldots}{A_1 + A_2 + \ldots}$$

where A_1, A_2, etc. are the areas with the U-values U_1, U_2, etc.

Worked Example 3.3

A three-bedroom semi-detached house has a total window area of 20 m² with a mixture of 16 m² double glazing and 4 m² single glazing. The U-values are 1.8 W/m² K for the double glazing and 2.8 W/m² K for the single glazing. Calculate the average U-value for the windows.

Step	Calculation
Known values	$U_1 = 1.8$ $A_1 = 16$
	$U_2 = 2.8$ $A_2 = 4$
Unknown value	U (average) = ?
Use area weighted formula	$U = \dfrac{A_1 U_1 + A_2 U_2}{A_1 + A_2}$
Substitute in formula	$U = \dfrac{(16 \times 1.8) + (4 \times 2.8) = \dfrac{28.8 + 11.2}{20}}{16 + 4}$ $U_1 = 2.00$
So average U-value of windows = 2.00 W/m² K	

Proportional area calculation

To calculate the effect of thermal bridges, a proportional method can also be used. Because the bridging might take place in one layer of the construction, then a proportional method can be applied to values of thermal resistance before a final U-value is calculated. An example is shown in Worked Example 3.4 (overleaf).

Combined method calculation

Some building regulations in the United Kingdom refer to an international standard (BS EN ISO 6946) that defines a 'combined method' of calculating an overall insulation value for areas of mixed construction. The method is suitable for building elements containing repeating thermal bridges, such as those found in timber joists between insulation, or mortar joints around lightweight concrete wall block. A U-value is calculated by the combined method using the following steps.

1. Calculation of the upper resistance limit: the total resistance, combined in parallel, of all possible paths of heat flow through the plane building element.
2. Calculation of the lower resistance limit: the total resistance, combined in parallel, of the paths of heat flow through each layer separately.
3. Calculation of total resistance using

$$R_T = \frac{R_{upper} + R_{lower}}{2}$$

4. Calculation of the U-value using $U = 1/R_T$.

Worked Example 3.4

An external wall has the following construction: outer leaf of brickwork; air cavity; inner leaf of ply sheathing, timber stud frame filled with insulation, plasterboard.

The inner leaf has a mixed construction with lower heat flow through the insulation areas and higher heat flow through the timber stud areas. Therefore the thermal resistance is calculated separately for each part of the inner leaf, as in the following example.

Resistance through section containing timber stud	m² K/W
Inside surface resistance	= 0.12
Resistance of plasterboard = 0.013/0.16	= 0.08
Resistance of timber stud = 0.09/0.14	= 0.64
Resistance of ply sheathing = 0.009/0.14	= 0.06
Half cavity resistance = 1/2 (std 0.18)	= 0.09
Total resistance through timber stud section R_1	= 0.99

Resistance through section containing insulation	m² K/W
Inside surface resistance	= 0.12
Resistance of plasterboard = 0.013/0.16	= 0.08
Resistance of timber stud = 0.09/0.14	= 2.25
Resistance of ply sheathing = 0.009/0.14	= 0.06
Half cavity resistance = 1/2 (std 0.18)	= 0.09
Total resistance through insulation section R2	= 2.60

If the timber studs occur after every 600 mm of insulation and the studs are 38 mm wide, then the fractional areas are:

For timber stud $\quad F_1 = 38/600 \quad\quad = 0.063$
For insulation $\quad\quad F_2 = 1 - 0.063 \quad = 0.937$

The resistance of the inner leaf is obtained from using the partial resistances and corresponding fractional areas in the following formula:

$$\frac{1}{R_3} = \frac{F_1}{R_1} + \frac{F_2}{R_2}$$

$$\frac{1}{R_3} = \frac{0.063}{0.99} + \frac{0.937}{2.6}$$

$$\frac{1}{R_3} = 0.064 + 0.36 = 0.424$$

$$\frac{1}{R_3} = \frac{1}{0.424} = 2.36$$

So resistance of the inner leaf is 2.36 m² K/W.

This resistance for the inner leaf (and half the cavity) can then be added to the resistance for the outer leaf (including half the cavity) to give the total resistance for the wall. For example, if the resistance of the outer leaf has a typical value of 0.30 m² K/W, then

$$R_{total} = R_{inner} + R_{outer}$$
$$= 2.36 + 0.30$$
$$= 2.66 \text{ m}^2 \text{ K/W}$$

so

$$U = 1/R = 1/2.66 = 0.38$$

So the U-value for the wall is **0.38 W/m² K.**

STRUCTURAL TEMPERATURES

The thermal insulation installed in a building affects the rate at which the building loses heat energy, which is measured by the U-value. The thermal performance of the building also depends on the thermal capacity of the insulating material, which affects the times taken to heat or cool the structure, and on the position of the insulation, which affects the temperatures in the structural elements.

Response times

It is possible for two types of wall to have the same thermal insulation, as measured by U-values, but to absorb or dissipate heat at different rates. As a result the

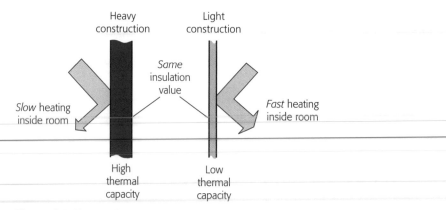

Figure 3.5 Thermal response

rooms inside such walls would take different times to warm or to cool, as indicated in Figure 3.5.

In general, lightweight structures respond more quickly to surrounding temperature changes than do heavyweight ones. This is because heavyweight materials have a higher thermal capacity and require more heat energy to produce given temperature changes. These effects are described further in Chapter 2.

Heavy materials: slow response *Light materials*: faster response

The storage properties and slow temperature changes of heavyweight materials, such as concrete and brick, can be useful if thoughtfully designed. Where quick heating is often required, in bedrooms for example, a lightweight structure may be more useful. Insulating the internal surfaces, for instance by carpeting a concrete floor, can also reduce the heating-up time of heavyweight construction.

Temperature gradients

A temperature difference between the inside and outside of a wall or roof causes a progressive change in temperature from the warm side to the cold side. This *temperature gradient* changes uniformly through each component, provided that the material is homogeneous and that the temperatures remain constant.

Steeper graph gradient = Larger temperature difference = Higher condensation risk

A structure made up of different materials, such as the wall shown in Figure 3.6, will have varying temperature gradients between the inside and outside. The layers with the highest thermal resistances will have the steepest gradients. This is because the best insulators must have the greatest temperature differences between their surfaces.

The boundary temperatures between layers in a structural element can be determined from the thermal resistances that make up the *U*-value of that element.

- The ratio of the temperature changes inside a structure is proportional to the ratio of the thermal resistances.

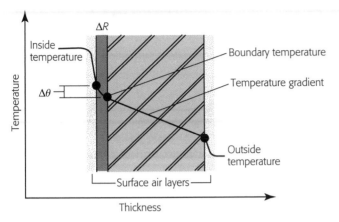

Figure 3.6 Temperature gradients through a wall

The most useful relationship is given by the following formula:

$$\frac{\Delta\theta}{\theta_T} = \frac{R}{R_T}$$

where
  $\Delta\theta$ = temperature difference across a particular layer
  R = resistance of that layer
  θ_T = total temperature difference across the structure
  R_T = total resistance of the structure.

From this relationship the temperatures at boundaries in a structure can be predicted for specified conditions. A calculation of all the boundary temperatures allows temperature gradients to be drawn on a scaled diagram. These diagrams can be used to predict zones of possible condensation, as shown in Chapter 5.

See also:

section on
*Condensation
conditions* in
Chapter 4

Worked Example 3.5

A room has an external wall with a U-value of 1.5 W/m K and contains air at 20°C when the outside temperature is 5°C. The internal surface resistance is 0.12 m² K/W. Calculate the boundary temperature on the internal surface of the wall.

Temperature drop across internal surface layer $\Delta\theta$ = ?

Total temperature drop across wall $\theta_T = 20 - 5 = 15$

Resistance of internal surface layer $R_{si} = 0.12$

Total resistance of the wall $R_T = 1/U$
$$= 1/1.5$$
$$= 0.667$$

Using the ratio formula and substituting

$$\frac{\Delta\theta}{\theta_T} = \frac{R}{R_T}$$

$$\frac{\Delta\theta}{15} = \frac{0.120}{0.667}$$

$$\Delta\theta = \frac{0.120}{0.667} \times 15 = 2.699$$

Temperature drop across internal surface layer $= 2.7°C$
So temperature on the inside surface $= 20 - 2.70 = \mathbf{17.3°C}$

HEAT ENERGY CALCULATIONS

Calculation of heat loss

Various methods are available for calculating the rate at which heat flows out of a building, and the quantity of heat lost in a given time. It is relatively difficult to calculate heat losses for unsteady or *cyclic* conditions where temperatures fluctuate with time. However, certain simplified calculations can be used for predicting heating requirements and the amount of fuel required. The results obtained by these calculations are found to give adequate agreement with the conditions that actually exist.

With *steady state* conditions the temperatures inside and outside the building do not change with time and the various flows of heat from the building occur at constant rates. Assuming steady state conditions, the heat losses from a building can be classed as either a 'fabric loss' or a 'ventilation loss', then calculated by the methods described below. Conditions not suited to the steady state assumption occur when there is intermittent heating, such as that given by electric storage heaters or by solar radiation. The thermal capacity of the structure is then significant and needs to be considered in the calculation, as explained in the section about *thermal admittance*.

The calculation procedures shown below have been kept in a format which makes the link between the principles of heat loss, discussed earlier, and the practical calculations as obvious as possible. Specialist calculations of energy balances, such as those within the SAP energy rating procedures (see page 38), can take account of more factors than those used here. These procedures, which usually involve the input of many items of data into complex tables or computer programs, produce refined results but also lose the visible links between each factor and the heat losses they are associated with.

Fabric heat loss

Fabric heat loss from a building is caused by the transmission of heat through the materials of walls, roofs and floors. Assuming steady state conditions, the heat loss for each element can be calculated by the following formula.

$$P_\mathrm{f} = U A \, \Delta t$$

where
P_f = rate of fabric heat loss = heat energy lost/time (W)
U = U-value of the element considered (W/m² K)
A = area of that element (m²)
Δt = difference between the temperatures assumed for the inside and outside environments (°C).

The heat loss per second is a form of power (energy divided by time) and therefore measured in watts (which are joules per second). The notation P is used here to represent this rate of heat energy.

To calculate daily heat losses, appropriate temperatures would be the internal environmental temperature and the outside environmental temperature, both averaged over 24 hours. For the calculation of maximum heat losses, such as when choosing the size of heating equipment, it is necessary to assume a lowest external design temperature, such as -1°C.

Ventilation loss

Ventilation heat loss from a building is caused by the loss of warm air and its replacement by air that is colder and has to be heated. The rate of heat loss by such ventilation or infiltration is given by the following formula:

$$P_v = \frac{c_v \, N \, V \, \Delta t}{3600}$$

where
P_v = rate of ventilation heat loss
= heat energy/time (W)
c_v = volumetric specific heat capacity of air
= specific heat capacity (density (J/m³ K)

N   = air infiltration rate for the room (the number of complete air changes
       per hour, ach)

V   = volume of the room (m³)

Δt = difference between the inside and outside air temperatures (°C).

Again, the heat loss per second, being a form of power (energy divided by time), is measured in watts (joules per second). The notation P is used here to represent this rate of heat energy. The specific heat capacity for a gas depends upon density and temperature but in practice a set figure is used, which leads to the following alternative formula.

Seconds in 1
hour = 3600

$$P_v = 0.33\, N\, V\, \Delta t$$

To calculate daily heat losses, appropriate temperatures would be the mean internal air temperature and the mean external air temperature, both averaged over 24 hours. For the calculation of maximum heat losses, for example when choosing the size of heating equipment, it is necessary to assume a lowest design temperature for the external air, such as -1°C.

External temperature

When designing heating systems for buildings it is necessary to assume a temperature for the outside environment. For winter heating an overcast sky is assumed and the outside air temperature can be used for design purposes. For heat transfer calculations in summer, it is necessary to take account of solar radiation as well as air temperature.

Worked Example 3.6

A window measuring 2 m by 1.25 m has an average U-value, including the frame, of 6.2 W/m² K. Calculate the rate of fabric heat loss through this window when the inside comfort temperature is 20°C and the outside air temperature is 4°C.

Known:
$U = 6.2$ W/m² K, $A = 2 \times 1.25 = 2.5$ m², $\Delta t = 20 - 4 = 16$°C.

Using formula for ventilation loss and substituting:

$$P_t = U A \Delta t$$
$$= 6.2 \times 2.5 \times 16 = 248$$

So fabric loss = **248 W**

Worked Example 3.7

A simple building is 4 m long by 3 m wide by 2.5 m high. In the walls there are two windows, each 1 m by 0.6 m, and there is one large door 1.75 m by 0.8 m. The construction has the following U-values in W/m² K: windows 5.6, door 2.0, walls 2.5, roof 3.0, floor 1.5. The inside environmental or comfort temperature is maintained at 18°C while the outside air temperature is 6°C. The volumetric specific heat capacity of the air is taken to be 1300 J/m K. There are 1.5 air changes per hour. Calculate the total rate of heat loss for the building under the above conditions.

Step 1

Sketch the building, with its dimensions, as in Figure 3.7. Calculate the areas and temperature differences.

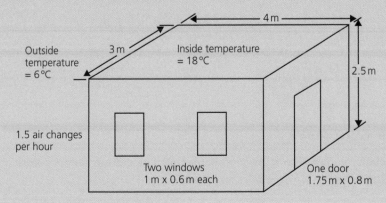

Outside temperature = 6 °C

3 m

Inside temperature = 18 °C

4 m

2.5 m

1.5 air changes per hour

Two windows 1 m x 0.6 m each

One door 1.75 m x 0.8 m

An actual building can be simplified into this type of 'shed' format to get a first estimate of heat losses

Figure 3.7 Sketch of building

Step 2

Tabulate the information and calculate the rate of fabric heat losses using $P_v = U A \Delta t$ (see table).

Element	U-value (W/m² K)	Area (m²)	Temperature difference (°C)	Rate of heat loss (W)
Windows	5.6	1.2	12	80.64
Door	2.0	1.4	12	33.6
Walls	2.5	3-5–2.6	12	972
Roof	3.0	12	12	432
Floor	1.5	12	12	216
Total rate of fabric heat loss				**1734.24 W**

Step 3

Calculate the ventilation heat loss.

$$c_v = 1\,300 \text{ J/m}^3 \text{ K}, \qquad\qquad N = 1.5 \text{ h}^{-1}$$

$$V = 4 \times 3 \times 2.5 = 30 \text{ m}^3, \qquad \Delta t = 18 - 6 = 12°\text{C}$$

Using

$$Pv = \frac{C_v\, N\, V\, \Delta t}{3600} = \frac{1\,300 \times 1.5 \times 30 \times 12}{3600}$$

So rate of ventilation heat loss = **195 W**

Step 4

Total rate of heat loss = fabric heat loss + ventilation heat loss

$$= 1\,734.24 + 195$$

$$= \textbf{1\,929.24 W}$$

Sol-air temperature t_{eo}

The sol-air temperature is an environmental temperature for the outside air which includes the effect of solar radiation.

- The rate of heat flow due to the *sol-air temperature* is equivalent to the rate of heat flow due to the actual air temperature combined with the effect of solar radiation.

Sol-air temperature varies with climate, time of day and incident radiation. Values can be calculated or found from tables of average values.

Non-steady conditions

The calculations so far have assumed that internal and external temperatures remain constant. For situations where this *steady state* assumption is invalid, it is necessary to consider the effects of cyclic (daily) variations in the outside temperature, variations in solar radiation, and changes in the internal heat input.

Thermal admittance and thermal mass

The thermal admittance value is one method of describing the effect of variations in temperature.

- **Thermal admittance** or Y-value describes the ability of a material or construction element to exchange heat when subject to variations in temperature, for example over a 24-hour period.

    Unit: W/m² K

> Diurnal variation = daily

The greater the admittance, the smaller the temperature swing inside the building. The factors that affect the admittance values of a particular material include thermal conductivity and specific heat capacity. Methods of calculation using admittance values are available. Dense heavyweight materials, such as concrete, have larger admittance values than the values for less dense materials, such as lightweight thermal insulating materials. Therefore, heavyweight structures have smaller temperature swings than lightweight structures. See Figure 3.8.

Although the unit for admittance value is the same as for *U*-value, it is possible to have elements of different construction with identical insulation measured in *U*-values but different thermal damping properties as measured in admittance values. Table 3.7 shows some examples. For very thin units, such as glass, the admittance becomes the same as the *U*-value.

> *Heavier materials give slower and lower temperature swings*

- **Thermal mass parameter**, PMT, is an overall figure for the thermal mass of a building that is derived from the admittance of individual elements.

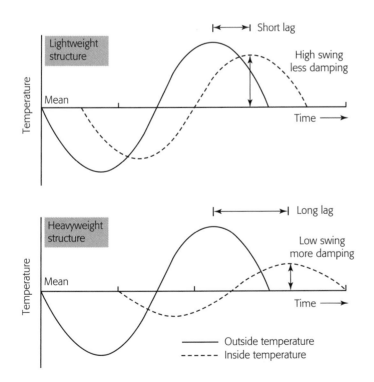

Figure 3.8 Thermal response with time

Table 3.7 Contrasting values of admittance and transmittance

Element	Y-value (W/m² K)	U-value (W/m² K)
Typical heavyweight wall (brick/blockwork with cavity insulation)	4.0	0.6
Typical lightweight wall (cladding, insulation, lining)	1.0	0.6

Typical values for PMT range from: 5 W/m² K for a lightweight building constructed with a suspended timber floor, and framed walls with plasterboard, to 19 W/m² K for a heavyweight building constructed with a solid floor and masonry walls with dense plaster.

Calculation of heat loss

As discussed in Chapter 2, the factors affecting heat gains in buildings can be considered and calculated under the following two broad categories.

- solar heat gains – from the sun
- casual heat gains – from occupants and equipment in the building.

The rate at which heat from the sun falls on a surface varies throughout the day and the year. Figure 3.9 gives an indication of the intensity of solar radiation on a vertical and horizontal surface at different times. The figures are for London (latitude 51.7°N) and assume cloudless days. The heat gains for other surfaces and locations can be calculated from appropriate tables and charts.

Most solar heat gain to buildings in the United Kingdom is by direct radiation through windows. The maximum gains through south-facing windows tend to occur in spring and autumn, when the lower angle of the sun causes radiation to fall more directly onto vertical surfaces. This heat gain via windows can, if used correctly, be useful for winter heating.

The fabric solar heat gains through walls and roofs are considered negligible for masonry buildings during the UK winter. Little solar heat reaches the interior of the buildings because the high thermal capacity of heavyweight construction tends to delay transmission of the heat until its direction of flow is reversed with the arrival of evening.

See also:

section on *Solar data* in Chapter 14

The solar heat gains for a particular building at a specific time are relatively complicated to calculate, although it is important to do so when predicting summer heat gains in commercial buildings. For winter calculations however, it is useful to consider the total solar gain over an average heating season. Typical figures for solar heat gain through the windows of typical buildings in Britain are given in Table 3.8.

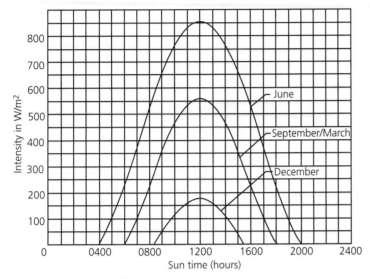

(a) Horizontal surface

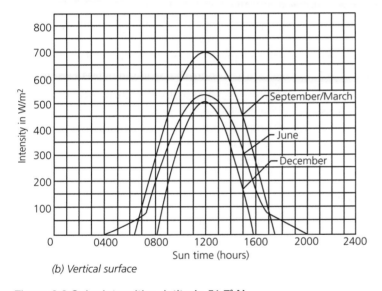

(b) Vertical surface

Figure 3.9 Solar intensities, latitude 51.7° N

Table 3.8 Seasonal solar gain through windows

Type of window (vertical unobstructed)	Average solar flux	Seasonal total heat gain
South-facing windows	72 W/m² glass	680 MJ/m² glass
East and west-facing windows	48 W/m² glass	410 MJ/m² glass
North-facing windows	29 W/m² glass	250 MJ/m² glass
Total for average semi-detached house		7500 MJ

Note: Seasonal figures are for 33 weeks.

Sun controls

Sun controls are parts of a building that help prevent excessive heat gain and glare caused by direct sunshine. The main types of device are described below:

- *External controls* are the most effective form of sun control because they minimise the radiant heat reaching the fabric of the building. Examples include external shutters, awnings, projecting eaves or floor slabs.
- *Internal controls* such as blinds give protection against glare and direct radiation. The system is less effective than controls outside the glass because the blind will absorb some solar heat and re-emit this heat into the room. Examples include curtains, blinds and internal shutters.
- *Special glasses* are available which prevent the transmission of most heat radiation with only some loss of light transmission. A similar effect is given by special film which sticks onto plain glass.

Casual heat gains

Casual heat gains take account of the heat given off by various activities and by equipment in a building that is not primarily designed to give heat. The major sources of such heat are:

- heat from people
- heat from lighting
- heat from cooking and water heating
- heat from machinery, refrigerators, electrical appliances.

In commercial or public buildings this type of heat gain must be allowed for in the design of the heating/cooling system. Where possible this heat should be used rather than wasted. Table 3.9 gives typical rates of heat output from various sources.

In houses the casual heat gain is useful in winter, and as dwellings become better insulated, it forms a higher proportion of the total heat needed. The combined heat output of the various sources varies from hour to hour, so it is

See also:

section on *Lighting design* in Chapter 6

Table 3.9 Heat emissions from casual sources

Type of source		Typical heat emission
Adult person (for 20°C surroundings)	Seated at rest	90 W
	Walking slowly	110 W
	Medium work	140 W
	Heavy work	190 W
Lighting	Fluorescent system giving 400 lux (e.g. class-room)	20 W/m² floor area
	Tungsten system giving 200 lux (e.g. domestic kitchen)	40 W/m² floor area
Equipment	Desktop computer	150 W
	Computer printer	100 W
	Photocopier	800 W
	Small Bunsen burner	600 W
	Hair dryer	800 W
	Gas cooker	3500 W per burner
	Domestic fridge-freezer	150 W
	Television or hi-fi unit	100 W

again useful to consider the total heat gain over a standard heating system. Table 3.10 quotes estimates of the heat gains in a typical semi-detached house, with a floor area of 100 m², totalled over a heating season of 33 weeks.

Table 3.10 Domestic seasonal heat gains

Source	Typical gain per heating season
Body heat (per person)	1000 MJ
Cooking (gas)	6500 MJ
Cooking (electric)	3500 MJ
Water heating	2000 MJ
Electricity (including lights)	3000 MJ

Notes:
Figures for semi-detached house with 100 m².
Seasonal results are for 33 weeks.

ENERGY BALANCE

The thermal comfort of humans requires that the inside temperature of a building is kept constant at a specified level. The storage of most goods also needs specified temperatures, and to provide these conditions the building will generally require heating or cooling. Both of these processes involve the consumption of energy.

Calculation of energy

Heat is defined as a form of energy. Power is defined as the rate (divided by time) at which energy is used. The rate at which heat energy is used, measured in watts, may therefore be called a power loss. These power losses in watts are sometimes, and less accurately, called 'heat losses'.

The true heat or energy use can only be determined when it is decided which period of time is being considered. The quantity of energy used over a given period depends upon both the power (rate of using energy) and upon the time involved.

By definition

$$\text{Energy} = \text{Power} \times \text{Time}$$

Written as a formula

$$E = Pt$$

Unit: joule (J), where 1 joule = 1 watt × 1 second

There are some other units of energy which may be found in use:

- kilowatt hour (kWh), where 1 kWh = 3.6 MJ
- British Thermal Unit (BTU), where 1 BTU = 1.055 kJ

Energy balance

When heat losses and heat gains have been determined it is possible to calculate the extra energy needed to 'balance' the losses and the gains and to give a constant temperature. The following is a general expression of balance which is true for summer and winter conditions:

$$\text{Fabric heat losses} + \text{Ventilation heat losses} = \text{Solar heat gains} + \text{Casual heat gains} + \text{Energy for heating or cooling}$$

The quantities can be measured as true heat energy in joules or as rate of energy use in watts, but not by a mixture of methods. For most buildings in winter the 'useful energy' which needs to be supplied by the heating plant is given by the following expression:

Energy needed = Heat losses – Heat gains

Seasonal energy requirements

The energy requirement of a building at any particular time depends on the state of the heat losses and the heat gains at that time. These factors vary but, as described in the previous sections, it is useful to consider the total effect over a standard heating season.

It is important to note that the calculation of seasonal heat losses and gains assumes average temperature conditions and cannot be used to predict the size of the heating or cooling plant required; such a prediction needs consideration of the coldest and hottest days. Seasonal heat calculations are, however, valid for calculating total energy consumption and can therefore be used to predict the quantity of fuel required in a season and how much it will cost.

Worked Example 3.8

Over a heating season of 33 weeks, the average rate of heat loss from a certain semi-detached house is 2500 W for the fabric loss and 1300 W for the ventilation loss. The windows have areas: 6 m² south-facing, 5 m² east-facing, 6 m² north-facing. The house is occupied by three people and cooking is by gas.

Use the values for seasonal heat gains given in Tables 3.9 and 3.11 and calculate:

(a)   the seasonal heat losses
(b)   the seasonal heat gains
(c)   the seasonal heat requirements.

(a)   Total rate of heat loss   = fabric loss + ventilation loss

$$= 2\,500\ W + 1\,300\ W$$

$$= 3\,800\ W$$

Heat energy lost = rate of heat loss × time taken

$$= 3\,800\ W \times (33 \times 24 \times 7 \times 60 \times 60)\ s$$

$$= 7.5842 \times 10^{10}$$

$$= 75.842\ GJ$$

So seasonal heat loss = **75 842 MJ**

(b) Heat gains

Solar window gain (Table 3.8)	MJ
south (680 MJ/m² × 6)	4080
east (410 MJ/m² × 5)	2050
north (250 MJ/m² × 6)	1500

Casual gains (Table 3.10)	MJ
body heat (1 000 MJ × 3)	3000
cooking (gas)	6500
water heating	2000
electrical	3000
	22130 MJ

So seasonal heat gain = **22 130 MJ**

(c) Seasonal heat requirements = Heat loss – Heat gain

= 75 842 – 22 130 = 53 712 MJ

= **53.712 GJ**

Efficiency

The heat energy required for buildings is commonly obtained from fuels such as coal, gas and oil, even if the energy is delivered in the form of electricity. Each type of fuel must be converted to heat in an appropriate piece of equipment and the heat distributed as required. The amount of heat finally obtained depends upon the original heat content of the fuel and the efficiency of the system in converting and distributing this heat.

Efficiency is a measure of the effectiveness of a system that converts energy from one form to another. The efficiency index or percentage is calculated by comparing the output with the input. The maximum value is 100 per cent although most practical processes are well below the maximum.

$$\frac{\text{Efficiency \%}}{100} = \frac{\text{Output energy}}{\text{Input energy}} = \frac{\text{Useful energy}}{\text{Delivered energy}}$$

or   Input energy = Output energy × 100 / Efficiency %

In general, the useful energy is the output energy from the system which is used to balance the heat losses and heat gains. The delivered energy is the input energy needed for the boiler, or other device, and it is the energy which is paid for.

Boiler efficiency

The efficiency of a boiler is dependent on the boiler load at a particular time. Boilers are generally more efficient when they are working at higher loads, but most installed boilers work at a load which is lower than their maximum design load. An average figure of boiler efficiency is used.

- SEDBUK (Seasonal Efficiency of Domestic Boilers in the UK) is the average annual efficiency of a boiler achieved in typical domestic conditions.

This figure makes standard assumptions about the pattern of usage, climate control and other influences. High-efficiency boilers, such as condensing boilers, are typically in a SEDBUK range of 80 to 90 per cent. The latest methodology for calculating the figure is indicated by a year in the title, such as *SEDBUK (2009)*.

System efficiency

The overall efficiency of a system depends upon how much of the heat is extracted from the fuel, how much heat is lost through the flue and how much heat is lost in the distribution system. The house efficiency is an approximate figure for domestic systems that takes all these effects into account. Some typical values are quoted in Table 3.11 and used in the simplified calculation of Worked Example 3.9. The SAP energy rating described in the next section combines various factors affecting services efficiency in the building to give a more sophisticated result.

Table 3.11 Domestic heating efficiencies

Type of system	House efficiency
Central heating (gas, oil, solid fuel)	60–80%
Gas radiant heater	50–60%
Gas convector heater	60–70%
Electric fire	100%

Note: Although the efficiency of an electrical heating appliance is 100% within the building, the overall efficiency of the generation and distribution system is about 30%.

Worked Example 3.9

The seasonal heat requirement of a house is 54 GJ, which is to be supplied by a heating system with an overall house efficiency of 67 per cent. The solid fuel used has a calorific value of 31 MJ/kg. Calculate the mass of fuel required for one heating season.

Know:

$$\text{Efficiency} = \frac{67}{100}, \text{ Output energy} = 54 \text{ GJ, Input energy} = ?$$

Using efficiency formula and substituting

$$\frac{\text{Efficiency \%}}{100} = \frac{\text{Useful energy (system output)}}{\text{Delivered energy (boiler input)}}$$

$$\frac{67}{100} = \frac{54}{\text{Input energy}}$$

$$\text{Input energy} = 54 \times \frac{100}{67} = 80.597 \text{ GJ} = 80{,}597 \text{ MJ}$$

$$\text{Mass of fuel needed} = \frac{\text{Energy required}}{\text{Calorific value}} = \frac{80{,}597 \text{ MJ}}{31 \text{ MJ/kg}} = 2599.9$$

$$= \mathbf{2600 \text{ kg}}$$

EXERCISES

Suggestion: Spreadsheets are a useful way to do the calculations in some of the exercises below. One option is to use the Worked Examples 3.1 and 3.2 as initial layouts for your spreadsheets.

1. Choose three different insulating materials used in modern buildings. List the physical properties of each material and use the principles of heat transfer to explain why the material acts as a good thermal insulator. Use construction sketches to help describe where and how the material is installed in a building.

2. Use definitions, units and examples to explain the difference between the following terms:
   (a) thermal conductivity
   (b) thermal resistance
   (c) thermal transmittance.

3. A cavity wall is constructed as follows: brickwork outer leaf 103 mm, air gap 50 mm, expanded insulation board 60 mm, high performance concrete block inner leaf 100 mm, lightweight plasterboard 10 mm. The relevant values of thermal conductivity, in W/m K, are: brickwork 0.84, polystyrene 0.025, concrete block 0.11, plasterboard 0.18. The

standard thermal resistances, in m² K/W are: outside surface 0.06, inside surface 0.12, air gap 0.18. Calculate the U-value of this wall.

4. The cavity wall of an existing house has outer and inner brickwork leaves each 105 mm with a 50 mm air gap between them, finished with a 16 mm layer of plaster inside. The relevant values of thermal conductivity, in W/m K are: brickwork 0.73, plaster 0.46. The standard thermal resistances, in m2 K/W, are: outside surface 0.055, inside surface 0.123, air gap 0.18.

    (a) Calculate the U-value of the existing wall.

    (b) Calculate the U-value of the wall if the cavity is completely filled with foamed urea formaldehyde (k = 0.026 W/m K).

5. Compare the U-values obtained in Questions 3 and 4 with the U-values required by current building regulations applicable to your area. Comment on the suitability of the walls for different purposes.

6. Compare the U-values of a single-glazed window made up of one sheet of 4 mm glass with a double-glazed window made up of two sheets of 4 mm glass which have a 5 mm airspace between them. The thermal conductivity of the glass is 1.022 W/m K. The standard thermal resistances, in m² K/W, are: outside surface 0.06, inside surface 0.12, airspace 0.11. Comment on the proportion of the thermal resistance provided by the glass layers. Comment on the effect of the window frames.

7. A blockwork wall measures 5 m × 2.8 m in overall length and height. The wall contains one window 1400 mm by 800 mm and one door 1900 mm by 750 mm. The U-values, in W/m² K, are: blockwork 0.58, window 5.6, door 3.4. Calculate the average U-value of this wall.

8. A wall panel is to have the following construction: outer metal sheeting, foamed polyurethane board, airgap, and 15 mm of lining board. The relevant values of thermal conductivity, in W/m K, are polyurethane board 0.025, and lining board 0.16 (the metal sheeting is ignored). The standard thermal resistances, in m² K/W are: outside surface 0.06, inside surface 0.12, airgap 0.18. Calculate the minimum thickness of polystyrene needed to give the wall panel a U-value of 0.35 W/m² K.

9. A domestic pitched roof of tiles on felt sacking, with a plasterboard ceiling, has an existing U-value of 1.9 W/m² K. Calculate the minimum thickness of fibreglass insulation in the roof space required to give the roof a new U-value of 0.25 W/m² K. The thermal conductivity of the fibreglass quilting used is 0.04 W/m K.

10. A wall has a U-value of 2.5 W/m² K. The thermal resistance of the inside surface layer is 0.123 m² K/W. The inside air temperature is 18°C and the outside air temperature is 0°C. Calculate the temperature on the inside surface of the wall.

11. The external wall of a room measures 4.8 m by 2.6 m and has an average U-value 1.8 W/m² K. The internal air temperature is 21°C, the mean radiant temperature is 18°C, and the external air temperature is 0°C.
    (a) Calculate the environmental temperature inside the room.
    (b) Use the environmental temperature to calculate the rate of heat loss through the wall.

12. A house has a floor area of 92 m² and a ceiling height of 2.5 m. The average inside air temperature is kept at 18°C, the outside air temperature is 6°C, and the average infiltration rate is 1.5 air changes per hour. The volumetric specific heat capacity of the air is 1300 J/m³ K.
    (a) Calculate the rate of ventilation heat loss.
    (b) Calculate the cost of the heat energy lost during 24 hours if the above conditions are maintained and replacement heat costs 12 pence per megajoule.

13. A sports pavilion has internal dimensions of 11 m × 4 m × 3 m high. 20 per cent of the wall area is glazed and the doors have a total area of 6 m². The U-values in W/m K are: walls 1.6, windows 5.5, doors 2.5, roof 1.5, floor 0.8. The inside air temperature is maintained at 18°C when the outside air temperature is -2°C. There are four air changes per hour and the volumetric specific heat capacity of air is 1300 J/m³ K. The heat gains total 2 200 W.
    (a) Calculate the net rate of heat loss from this building.
    (b) Calculate the surface area of the radiators required to maintain the internal temperature under the above conditions. The output of the radiators is 440 W/m² of radiating surface area.

14. A room has 7.5 m² area of single-glazed windows, which have a U-value of 5.6 W/m² K. It is proposed to double-glaze the windows and reduce the U-value to 3.0 W/m² °C. During a 33-week heating season, the average temperature difference across the windows is 7°C.
    (a) For both types of glazing, calculate the total heat lost during the heating season.
    (b) Obtain current figures for the cost of electrical energy and the approximate cost of double glazing such windows. Estimate the number of years required for the annual fuel saving to pay for the cost of the double glazing.

15. The average rates of heat loss for a particular house are 1580 W total fabric loss and 870 W ventilation loss. The seasonal heat gains of the house total 27 500 MJ. The fuel used has a calorific value of 32 MJ/kg and the heating system has an overall efficiency of 75 per cent.

   (a) Calculate the input heat required during a heating season of 33 weeks.

   (b) Calculate the mass of fuel required to supply one season's heating.

Answers are on page 325.

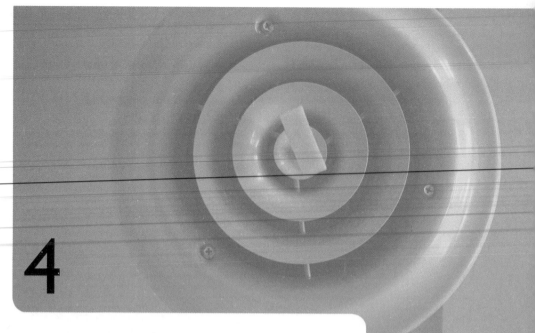

4

Air Control in Buildings

CHAPTER OUTLINE

To feel comfortable inside our buildings we need fresh air which is at the correct temperature, contains the right amount of moisture and is free from odours and pollutants. The air supply for a building also affects the thermal design and energy performance of buildings, so balances have to be found between discarding stale air and keeping energy in the building.

The properties and principles of heat, gases and vapours are described in the Resource sections of the book and make a useful reference if you want further information about the underlying science or want to investigate topics further. Much of this chapter is concerned with the effects of moisture in the air, which plays a large role in human comfort and in the performance of buildings. The contents of this chapter allow you to:

▶ understand the nature and effects of moisture in the air
▶ work with the terms and measurements that describe humidity
▶ use psychrometric charts to predict states of air
▶ understand the principles of condensation in buildings
▶ understand the practical causes and the control of condensation
▶ calculate the risk of condensation in walls and roofs
▶ understand the principles of refrigeration and heating cycles
▶ appreciate the role of refrigerants and their changing nature
▶ understand the principles behind applications such as refrigerators and heat pumps
▶ understand the reasons and requirements for ventilation

→

▶ specify and calculate appropriate rates of ventilation for particular environments
▶ understand the principles and applications of natural ventilation
▶ understand the nature of mechanical ventilation and air-conditioning installations.

The supply and control of air in buildings has important links with the topics of energy conservation and carbon management discussed in Chapter 2 and with the overall performance of buildings discussed in Chapter 14.

AIR SUPPLIES

Ventilation in buildings is the process of changing the air in a room or in some other internal space. This process should be continuous with new air taken from a clean source. Although we require oxygen for life, a build-up of carbon dioxide is more life-threatening and a general built-up of odours will be more critical long before there is danger to life. In addition to the comfort of the occupants the ventilation of a building has other objectives, such as those listed below:

> **See also:**
>
> Chapter 2 on *Energy use in buildings* and Chapter 14 on *Green buildings*

- supply of oxygen for human breathing
- removal of carbon dioxide from human breathing
- control of humidity for human comfort or conditioning of materials
- control of air velocity for human comfort
- removal of odours
- removal of micro-organisms, mites, moulds, fungi
- removal of excessive heat energy
- removal of water vapour to help prevent condensation
- removal of particles such as smoke and dust
- removal of volatile organic compounds (VOCs) from sources such as cleaning solvents, carpet, furniture and building products
- removal of combustion products from heating and cooking
- removal of ozone gas from photocopiers and laser printers
- removal of methane gas and decay products from ground conditions.

The 'old' air being replaced has often been heated and, to conserve energy, the rate of ventilation may be limited or heat may be recovered from the extracted air. Other factors to consider in the provision of ventilation include the following:

> 'Every building must be designed and constructed in such a way that the air quality inside the building is not a threat to the health of the occupants or the capability of the building to resist moisture, decay or infestation'
> *(Scottish Building Standards)*

- control of fire
- conservation of energy
- noise from the system.

Ventilation requirements

Adequate ventilation of buildings and spaces is required by several sources of regulations in the United Kingdom and Ireland such as the Building Regulations,

Table 4.1 Typical ventilation rates

Element	Rate
Commercial kitchens	20–40 air changes per hour
Restaurants (with smoking)	10–15 air changes per hour
Classrooms	3–4 air changes per hour
Offices	2–6 air changes per hour
Domestic rooms	1 air change per hour
General occupied rooms	8 litres/second fresh air per occupant

Workplace Regulations, Housing Acts, and the Health and Safety at Work legislation. Various authorities produce technical guidance for providing adequate ventilation, and important sources for the UK include British Standards, Building Research Establishment (BRE) and the Chartered Institution of Building Services Engineers (CIBSE).

The method of specifying the quantity and rate of air renewal varies and includes the following:

- air supply to a space – such as 1.5 room air changes per hour
- air supply to a person – such as 8 litres per second per person.

In the simplest case it can be assumed that if air is actively extracted, say by a fan, then air will flow in to replace the extracted air. So the supply rate will match the extraction rate, although the source of the supply may need to be considered in the ventilation design. Other methods of ensuring minimum ventilation, as used by Building Regulations, include the following:

Table 4.2 Typical ventilation standards for dwellings

Type of room	Rapid ventilation e.g. opening windows	Background ventilation e.g. trickle vents	Extract ventilation e.g. fan rates
Habitable room	1/30th of floor area	8000 mm²	–
Kitchen	opening window (no minimum size)	4000 mm²	60 litres/second or PSV
Utility room	opening window (no minimum size)	4000 mm²	30 litres/second or PSV
Bathroom (with or without WC)	opening window (no minimum size)	4000 mm²	15 litres/second or PSV*
WC (separate from bathroom)	1/30th of floor area	4000 mm²	6 litres/second minimum

Note:
*   PSV = passive stack ventilation, which uses a duct between ceiling and roof terminal.

- number or area of open windows to provide *rapid ventilation*
- area of opening to provide *trickle ventilation*
- performance of fan or stack to provide *extract ventilation*.

Table 4.2 summarises typical ventilation requirements for dwellings.

Table 4.3 Useful ventilation rate conversions

1 000 litres	= 1 m³
1 litre/sec	= 3 600 litres/hour
1 litre/sec	= 3.6 m³/hour

Air supply rates

Useful conversion between systems of specifying ventilation are given in Table 4.3 and Worked Example 4.1. For example, a common standard per person is 8 litres/sec = 30 m³/hour.

Worked Example 4.1

An extract fan is required for a room with sanitary accommodation (WC or urinal) with the following details:

target air change per hour = 3 ach
volume of room 2 m × 3 m × 2.5 m = 15 m³

So

3 ach = 3 × 15 = 45 m³/hour = 45,000 litres/hour
45,000 litres/hour = 45,000/60 × 60 = 15.5 litres/second

Therefore the extract fan needs to have an extraction rate of at least 15.5 litres/second.

HUMIDITY

The atmosphere surrounding our built environment contains moisture in the form of water vapour. This water is only about 5 per cent by weight of the total gases contained in air, yet this relatively small proportion of moisture creates huge global effects such as the weather systems, the water supplies and our ability to grow food crops. This condition of moisture in the air, or *humidity*, also influences the heat balance and comfort of the human body, the durability of building materials and the rate at which materials dry, and it also causes condensation in buildings,

See also:

section on *Gases and vapours* in Resource 2

Most of the moisture in the atmosphere is a result of evaporation from the sea, which covers more than two-thirds of the earth's surface. At any particular place, natural humidity is dependent on local weather conditions and, inside a building, humidity is further affected by the thermal properties and the use of the building.

Water vapour

A vapour can be defined as a substance in the gaseous state which may be liquefied by compression. Water vapour, for example, is formed naturally in the space above liquid water which is left open to the air. This process of *evaporation* occurs because some liquid molecules gain enough energy, from chance collisions with other molecules, to escape from the liquid surface and become gas molecules. The latent heat required for this change of state is taken from the other molecules of the liquid, which therefore becomes cooled. The rate of evaporation increases if there is a movement of air above the liquid.

Water vapour is invisible. Steam and mist, which can be seen, are actually suspended droplets of water liquid, not water vapour. The molecules of water vapour rapidly occupy any given space and exert a vapour pressure on the sides of any surface that they are in contact with. This pressure behaves independently of the other gases in the air – an example of Dalton's Law of partial pressures described in Resource 2.

Water vapour is invisible

Saturation

If the air space above a liquid is enclosed then the evaporated vapour molecules collect in the space and the vapour pressure increases. Some molecules are continually returning to the liquid state and eventually the number of molecules evaporating is equal to the number of molecules condensing. The air in the space is then said to be saturated. Saturated air is a sample that contains the maximum amount of water vapour possible at that temperature.

Vapour pressure increases as the amount of water vapour increases and at saturation the vapour pressure reaches a steady value called the saturated vapour pressure.

- **Saturated vapour pressure** (SVP) is the vapour pressure of the water vapour in an air sample that contains the maximum amount of vapour possible at that temperature.

For any fixed volume of air the saturated vapour pressure is found to increase with increase in temperature. At higher temperatures three more energy available for molecules to evaporate from the liquid and become vapour molecules. Figure 4.1 shows this effect in a diagrammatic way. The saturated vapour pressure of water vapour is still a limit, but it is a limit which changes with temperature. The change with temperature is a non-linear relationship, as shown by the graph in Figure 4.2.

Increased vapour pressure indicates an increased moisture content; therefore

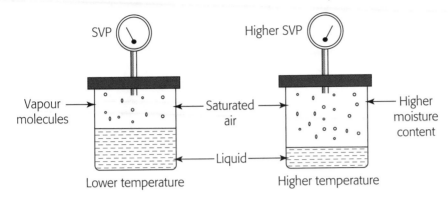

Figure 4.1 Saturated air at different temperatures

the saturation of a fixed sample of air is delayed if the temperature of the air is raised. This property gives rise to the following general principle.

- Warm air can hold more moisture than cold air.

If an unsaturated sample of moist air is cooled then, at a certain temperature, the sample will become saturated. If the sample is further cooled below this dew-point temperature then some of the water vapour must condense to liquid. This condensation may occur on surfaces, inside materials, or around dust particles in the form of cloud or fog.

> 1 atmosphere of pressure
> = 101.325 kPa or
> = 760 mm of mercury

Specification of humidity

Some of the properties of water vapour can be used to specify the amount of moisture in a sample of air. The different variables and their applications are listed below and described in the sections that follow:

- moisture content
- vapour pressure
- dew-point
- relative humidity.

Moisture content

Moisture content is a measure of absolute humidity – the actual quantity of water vapour present in the air.

$$\textbf{Moisture content} = \frac{\text{mass of water vapour in air sample}}{\text{mass of air sample when dry}}$$

Unit: kg/kg of dry air (or g/kg)

Moisture content is not usually measured directly but it can be obtained from other types of measurement. Moisture content values are needed, for example, in determining what quantity of water an air-conditioning plant needs to add or to extract from a sample of air.

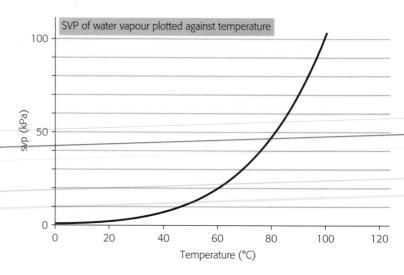

Figure 4.2 Saturated vapour pressures of water

Vapour pressure

The molecules of water vapour in the air exert a pressure that increases as the amount of water vapour increases.

Increased
moisture content
means higher
vapour pressure

- **Vapour pressure** is the partial pressure exerted by the molecules of a vapour.

    Unit: pascal (Pa)

Other units which may be found in use include:

- millibars (mb), where 1 mb = 100 000 Pa
- mm of mercury, where 1 mm = 133 Pa.

The vapour pressure of moisture in the air behaves independently of the other gases in the air. Vapour pressure values are usually derived from other measurements, and one of the uses of vapour pressure is to determine the rate at which water vapour moves through materials.

Dew-point

Higher dew-point
means increased
condensation risk

If moist air is cooled then, at a certain temperature, the air becomes saturated with water vapour. If this air is in contact with a surface that is at or below this temperature then a thin film of liquid will form. This effect is known as dew or condensation.

- The **dew-point** is the temperature at which a fixed sample of air becomes saturated.

    Unit: °C or K

A sample of air with a fixed moisture content has a constant dew-point, even if the air temperature changes. The dew-point is particularly relevant in the study of condensation in buildings. Dew-point values can be measured directly or derived from other measurements.

Relative humidity

The relative humidity (RH) of a sample of air compares the actual amount of moisture in the air with the maximum amount of moisture the air can contain at that temperature. The strict definition of relative humidity is:

$$\textbf{Relative humidity} = \frac{\text{vapour pressure of sample}}{\text{SVP of sample at same temperature}} \times 100$$

Unit: per cent RH, at a specified temperature

It is also common practice to describe humidity in terms of *percentage saturation*, which is defined by the following formula:

$$\textbf{\% saturation} = \frac{\text{mass of water vapour in air sample}}{\begin{array}{c}\text{mass of water vapour required to}\\\text{saturate sample at same temperature}\end{array}} \times 100$$

Percentage saturation and relative humidity are identical only when air is perfectly dry (0 per cent) or fully saturated (100 per cent). For temperatures in the range 0 to 25°C, the difference between relative humidity and percentage saturation is small.

A convenient alternative formula for relative humidity is:

$$\textbf{Relative humidity} = \frac{\text{SVP at dew-point}}{\text{SVP at room temperature}} \times 100$$

An RH of 100 per cent represents fully saturated air, such as occurs in condensation on a cold surface or in a mist or fog. An RH of 0 per cent represents perfectly dry air, a condition that is approached in some desert regions and in sub-zero temperatures when water is frozen solid.

The SVP or saturated moisture content varies with temperature; therefore RH changes as the temperature of the air changes. Despite this dependence on temperature, RH values are a good measurement of how humidity affects human comfort and drying processes.

Because warm air can hold more moisture than cool air, raising the temperature increases the SVP or the saturation moisture content. The denominator on the bottom line of the RH fraction then increases and so the RH value decreases. This property gives rise to the following general effects for any sample of air:

- Heating the air lowers the relative humidity.
- Cooling the air increases the relative humidity.

Worked Example 4.2

A sample of air has an RH of 40 per cent at a temperature of 20°C. Calculate the vapour pressure of the air (given: SVP of water vapour = 2340 Pa at 20°C).

RH = 40 per cent $vp = ?$ SVP = 2340 Pa

Using the formula for RH and substituting values

$$RH = \frac{vp}{SVP} \times 100$$

$$40 = \frac{vp}{2340} \times 100$$

$$vp = \frac{2340}{100} \times 40 = 936$$

Vapour pressure = **936 Pa**

Hygrometers

Hygrometry, also called psychrometry, is the measurement of humidity. The absolute humidity (moisture content) of a sample of air could be measured by carefully weighing and drying it in a laboratory. Usually however, other properties of the air are measured and these properties are then used to calculate the value of the RH.

- **Hygrometers**, or **psychrometers**, are instruments which measure the humidity of air.

Hair and paper hygrometers

The hair hygrometer and the paper hygrometer make use of the fact that animal hair or paper change their dimensions with changes in moisture content. These instruments can be made to give a direct reading of RH on their dials, but they need calibration against another instrument.

Dew-point hygrometer

The temperature of the dew-point is a property that can be observed and measured by cooling a surface until water vapour condenses upon it. The dew-point temperature and the room temperature can then be used to obtain an RH value from tables or charts.

The *Regnault hygrometer* is one form of dew-point hygrometer. A gentle passage of air is bubbled through liquid ether in a container and causes the ether to evaporate. The latent heat required for this evaporation is taken from the liquid ether and from the air immediately surrounding the container, which therefore becomes cooled. The polished outside surface of the container is observed and the moment a thin film of mist appears the temperature is noted from the thermometer in the ether. When the instrument is used carefully, this observed temperature is also the dew-point temperature of the surrounding air.

Modern forms of dew-point hygrometer use a variety of methods to cool a mirror surface and various forms of photoelectrical detection to determine the temperature at which condensation occurs on the mirror.

Wet-and-dry-bulb hygrometer

If the bulb of a thermometer is wrapped in a wetted fabric the evaporation of the moisture absorbs latent heat and will cool the bulb. This wet-bulb thermometer will record a lower temperature than an adjacent dry-bulb thermometer (an ordinary thermometer). Dry air with a low RH causes rapid evaporation and produces a greater wet-bulb depression than moist air. Saturated air at 100 per cent RH causes no net evaporation and the dry-bulb and wet-bulb thermometers then record the same temperature. This changing difference between the wet and the dry thermometers – the 'depression' – can therefore be used as a indicator of relative humidity.

Drier air = greater wet-bulb depression

The two thermometers are mounted side by side, and a water supply and fabric wick are connected to the wet bulb. The *whirling sling hygrometer*, shown in Figure 4.3, is rapidly rotated by means of a handle before reading, to circulate fresh air around the bulbs. Each type of hygrometer has calibrated tables which give the RH value for a particular pair of wet-bulb and dry-bulb temperatures.

Electronic hygrometer

Practical hygrometers, for use in the field, usually detect humidity by electronic sensors which then show the results on an instrument display, or pass them as data to other systems. *Capacitive sensors* apply an AC signal between two plates and measure the change of capacitance caused by the water vapour present. *Resistive sensors* use a polymer membrane whose resistance changes with water absorbed from the air. The calibration of electronic instruments is important and usually varies with temperature.

Psychrometric chart

The different variables used to specify the amount of water vapour in the air are related to one another. These relationships between different types of measurement can be expressed in the form of tables of values, or as graphs.

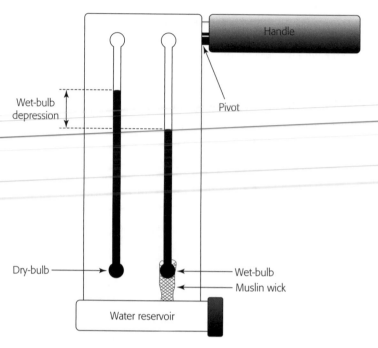

Figure 4.3 Whirling sling hygrometer

- A **psychrometric chart** is a set of graphs which are combined so that they plot the relationships between the different variables used to specify humidity.

Figure 4.4 is one form of psychrometric chart and it displays the following measurements.

- **Dry-bulb temperature**: the ordinary air temperature, is read from a horizontal scale on the base line.
- **Moisture content**: the mixing ratio, is read from one of the right-hand vertical scales.
- **Vapour pressure**: read from one of the right-hand vertical scales.
- **Wet-bulb temperature**: read from the sloping straight lines running from the saturation line.
- **Dew-point temperature**: read from the horizontal lines running from the saturation line.
- **Relative humidity**: read from the series of curves running from the left-hand vertical scale.

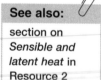

See also:

section on
*Sensible and
latent heat* in
Resource 2

Notice that the saturation curve represents 100 per cent RH and that, at this saturated condition, the dry-bulb, the wet-bulb and the dew-point temperatures all have the same value.

A psychrometric chart is strictly valid for one value of atmospheric pressure, the sea-level pressure of 101.3 kPa being a common standard. Some versions of

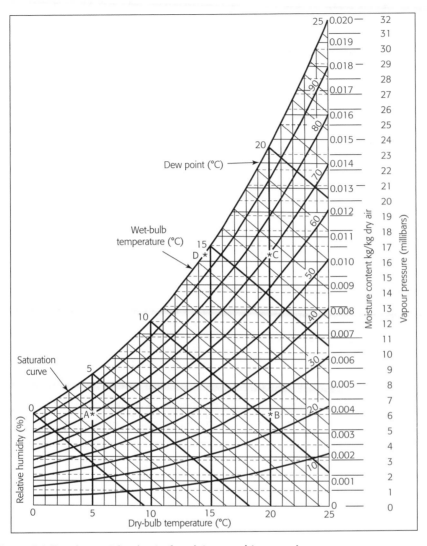

Figure 4.4 Psychrometric chart of moisture and temperature

the psychrometric chart contain additional information, such as *sensible heat* and *latent heat* contents, which are of particular use to building services engineers.

Use of the psychrometric chart

The temperature and moisture content of a particular sample of air is described by a pair of values. These two values also act as coordinates on the chart and define a single *state point* for the air. Moving to any other state point on the chart represents a change in the condition of the air.

The direction of movement on the chart also represents specific types of air-conditioning change. For example, horizontal movements mean that the moisture content of the air remains constant, as happens when air is simply heated without humidification. The relative humidity, however, does change during horizontal movements.

The following air conditions and changes are marked on Figure 4.5.

- **Point A** represents air of 5°C and 70 per cent RH. From the scales it can also be read that the moisture content is 0.0038 kg/kg and the vapour pressure is 6 mb.
- **Point B** represents air where the dry bulb temperature has been raised to 20°C and so the RH has decreased to 25 per cent. The moisture content and the vapour pressure are unchanged.
- **Point C** represents air at the same dry bulb temperature but with an increase in moisture content to 0.0103 kg/kg. The vapour pressure has also increased. The RH has been restored to 70 per cent. The wet bulb at this state is 16.5°C.
- **Point D** represents the dew-point of this same air at 14.5°C. This is the temperature at which the air is saturated and has 100 per cent RH.

Worked Example 4.3

External air at 0°C and 80 per cent RH is heated to 18°C. Use the psychrometric chart to determine the following information:

(a) The RH of the heated air.
(b) The RH of the heated air if 0.005 kg/kg of moisture is added.
(c) The temperature at which this moistened air would first condense.

Initial conditions: dry bulb = 0°C, RH = 80 per cent, so moisture content = 0.003 kg/kg

(a) For the heated air
        moisture content = 0.003 kg/kg
        dry bulb = 18°C
    so, reading from chart
        RH = 23 per cent

(b) For the moistened air
        moisture content = 0.003 + 0.005 = 0.008 kg/kg
        dry bulb = 18°C
    so, reading from chart
        RH = 62 per cent

(c) For condensation
        RH = 100 per cent
        moisture content = 0.008
    so, reading from chart
        dew-point = 10.8°C gives condensation

CONDENSATION IN BUILDINGS

Condensation in buildings is a form of dampness caused by water vapour in the air. Among the effects of condensation are misting of windows, beads of water on non-absorbent surfaces, dampness of absorbent materials and mould growth.

Condensation is not likely to be a problem in a building where it has been anticipated and designed for, as in a tiled bathroom or at an indoor swimming pool. Unwanted condensation, however, is a problem when it causes unhealthy living conditions; damage to materials, to structures and to decorations; or general concern to people.

Condensation as a problem is a relatively recent concern and one that has been increasing. It is affected by the design of modern buildings and by the way in which buildings are heated, ventilated and occupied. These factors are considered in the sections on the causes of condensation and the remedies for it.

> Water vapour is invisible; and moves quickly

Principles of condensation

Warm air can hold more moisture than cold air. If air in a building acquires additional moisture, this increased moisture content will not be seen in places where the air is also warmed. But if this moist air comes into contact with colder air, or with a cold surface, then the air is likely to be cooled to its dew-point. At this temperature the sample of air becomes saturated, it can no longer contain the same amount of water vapour as before, and the excess water vapour condenses to liquid.

- **Condensation in buildings** occurs whenever warm moist air meets surfaces that are at or are below the dew-point of that air.

> Cooled air can hold less water vapour

It is convenient to classify the effects of condensation into the following two main types:

- surface condensation
- interstitial condensation.

Surface condensation

Surface condensation occurs on the surfaces of walls, windows, ceilings and floors. It appears as a film of moisture or as beads of water on the surface and is most obvious on the harder, more impervious surfaces. An absorbent surface may not show condensation at first, although persistent condensation will eventually cause dampness.

Interstitial condensation

Interstitial condensation occurs within the construction of a building as shown in Figure 4.5. Most building materials are, to some extent, permeable to water

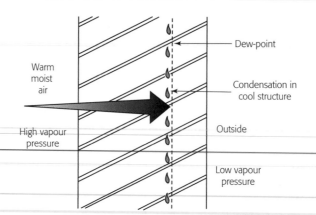

Figure 4.5 Interstitial condensation

vapour (that is, they allow the passage of air containing moisture). If this air cools as it passes through the structure then, at the dew-point temperature, condensation will begin to occur inside the structure. The dampness caused by this interstitial condensation can damage important structural materials such as steelwork, and can make insulating materials less effective.

Causes of condensation

The general requirements for condensation in buildings are moist air and cold structures. The factors that influence the production of condensation can be considered under the following headings:

- moisture sources
- ventilation
- temperatures
- use of buildings.

Moisture sources

The moisture present in any sample of air comes from a source of water. Local weather conditions determine the initial moisture state of the air but if the outside air is cool then it has a low moisture content and does not cause condensation when it is inside a building. Warm humid weather can cause condensation, especially if there is a sudden change from cool weather to warm moist conditions and the room surfaces are slow to heat up. In temperate climates like Britain the amount of rainfall does not have a big influence on condensation inside buildings.

Most of the moisture content of the air inside dwellings comes from the occupants and their activities. An average family produces 10 to 20 kg of moisture per day by the activities of normal breathing, cooking and washing. Flueless devices, such as gas cookers or bottle-gas heaters release large quantities of water vapour and increase the risk of condensation. The moisture produced in kitchens and bathrooms can quickly spread into other rooms and cause condensation far away from the original source of the moisture.

> Condensation is *not* usually caused by damp weather

The construction processes involved in a traditionally-built brick house add about 5000 kg of water to the structure – mainly from the water used for mixing concrete, mortar and plaster, and by exposure to weather. This water dries out during the first months of occupancy and adds to the moisture content of the air inside the building.

> 1 kg of water is 1 litre, the same quantity as a typical bottle of soft drink

Temperatures

Warm air can hold more moisture than cool air and is therefore less likely to cause condensation, but the extra moisture contained in the warm air can cause condensation when that air is cooled. For example, the kitchen itself may be warm enough to be free from condensation but the same moisturised air may cause condensation when it diffuses to a cooler part of the house.

Condensation occurs on those surfaces, or in those materials, that have a temperature at or below the dew-point of the moist air. If moisture is progressively added to the air in a room then condensation will first occur on the coldest surfaces, on windows and cold pipes for example. The temperatures of room surfaces are lowest where the thermal insulation is lowest. Condensation can indicate the areas of worst insulation, particularly those caused by 'cold bridges' such as lintels above windows.

> Cooled air can hold less water vapour

The speed at which structures change their temperature affects condensation. Heavyweight construction, with a high thermal capacity, will be slower to respond to heating than a lightweight construction. A brick wall or a concrete floor, for example, is slow to increase its temperature when heated and condensation may occur for a period while it is heating. The positioning of insulation on the inside of heavyweight structures will allow the surface temperatures to rise more rapidly.

Ventilation

The air outside a building usually has a much lower moisture content than the air inside, because cool air cannot hold as much moisture as warm air. Ventilating a building therefore lowers the moisture content inside and reduces the risk of condensation. It is theoretically possible to avoid all condensation by adequate ventilation, but as the ventilation rate increases the heat loss in the discarded air also increases.

The natural ventilation rates in dwellings have been decreasing over the years, and this decrease greatly contributes to the incidence of condensation in buildings. Chimneys have been eliminated in many modern houses and blocked-up in older ones. The construction of modern doors, windows and floors usually provides better seals against the entry of outside air than in the past. As a result, the natural ventilation in dwellings has fallen from typical rates of four air changes per hour to less than one.

Even with higher ventilation rates it is possible to have stagnant volumes of moist air where condensation still occurs: behind furniture and inside wardrobes, for example.

Use of buildings

Over the years there have been changes in the design of buildings and changes in the way that we live in these buildings. These changes have tended to increase the risk of condensation.

Methods of heating buildings have changed and, at the same time, people expect higher standards of thermal comfort and cleanliness than they did in the past. The presence of chimneys in older houses aided natural ventilation and the use of a fire increased the ventilation by drawing air for combustion. The cool draughts that resulted were offset to some extent by the direct warming effect of radiant heat from the fire.

Thermal comfort by means of warm air from modern convective heating requires low levels of air movement and this leads to the draught-proofing of houses. Public campaigns for better thermal insulation also encourage lower ventilation rates and these measures can increase the risk of condensation unless care is taken.

Factors linked to condensation: no open fires sealed buildings laundry in house working households peaks of activity

Moisture-making activities such as personal bathing and washing laundry have increased because of improved facilities and changing social customs. The laundry is often dried inside the building, sometimes by a tumble-drier that is not vented to the outside.

Many dwellings are only occupied during the evening and the night. This pattern of living tends to compress the cooking and washing activities of a household into a short period, at a time when windows are unlikely to be opened. During the day, when the dwelling is unoccupied, the windows may be left closed for security reasons and heating is turned off, allowing the structure to cool.

Remedies for condensation

The causes of condensation, discussed above, also link to the methods for preventing condensation. Remedies for persistent condensation inside buildings can be classified under the following three types:

- ventilation
- heating
- insulation.

Condensation is best prevented by a combination of approaches

A combination of these three remedies is usually necessary. The use of vapour barriers as a measure against interstitial condensation is discussed under a separate heading. *Anti-condensation paint* can be useful for absorbing temporary condensation but it is *not* a permanent remedy for condensation.

Ventilation

Ventilation helps to remove the moist air, which might otherwise condense if it was cooled inside the building. Ventilation is most effective if it is used near the source of moisture (for example, an extractor fan in the kitchen). Care should be taken that natural ventilation from open windows does not blow water vapour into other rooms; kitchen and bathroom doors should be self-closing.

Some windows in each room should have a ventilator, which can give a small controlled rate of ventilation without loss of security, loss of comfort, or loss of significant amounts of heat energy. The cost of the heat lost in necessary ventilation is small and it should decrease as techniques of heat recovery from exhaust air are applied to houses.

Heating

Heating a building raises the temperature of the room surfaces and helps keep them above the dew-point of the air inside the building. Heated air also has the ability to hold more moisture, which can then be removed by ventilation before it has a chance to condense upon a cold surface.

The level and timing of the heating used in a building affects condensation and, in general, long periods of low heating are better than short periods at high temperature. Heavyweight construction, such as concrete and brickwork, should not be allowed to cool completely, and a continuous level of background heating helps maintain temperatures. Heating devices without flues should not be used if there is a risk of condensation. A bottle gas or paraffin heater, for example, gives off about 1 litre of water for each litre of fuel used.

Insulation

Thermal insulation reduces the rate at which heat is lost through a structure and will help in keeping inside surfaces warm, although insulation by itself cannot keep a room warm if no heat is supplied. Insulation placed on the inside surface of a heavyweight construction, such as a brick wall, helps to raise the surface temperature more quickly when the room is heated. However, with the insulation near the inside of a room, the outer part of the wall will remain cool and a vapour barrier is needed to prevent interstitial condensation, as discussed in the next section.

CONDENSATION CONDITIONS

The risk of condensation occurring on or in a building material depends upon the temperature and the humidity of the air on both sides of the structure, and also upon the resistance of the material to the passage of heat and vapour. The thermal resistance of materials was used to calculate U-values in Chapter 2. This section introduces the similar idea of vapour resistance and uses it to calculate when and where condensation may occur in a structure.

Vapour transfer

The vapour pressure of a sample of air increases when the moisture content increases. Because the occupants of a building and their activities add moisture to the air, the vapour pressure of the inside air is usually greater than that of the cooler outside air. This pressure difference results in the following general rule:

● Water vapour passes through structures from inside to outside.

The rate at which water vapour passes through a structure depends upon the vapour *permeability* of the various building materials present – that is, the ease with which they permit the diffusion of water vapour. This property of a material concerning the behaviour of water vapour is often totally different to the behaviour of that same material concerning water liquid. It is relatively easy to make a material waterproof against liquid molecules but harder to make the material vapour proof against the much smaller and more energetic molecules of a gas. The permeability of materials to water vapour can be expressed in a number of ways, as described below.

Vapour resistivity

● **Vapour resistivity** (r_v) is a measure of the resistance to the flow of water vapour offered by unit thickness of a particular material under standardised conditions.

        Unit: GN s/kg m

> Resisting liquid water is easy compared with resisting water vapour

This property of a material is sometimes alternatively expressed as a *vapour diffusivity* or *vapour permeability* value, which is the reciprocal of vapour resistivity. Table 4.4 lists typical values for the vapour resistivity or vapour resistance of various building materials.

Vapour resistance

● **Vapour resistance** (R_v) describes the resistance of a specific thickness of material.

Vapour resistance is calculated by the following formula.

$$R_v = r_v L$$

where
    R_v = vapour resistance of that material (GN s/kg)
    L = thickness of the material (m)
    r_v = vapour resistivity of the material (GN s/kg m)

A vapour resistance value is usually quoted for thin membranes, such as aluminium foil or polythene sheet, and some typical figures are given in Table 4.4.
    Total vapour resistance (R_{vT}) of a compound structure is the sum of the vapour resistances of all the separate components.

$$R_{vT} = R_{v1} + R_{v2} + R_{v3} + ... \text{ etc.}$$

Table 4.4 Vapour transfer properties of materials

Material	Vapour resistivity (MN s/g m)
Brickwork	25–100
Concrete	30–100
Fibre insulating board	15–16
Hardboard	450–750
Mineral wool	5
Plaster	60
Plasterboard	45–60

Plastics
  expanded polystyrene
  foamed polyurethane
  foamed urea-formaldehyde
	100–600
	30–1 000
	20–30

Plywood	1 500–6 000
Timber	45–75
Stone	150–450
Strawboard, compressed	45–75
Wood wool	15–40

Membrane	Vapour resistance (GN s/kg)
Aluminium foil	4 000 +
Bitumenised paper	11
Polythene sheet (0.06 mm)	125
Paint gloss (average)	6–20
Vinyl wallpaper (average)	6–10

Dew-point gradients

The changes in temperature inside a structure such as a wall were calculated in Chapter 2. The temperature change across any particular component is given by the formula:

$$\Delta\theta = \frac{R \times \theta_T}{R_T}$$

where

$\Delta\theta$ = temperature difference across a particular layer
R = resistance of that layer
$\Delta\theta_T$ = total temperature difference across the structure
R_T = total resistance of the structure

The temperature drop across each component can be plotted onto a scale drawing of the structure to produce the temperature gradients, as shown in Figure 4.5.

The *vapour pressure* drop across a component can be obtained in a similar manner from the following formula:

$$\Delta P = \frac{R_v}{R_{vT}} \times P_T$$

where
    ΔP = vapour pressure drop across a particular layer
    R_v = vapour resistance of that layer
    P_T = total vapour pressure drop across the structure
    R_{vT} = total vapour resistance of the structure

Vapour pressure changes can be plotted as gradients, but they are usually converted to dew-point readings and plotted as dew-point gradients. The dew-point at each boundary in the structure is obtained from a psychrometric chart by using the corresponding structural temperature and vapour pressure.

Condensation risk

Condensation occurs on room surfaces, or within structures, if moist air meets an environment that is at or below the dew-point temperature of that air. The prediction of such condensation risk is an important design technique, for which assumptions are made about the air conditions expected inside and outside the building. The air temperatures and the dew-points at any particular position within the structure can then be calculated and plotted.

Worked Example 4.4

An external wall is constructed with an inside lining of plasterboard 10 mm, then expanded polystyrene board (EPS) 25 mm, then dense concrete 150 mm. The thermal resistances of the components, in m² K/W, are: internal surface resistance 0.123, plasterboard 0.06, EPS 0.75, concrete 0.105, and external surface resistance 0.055. The vapour resistivities of the components, in MN s/g m, are: plasterboard 50, EPS 100, and concrete 30. The inside air is at 20°C and 59 per cent RH; the outside air is at 0°C and saturated. Use a scaled cross-section diagram of the wall to plot the structural temperature gradients and the dew-point gradients.

Step 1

Calculate the total temperature drop

$$20° - 0° = 20°C$$

Step 2

Use thermal resistances to calculate the temperature drops across each layer and the temperature at each boundary. Tabulate the information.

Layer	Thermal resistance m^2 K/W	Temperature drop $\Delta\theta = \dfrac{R}{R_T} \times \theta_T$	Boundary temperature °C
Inside air	–	–	20
Internal surface	0.123	$\dfrac{0.123}{1.093} \times 20 = 2.3$	
Boundary	–	–	17.7
Plaster	0.06	$\dfrac{0.06}{1.093} \times 20 = 1.1$	
Boundary	–	–	16.6
EPS	0.75	$\dfrac{0.75}{1.093} \times 20 = 13.7$	
Boundary	–	–	2.9
Concrete	0.105	$\dfrac{0.105}{1.093} \times 20 = 1.9$	
Boundary	–	–	1.0
External surface	0.055	$\dfrac{0.055}{1.093} \times 20 = 1.1$	
Outside air	–	–	0.0
$R_T = 1.093$			

Step 3

Plot the boundary temperatures on a scaled section of the wall and join the points to produce temperature gradients, as in Figure 4.6.

Step 4

Use vapour resistances to calculate the vapour pressure drops across each of the layers then, using the psychrometric chart, find the dew-point temperature at each boundary.

$$\left.\begin{array}{l} \text{Inside vapour pressure} \ \ = \ 1\,400 \\ \text{Outside vapour pressure} = \ \ \ 600 \end{array}\right\} \text{from psychrometric chart}$$

Total vapour pressure drop = 1 400 – 600 = 800 Pa

Layer	Thickness L(m)	Vapour resistivity r_v	Vapour resistance $R_v = r_v L$	vp drop $\Delta P = \dfrac{R_v}{R_{vT}} \times P_T$	vp at boundary (Pa)	Dew-pt at boundary
Internal surface	–	–	neg.	–	–	–
Boundary	–	–	–	–	1 400	12°C
Plaster	0.010	50	0.5	$\dfrac{0.5}{7.5} \times 800 = 53$	–	–
Boundary	–	–	–	–	1 347	11.5°C
EPS	0.025	100	2.5	$\dfrac{2.5}{7.5} \times 800 = 267$	–	–
Boundary	–	–	–	–	1 080	7.4°C
Concrete	0.150	30	4.5	$\dfrac{4.5}{7.5} \times 800 = 480$	–	–
Boundary	–	–	–		600	0°C
External surface	–	neg	–	–	–	–
$R_{vT} = 7.5$						

Step 5

Plot the dew-point temperatures on the scaled section diagram and produce dew-point gradients, as in Figure 4.6.

Note: Diagrams should be as large and as accurate as possible so they can be used to visually predict intermediate values of dew-point at any place within the wall or roof.

Surface condensation

Alongside the surface of any partition in a building there is a layer of stationary air. The thermal resistance provided by this air layer – the internal surface resistance – causes the temperature of the surface to be lower than the air temperature in the room. Consequently, air coming into contact with the surface will be cooled and may condense on the inside surface.

The surface temperature can be calculated using the following formula:

$$\Delta\theta = \frac{R}{R_T} \times \theta_T$$

In Worked Example 4.4, for example, the surface temperature of 17.7°C is above the dew-point of 12°C, so surface condensation would not occur under these conditions.

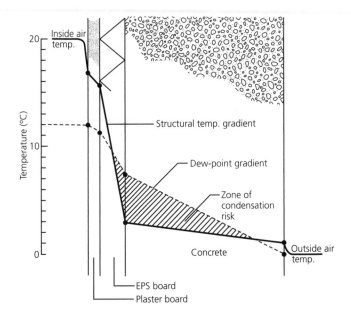

Figure 4.6 Prediction of condensation

Interstitial condensation

If moist air is able to permeate through a structure from inside to outside, then it is usually cooled and follows the gradient of structural temperature change. The dew-point temperature also falls and follows a gradient determined by the drop in vapour pressure across each material.

For specified air conditions, condensation will occur in any zone where the structural temperature gradient falls below the dew-point gradient. These zones can be predicted by scaled diagrams, such as Figure 4.6 where condensation will occur in the EPS layer and in the concrete.

> Higher dew-point means increased condensation risk

Vapour barriers, vapour checks

The risk of interstitial condensation in structures is reduced if moist air is deterred from permeating through the materials of the structure.

- A **vapour barrier** or **vapour check** is a layer of building material which has a high resistance to the passage of water vapour.

It can be seen from Table 4.4 of vapour resistance properties that no material is a perfect 'barrier' to the transfer of water vapour, but some do offer an acceptably high resistance. Vapour barriers or checks can be broadly classified by the forms in which they are applied.

- **Liquid films**. Examples: bituminous solutions, rubberised or siliconised paints, gloss paints.

- **Pre-formed membranes**. Examples: aluminium-foil board, polythene-backed board, polythene sheet, bituminous felt, vinyl paper.

Vapour barriers need to be installed when there is a danger of interstitial condensation causing damage to the structure or to the insulation. Some types of interstitial condensation may be predicted and tolerated, for example when it occurs in the outer brick leaf of a cavity wall. For a vapour barrier to be effective, it must block the passage of water vapour before the vapour meets an environment below the dew-point temperature.

- Vapour barriers must be installed on the warm side of the insulation layer.

> A vapour barrier is totally defective with just one hole; like the *Titanic*

Poor installation can cause the real performance of a vapour barrier to be much worse than theory predicts. A small defect in one area of a vapour barrier degrades the performance of the entire vapour barrier, just like a 'small' leak in a gas mask. Incomplete seals at junctions, such as those between walls and ceilings, and punctures by pipes or electrical fittings are common defects.

Wall vapour barriers

Vapour barriers should be installed on the warm side of the insulation layer in a wall, as shown in Figure 4.7. It is important that any water vapour inside the structure is able to escape and the outside surface should normally be permeable to water vapour, even though it may also need to be weatherproof. Various materials and constructions can be waterproof but still allow water vapour to

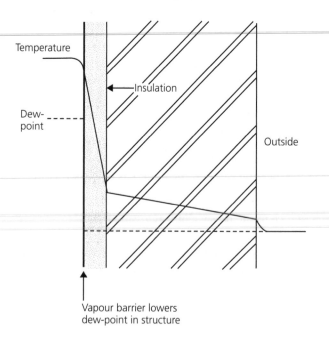

Figure 4.7 Vapour barrier in wall

pass, because the molecules of liquid water are much larger than those of water vapour.

Roof vapour barriers

Roof cavities can easily suffer from severe condensation, and appropriate ventilation and vapour barriers must be carefully provided.

Cold roofs

If thermal insulation is installed at ceiling level then the remainder of the roof space will stay at a low temperature, which is a sign of effective insulation. However, this *cold roof* structure runs the risk of interstitial condensation because any water vapour passing through the roof will be cooled below its dew-point temperature and condense into water liquid.

A vapour barrier must be installed on the warm side of the insulation in a cold roof, as shown in Figure 4.8. As a further precaution, the roof space must be ventilated to get rid of any vapour which does reach the roof space. In a flat roof, this ventilation must be carefully designed to achieve reliable flows of air and it may be easier to consider a warm roof design.

When the thermal insulation of a traditional pitched tile roof is upgraded by placing insulating material on top of the ceiling then the structure becomes a cold roof. There is usually enough accidental ventilation through the tiles and eaves to avoid condensation problems.

Warm roofs

A *warm roof* has its thermal insulation placed immediately beneath the water-proof covering and then protected with a vapour barrier, as shown in Figure 4.8.

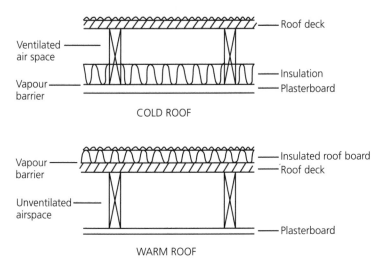

Figure 4.8 Vapour barriers in roofs

The roof deck and the roof space therefore remain on the warm side of the insulation and will not become cool enough to suffer condensation. Any ceiling that is installed in a warm roof should have minimal thermal insulation so that the temperature of the roof structure remains close to room temperature.

An inverted roof is a variation of warm roof design resulting from the installation of thermal insulating boards above the waterproof layer of the roof. Because all or part of the rainwater is designed to drain beneath the insulating boards this type of roof is also known as an 'upside-down roof'.

Cold roof: insulation is close to ceiling
Warm roof: insulation is close to sky

REFRIGERATORS AND HEAT PUMPS

A refrigeration cycle causes heat energy to be transferred from a cooler region to a warmer region. This is against the natural direction of heat flow and can only be achieved by supplying energy to the cycle. This movement of heat can be used for cooling, as in a refrigerator, or for heating, as in a heat pump. A refrigeration cycle can also be the basis of a single air-conditioning plant capable of either heating or cooling a building, according to needs.

The refrigeration cycle

A refrigeration is used for both cooling and heating

A refrigeration system is a process that absorbs heat, transfers that heat to a new position, releases the heat and is then ready to repeat the cycle. This process can also be given the general name of a 'heat cycle'. The most common forms of refrigeration make use of the following two physical mechanisms, which are described in the section on Gases and vapours in Resource 2:

- latent heat changes that occur with changes of state
- the behaviour of vapours when compressed and expanded.

See also:

section on *Gases and vapours* in Resource 2

A refrigeration system employs a volatile liquid called a refrigerant which undergoes the cycle described below and is illustrated in Figures 4.9 and 4.10.

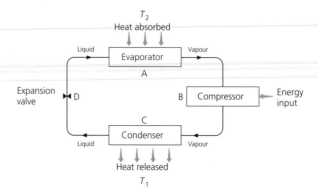

Figure 4.9 Compression refrigerator and heat pump cycle

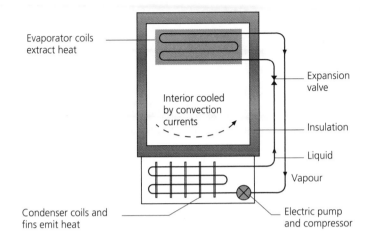

Figure 4.10 Compression refrigerator layout

Stage A

The liquid refrigerant is allowed to evaporate. The latent heat required for the evaporation is extracted from the surroundings of the evaporator, which acts as a cooler.

Stage B

The pressure of the refrigerant vapour is increased by means of a compressor or some other mechanism. This pressurisation raises the temperature of the refrigerant and requires an input of energy into the system.

Stage C

The refrigerant vapour condenses to liquid when it cools below its boiling point. The latent heat released is emitted by the condenser, which acts as a heater.

Stage D

The liquid refrigerant is passed through an expansion valve and its pressure drops. The refrigerant then evaporates and is ready to repeat the cycle.

Refrigerants

Not every gas behaves in a way that is suitable for the cycle of vapour–liquid–vapour changes used in refrigeration. Air, for example, does not normally turn to liquid when it is condensed inside a car tyre. A refrigerant must therefore be below its critical temperature, or it will not liquefy; and it must be above its melting point, or else it will solidify. In general a useful refrigerant should possess the following properties:

- low boiling point

- high latent heat of vaporisation
- easy liquefaction by compression
- being stable, non-toxic and non-corrosive
- environmental safety.

From the time of the first commercial refrigeration equipment, in the late 1800s, the substances used as refrigerants have both evolved and also returned to their origins. Early equipment was large and used 'natural' substances such as ammonia and carbon dioxide. From the 1930s smaller devices, such as domestic refrigerators and air-conditioning units, used synthetic compounds such as chlorofluorocarbons (CFCs). CFCs had very good properties as refrigerants but unfortunately they were identified as helping to cause a reduction of the ozone layer. The ozone in the upper atmosphere of the earth absorbs much of the harmful UVB ultra-violet radiation from the sun and helps prevent biological damage such as skin cancer. International meetings in the 1980s agreed to discontinue the production and use of CFCs. Safer alternatives, described below, were developed and most countries have ceased the use of CFCs. However, all countries have millions of existing refrigerators whose CFC refrigerant needs to be safely disposed of as they are 'retired'.

Modern large-scale refrigeration and air-conditioning plants do not necessarily circulate the refrigerant throughout the whole of the equipment. The evaporator and condenser components can be kept compact and their heating/cooling loads transferred to another medium, such as water/brine, for circulation throughout a large area. In this way the quantity of refrigerant used is reduced and kept contained in a controlled place. The properties of the more common refrigerants are outlined below.

- **Chlorofluorocarbons** (CFCs) are a mixture of organic compounds containing carbon, chlorine and fluorine which have been available in various commercial formulations such as freon. These compounds have been widely used in air-conditioning and refrigeration systems, including domestic refrigerators. They have desirable properties as refrigerants, being colourless, non-flammable, non-toxic and non-corrosive but, because of their link to ozone depletion, the use of CFCs has been phased out. They have been replaced with alternate refrigerants, such as HCs and HFCs described below.
- **Hydrocarbons** (HCs) such as propane and butane are now widely used for refrigeration applications, including the majority of domestic refrigerators made in Europe. Similar to the gases used in camping stoves, HCs are flammable and potentially explosive at certain concentrations in air. But as refrigerants they are contained within factory-sealed units and their use has proved to be safe, efficient, reliable and economic.
- **Hydrofluorocarbons** (HFCs) are another refrigerant manufactured to replace CFCs and have similar operating properties to HCs.
- **Ammonia** (NH_3) is a pungent clear gas that dissolves readily in water and is an efficient refrigerant. It was used in early refrigeration equipment of the late

1800s, but its toxicity and flammability require special consideration. The use of ammonia has had a revival in modern commercial installations, such as large air-conditioning plants at airports, where modern techniques minimise the qualities of gas used. It is used in the absorption refrigeration cycle explained in a later section.

- **Carbon dioxide** (CO_2) is another early refrigerant whose use has been revived for large refrigeration plants and low temperatures. The equipment runs at high operating pressures but has lower energy costs than other systems. The carbon dioxide gas is contained within the equipment and so does not have an impact on the ozone layer.

Refrigerators and heat pumps

When designing and building a practical working refrigerator there is a choice of using the compression cycle or the absorption cycle, described in the following sections.

Compression refrigeration cycle

The compression refrigeration cycle is driven by an electric compressor which also circulates the refrigerant. The operating principles of the compression cycle are shown schematically in Figure 4.9.

A domestic refrigerator usually uses the compression cycle and a layout of a practical refrigerator is shown in Figure 4.10. The evaporator coils are situated at the top of the compartment where natural convection currents aid the flow of heat. The heat is extracted and transferred to the condenser coils, which release the heat outside the refrigerator with the help of cooling fins. The temperature of the interior is regulated by a thermostat which switches the compressor motor on and off.

Absorption refrigeration cycle

The absorption refrigeration cycle, shown in Figure 4.11, operates without a compressor and needs no moving parts. The principle of evaporative cooling is the same as for the compression cycle but the driving mechanism is less obvious.

The refrigerant is pressurised and circulated by a generator or 'boiler', which can be heated by any source. A concentrated solution of ammonia in water is heated by the generator and the ammonia is driven off as a vapour, leaving the water. The pressure of the ammonia vapour increases as it passes through the condenser and evaporator, so acting as a refrigerant. In the absorber the ammonia is redissolved by water flowing from the generator and the cycle then repeats.

The absorption cycle is used in domestic gas-operated refrigerators and for commercial plants, particularly where there is a source of waste heat that can be used to drive the cycle.

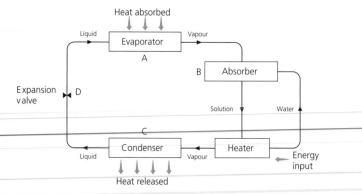

Figure 4.11 Absorption refrigeration cycle

Heat pumps

Every object in the environment is a potential source of heat, even ground that is frozen solid. The quantity and the temperature of this 'low grade' heat is usually too low for it to be useful. With some sources however, it can be profitable to 'pump' its heat to a higher temperature.

- **A heat pump** is a device that extracts heat from a low-temperature source and upgrades it to a higher temperature.

The heat pump employs a standard refrigeration cycle. The compression system, shown in Figure 4.9, is the commonest type and is outlined below.

Stage A

The evaporator coils are placed in the low-temperature heat source and heat is absorbed by the evaporating refrigerant.

Stage B

The temperature of the refrigerant is increased as it is compressed, the extra energy being supplied by the electric motor.

Stage C

The refrigerant condenses and releases heat energy at a useful temperature. The condenser coils are used to heat air or some other medium.

Stage D

The refrigerant is expanded and it evaporates for a repeat of the cycle.

This heat pump cycle is the same as that used in a domestic refrigerator which gives off heat at the rear: heat that has been 'extracted' from your food.

Air-conditioning units use the same refrigeration cycle in order to chill air and lower its humidity. An air-conditioning unit can therefore be designed to run in reverse as a heat pump and make good use of the heat output from the cycle, providing that there is a constant source of low-grade heat available.

Large air-conditioning installations have long made use of the reverse cycle to provide efficient heating. A more recent and welcome trend has been the introduction, at reasonable cost, of small compact units for homes and small offices that act as both heat pump and air conditioner.

Heat pump efficiency

The main feature of the heat pump is that it produces more usable energy than it consumes. This efficiency is measured by a coefficient of performance for heating.

- The **coefficient of performance** (COP_H) of a heat pump is the ratio of heat output to the energy needed to operate the pump.

$$COP_H = \frac{\text{Heat energy output}}{\text{Pump energy input}} = \frac{T_1}{T_1 - T_2}$$

Reverse cycle air conditioning is another term for heat pump

where
T_1 = absolute temperature of heat output (K)
T_2 = absolute temperature of heat source (K).

The above expressions show that the COP_H is always greater than unity, so that a heat pump always gives out more energy than is directly supplied to the pump. It therefore uses electricity more efficiently than any other system. The temperature equation also shows that the COP_H decreases as the temperature of the heat source decreases, which unfortunately means lower efficiency in colder weather.

The COP_H values achieved by heat pumps under practical conditions are between 2 and 5. These values take account of the extra energy needed to run circulatory fans and pumps, to defrost the heat extraction coils, and to supply supplementary heating if necessary. In terms of global energy, a heat pump does not actually give energy for nothing, as might be suggested by the coefficient of performance. The bottom line of the COP_H only takes account of the input energy from the user and ignores the energy taken from the source.

However, we only pay for the energy directly supplied, so using a heat pump provides big savings. For example, a heat pump may give you 8 kW of heat power into a room while you only pay for 3 kW of electrical power.

Heat sources for heat pumps

The extraction coils (evaporator) of a heat pump should ideally be placed in a heat source that remains at a constant temperature.

Air

Air is the most commonly used heat source for heat pumps but has the disadvantage of low thermal capacity. Outside air also has a variable temperature and when the air temperature falls the COP_H also falls. The exhaust air of a ventilation system is a useful source of heat at constant temperature.

Earth

The ground can be a useful source of low-grade heat in which to bury the extraction coils of a heat pump. Ground temperatures remain relatively constant but large areas may need to be used, otherwise the heat pump will cause the ground to freeze.

Water

Water is a good heat source for heat pumps provided that the supply does not freeze under operating conditions. Rivers, lakes, the sea and supplies of waste water can be used.

VENTILATION INSTALLATIONS

The previous sections of this chapter have revealed the importance of air control in buildings for the comfort and health of the inhabitants, for the optimum performance of the building, and to prevent deterioration in the building by interstitial condensation. The design of ventilation systems for buildings therefore needs to take account of the factors listed below.

- volume of air
- movement of air
- distribution of air
- filtration

- temperature change
- humidity change
- energy conservation
- feedback and control.

The common systems used to control some or all of these factors can be considered as two broad types, natural ventilation and mechanical ventilation. The following sections outline the principles and applications of ventilation systems.

Natural ventilation

See also:

section on *Fluid flow* in Resource 5

Natural ventilation is provided by the following two broad mechanisms.

- **Air pressure difference**: such as that caused by the wind movement over and around a building.
- **Stack effect**: caused by the natural movement of warm air rising within a building.

A simple example of using air pressure differences is when windows are opened on either side of a building. The stack effect is particularly noticeable when there is a considerable difference in height, for instance in the stairwell of a tall building. In a stairwell the stack effect creates large, uncomfortable airflows and a fire danger, but the stack effect can also be put to good use in a ventilation system. Some modern environmentally friendly buildings, such as the modern extension at the Houses of Parliament in Westminster, use the stack effect to minimise the need for mechanical ventilation.

Passive stack ventilation (PSV) systems use a combination of effects such as cross ventilation by air pressure differences, the stack effect of warm air rising because it is less dense and is displaced by cooler air, and the venturi effect of wind across an opening creating suction. In large buildings PSV can make use of stairwells and atria rising through the middle of the buildings. PSV in domestic buildings usually involves a duct linking the ceiling of a room, such as a bathroom, with a terminal on the rooftop or a special ridge tile. The flow of air through PSV systems can be regulated by shutters which may be linked to timers and control systems.

Trickle ventilators provide background ventilation by means of small ventilation openings, such as a slot in the top of a window frame. They normally have a controllable shutter.

Mechanical ventilation

The use of mechanical ventilation makes it possible to use spaces, such as those deep within buildings, that cannot be easily ventilated by natural means. The mechanical fans depend on a supply of energy, usually electricity.

A simple extract or input mechanical system provides movement of air but only a limited degree of control. A *plenum* system gives better control of air by using ductwork and usually has the ability to heat the air if necessary. The features of a typical mechanical ventilation system are shown in Table 4.5 and in Figure 4.12.

> Modern internal bathrooms *need* mechanical ventilation

Air conditioning

An air-conditioning installation has the aim of producing and maintaining a designed internal air environment, despite the variations in external air conditions.

Table 4.5 Features of mechanical ventilation installation

Installation feature	Purpose and features
Fresh air intake	Must be carefully situated
Recirculated air	Can be mixed with fresh air
Air filter	Removes particles and contaminants
Heater	Elements or coils over which air passes
Fan	Provides air movement for intake and distribution
Ductwork	Guides distribution of air
Diffusers	Control distribution of air into room

Figure 4.12 Schematic diagram of ventilation system

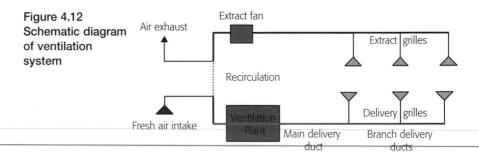

The equipment therefore has to be able to heat and cool the air, to humidify and dehumidify it, and to respond automatically to changes in the external air.

Typical air-conditioning equipment is described in Table 4.6 and in Figure 4.13. The removal of water from the air, to lower the humidity, is usually achieved by cooling the air with refrigeration plant and then reheating the air before passing it into the distribution system. More information about changes in humidity is given in an earlier section of this chapter, but the feature to note is that air-conditioning plant uses significant amounts of energy.

Table 4.6 Features of air-conditioning installation

Installation feature	Purpose and features
Fresh air intake	Must be carefully situated
Recirculated air	Can be mixed with fresh air
Air filter	Removes particles and contaminants
Heater	Elements or coils over which air passes
Preheater	To provide initial heat energy if needed
Chiller	Lowers temperature and therefore removes moisture
Humidifier	Adds moisture to the air if needed
Reheater	Adjusts final temperature of air if necessary
Fan	Provides air movement for intake and distribution
Ductwork	Guides distribution of air
Diffusers	Control distribution of air into room

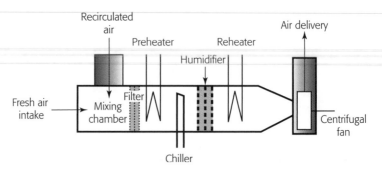

Figure 4.13 Schematic diagram of air-conditioning installation

EXERCISES

1. A sample of air at 1°C has a vapour pressure of 540 Pa. Calculate the RH of this air, given that the SVP of water is 1 230 Pa at 10°C.

2. The air temperature in a room is 20°C and the dew-point is 15°C. Calculate the RH of this air, given that the SVP of water is 2 340 Pa at 20°C and 1 700 Pa at 15°C.

3. Use the psychrometric chart for the following problems:
   (a) If air has a dry-bulb temperature of 20°C and a wet-bulb temperature of 15°C, then find the RH and the dew-point.
   (b) If air at 16°C dry-bulb temperature has an RH of 70 per cent, then find the moisture content.
   (c) If air at 20°C dry-bulb temperature has a vapour pressure of 7 mb, then find the RH.

4. External air at 4°C and 80 per cent RH is heated to 20°C. Use the psychrometric chart to solve the following conditions:
   (a) The RH of the heated air.
   (b) The RH of the heated air if 0.006 kg/kg of moisture is added while heating.
   (c) The increase in vapour pressure between the two states.

5. The surface of a wall has a temperature of 11°C when air at 14°C first begins to condense upon the wall surface. Use the psychrometric chart to find the following.
   (a) The RH of the air.
   (b) The reduction in moisture content necessary to lower the RH of the air to 50 per cent.

6. A 102 mm thick brick wall is insulated on the inside surface by the addition of 40 mm of mineral wool covered with 10 mm of plasterboard. The thermal resistances, in m² K/W, are: external surface 0.055, brickwork 0.133, mineral wool 0.4, and internal surface 0.123. The vapour resistivities, in GN s/kg m, are: brickwork 60, mineral wool 5, and plasterboard 50. The inside air is at 20°C and 59 per cent RH; the outside air is at 0°C and 100 per cent RH.
   (a) Calculate the boundary values of structural temperatures and dew-points.
   (b) Plot a structural temperature profile and a dew-point profile on the same scaled cross-section diagram of the wall.
   (c) Comment upon the above results.

Answers are on page 325.

5

Properties of Lighting

CHAPTER OUTLINE

Vision is one of our most important human senses. We use sight to understand the environment, to move about and to carry out tasks. Our vision also plays a major role in leisure activities and in our sense of well-being. Our ability to see depends upon the presence of light and the various effects light can produce. Being able to provide and to control light from natural and artificial light sources is therefore an important feature of the built environment and our comfort.

Aspects of lighting science, such as measurement of light, are oriented towards human perceptions because they depend on the working of the human eye and other parts of the 'visual system'. This chapter introduces the properties of light relating to the built environment so that you can:

▶ appreciate the principles for measuring light and its effects
▶ make use of the technical measurements and units associated with lighting
▶ understand how illumination on a surface changes with distance from the light source
▶ understand how illumination on a surface changes with the direction of the light
▶ understand the effects produced by reflection of light
▶ be familiar with typical values for lighting measurement and relate them to practical situations
▶ calculate practical quantities involved in lighting design
▶ understand the nature of colour and the ways it can be measured

▶ appreciate the interactions between spectral colour and human vision
▶ understand the principles of colour reproduction and perception
▶ predict the practical effects of colour in design.

The following chapters apply this knowledge to lighting design using both artificial and natural lighting. Resource 3 at the end of the book has useful information that explains the physical nature of light and the operation of the eye and visual system.

MEASUREMENT OF LIGHTING

The nature of electromagnetic radiation and energy is explained in the Resource sections of this book which also describe the nature of human vision. We define 'light' as that energy in the form of particular wavelengths of electromagnetic radiation which can be detected by the human sense of sight. Because light is a form of energy, the standard units of energy could be used to measure it. But the effect of light on the human environment also depends upon the sensitivity of the eye, and a special set of units has therefore been developed for the measurement of light and its effects on humans.

See also:

Chapters 6 and 7 on *Artificial lighting* and on *Natural lighting,* and Resource 3

See also:

section on *Nature of light* in Resource 3

Solid angle

As light can radiate in all three dimensions it is necessary to measure the way in which the space around a point can be divided into 'solid angles'. The standard SI unit of solid angle is the steradian, illustrated in Figure 5.1.

• **One steradian (ω)** is that solid angle at the centre of a sphere which cuts an area on the surface of the sphere equal to the size of the radius squared.

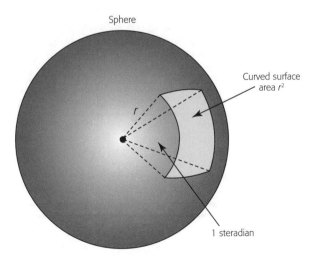

Sphere

Curved surface area r^2

r

1 steradian

Figure 5.1 The steradian

The size of a solid angle does not depend upon the radius of the sphere or upon the shape of the solid angle. The total amount of solid angle contained around a point at the centre of a sphere is equal to the number of areas, each of size radius squared, which can fit onto the total surface area of a sphere. That is:

$$\text{Total solid angle around a point} = \frac{\text{surface area at sphere}}{\text{area giving 1 steradian}} = \frac{4\pi r^2}{r^2} = 4\pi$$

Therefore a complete sphere contains a total of 4π steradians.

Luminous intensity

The concept of luminous intensity is used to compare different light sources and measure their 'strength'.

- **Luminous intensity** (I) is the power of a light source, or illuminated surface, to emit light in a particular direction.

    Unit: candela (cd)

Typical luminous intensity values:
candle: 1 cd
electric lamp:
 100 cd
lighthouse:
 10^5 cd

The candela is one of the base units in the Sl system. One candela is defined as the luminous intensity in a given direction of a source that emits monochromatic radiation of frequency 540×10^{12} Hz and of which the radiant intensity in that direction is $1/683$ W/ω.

The effect of one candela is still approximately the same as the original idea of one candlepower, and the mean spherical intensity (MSI) of a 100 W traditional light bulb, for example, is about 100 cd.

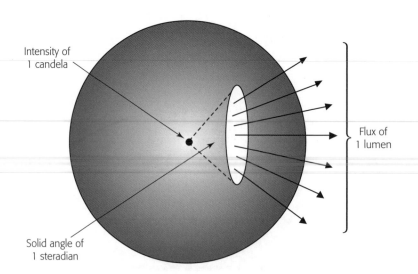

Figure 5.2 Definition of the lumen

Luminous flux

The rate of flow of any electromagnetic energy can be expressed in terms of power, but light energy is also measured by luminous flux.
- **Luminous flux** (F) is the rate of flow of light energy.

> Unit: lumen (lm)

By definition, one lumen is the luminous flux emitted within one steradian by a point source of light of one candela, as shown in Figure 5.2.

In general, luminous flux and luminous intensity are related by the following formula.

$$I = \frac{F}{\omega}$$

where
  I = mean spherical intensity of the source (cd)
  F = luminous flux emitted by the source (lm)
  ω = solid angle containing the flux (sterad).

In the common case of a point source emitting flux in all directions, the total solid angle around the point is 4π steradians.

Substituting the value of 4π and rearranging the formula

$$I = \frac{F}{\omega} = \frac{T}{4\pi}$$

$$F = I \times 4\pi$$

gives the following useful formula:

$$F = I \times 4\pi$$

Illuminance

When luminous flux falls on a surface it illuminates that surface. The lighting effect is termed illuminance.

- **Illuminance** (E) is the density of luminous flux reaching a surface.

> Unit: lux (lx) where 1 lux = 1 lumen/(metre)^2

Common luminance levels range from 50 lux for low domestic lighting to 50,000 lux for bright daylight. Recommended lighting levels are specified in terms of illuminance, and examples of standards are given in the section on lighting design.

Typical lighting values in lux
moonlight: 0.1 lx
street lighting: 10 lx
office lighting: 300 lx
overcast day: 1000 lx
direct sunlight: 100 000 lx

If light is falling on a surface at right angles to the surface then the illuminance is given by the following formula:

$$E = \frac{F}{A}$$

where

E = illuminance on surface (lx)

F = total flux reaching surface (lm)

A = area of the surface (m²)

Footcandle:
an older unit
where
1fc = 10.76 lx

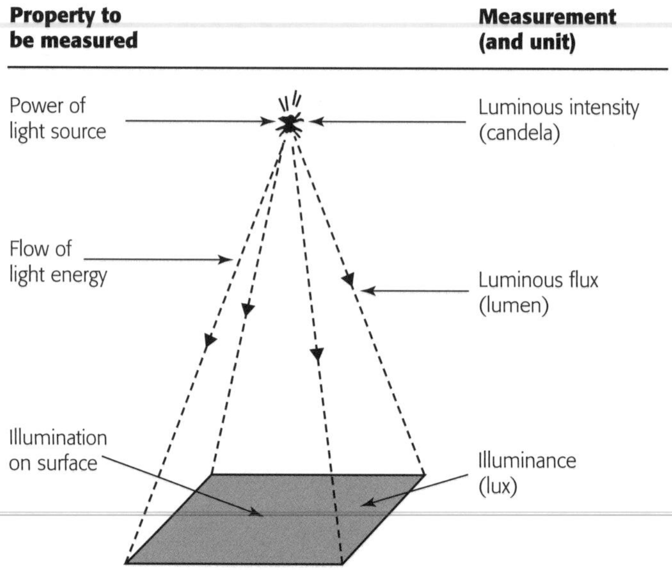

Property to be measured		Measurement (and unit)
Power of light source		Luminous intensity (candela)
Flow of light energy		Luminous flux (lumen)
Illumination on surface		Illuminance (lux)

Figure 5.3 Summary of lighting measurements

Worked Example 5.1

A small source of light has a mean spherical intensity of 100 cd. One-quarter of the total flux emitted from the source falls at right angles onto a surface measuring 3 m by 0.7 m. Calculate:

(a)   the total luminous flux given out by the source; and
(b)   the illuminance produced on the surface.

(a)   Know

$$I = 100 \text{ cd}, \quad \varpi = 4\pi, \quad F = ?$$

Using the formula for intensity and substituting

$$I \quad = \frac{F}{\omega}$$

$$100 \quad = \frac{F}{4\pi}$$

or $F = 4\pi \times 100$

$\qquad = 1\,256.64\ \text{lm}$

Total flux = **1256.64 lm**

(b)   Know $F = 1\,256.64 \times 0.25 = 314.16\ \text{lm}$

$\qquad\qquad A = 0.7 \times 3 = 2.1\ \text{m}^2$

$\qquad\qquad E = ?$

Using the formula for illuminance and substituting

$$E = \frac{F}{A}$$

$$\quad = \frac{314.16}{2.1} = 149.6$$

So illuminance = **150 lx**

Inverse square law of illumination

As the luminous flux emitted by a point source of light travels away from the source, the area over which the flux can spread increases. Therefore, the luminous flux per unit area (i.e. the illuminance) must decrease. This relationship is expressed by the inverse square law, as illustrated in Figure 5.4.

Inverse square law

- The illuminance produced by a point source of light decreases in inverse proportion to the square of the distance from the source.

Effects of
distance:
2 x distance gives
1/4 lux
3 x distance gives
1/9 lux
4 x distance gives
1/16 lux etc.

In SI units this law may be expressed mathematically by the following formula.

$$E = \frac{I}{d^2}$$

where I = intensity of a point source (cd)

$\qquad d$ = distance between source and surface (m)

$\qquad E$ = illuminance on that surface (lx)

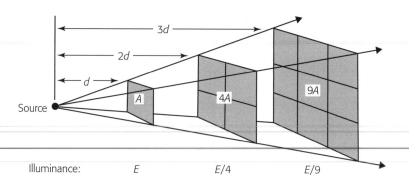

Illuminance:　　　　E　　　　　$E/4$　　　　$E/9$

Figure 5.4 Inverse square law of illumination

An important consequence of the inverse square law is that changes in the position of light sources produce relatively large changes in lighting effect. For example, doubling the distance between a lamp and a surface causes the illuminance on that surface to decrease to one-quarter of its original value.

Worked Example 5.2

A lamp has a luminous intensity of 1200 cd and acts as a point source. Calculate the illuminance produced on surfaces at the following positions.

(a)　At 2 m distance from the lamp, and
(b)　at 6 m distance from the lamp.

Know: I = 1200 cd, d_1 = 2 m and d_2 = 6 m, E = ?

Using the inverse square formula for illuminance and substituting

(a)　for a distance of 2 m

$$E = \frac{I}{d^2} = \frac{1200}{2^2} = 300$$

So illuminance at 2 m = **300 lx**

(b)　for a distance of 6m

$$E = \frac{I}{d^2} = \frac{1200}{6^2} = 33.33$$

So illuminance at 6 m = **33.33 lx**

Cosine law of illumination

When the luminous flux from a point source reaches the surface of a surrounding sphere, the direction of the light is always at right angles to that surface. However, light strikes many surfaces at an inclined angle and therefore illuminates larger areas than when it strikes at a right angle. The geometrical effect is shown in Figure 5.5. If the luminous flux is kept constant but spread over a larger area, then the illuminance at any point on that area must decrease.

Figures 5.5 and 5.6 also show how the area illuminated increases by a factor of 1/cosine θ. Because illuminance is equal to luminous flux divided by area, the illuminance decreases by a factor of cosine θ. This relationship is sometimes termed *Lambert's Cosine Rule* and can be expressed by a general formula combining the factors affecting illumination:

> Larger angle =
> Larger area =
> Less density of
> light

$$E = \frac{I}{d^2} \cos \theta$$

where
    E = illuminance on surface (lx)
    I = intensity of source (cd)
    d = distance between source and surface (m)
    θ = angle between direction of flux and the normal
Note: when $\theta = 0°$ then $\cos \theta = 1$.

An alternative formula using the perpendicular distance h between the source and the illuminated surface is:

$$E = \frac{I}{h^2} \cos^3 \theta$$

The cosine law affects most practical lighting arrangements as it is usually difficult for all surfaces to receive light at right angles. For example, when a lamp on

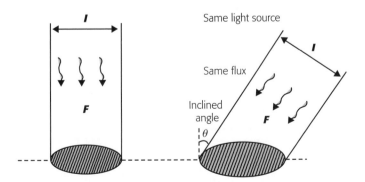

Figure 5.5 Cosine law of illumination

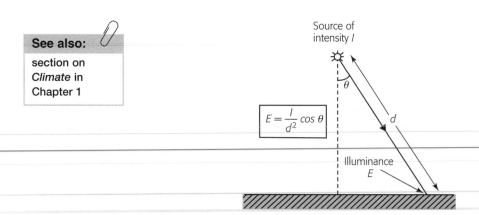

See also:

section on
Climate in
Chapter 1

Figure 5.6 Laws of illumination

the ceiling illuminates a floor, only the point directly below the light fitting will receive luminous flux at right angles; the light will strike all other parts of the floor at varying angles of inclination.

The cosine law also has a large effect on *solar radiation* received by different locations and buildings. The radiation from the Sun strikes the surface of the Earth at different angles of incidence, which vary from the ideal right angle, especially at higher latitudes away from the equator. The intensity of light and heat received on flat ground decreases in accordance with the cosine law but can be increased by arranging surfaces at right angles to the incident radiation. The walls and windows of buildings therefore have higher solar gains than the roofs when the Sun is low in the sky.

Worked Example 5.3

A lamp acts as a point source with a mean spherical intensity of 1500 cd. It is fixed 2 m above the centre of a circular table which has a radius of 1.5 m. Calculate the illuminance provided at the edge of the table, ignoring reflected light.

Know
$$I = 1500 \text{ cd}, \quad E = ?$$

Using triangle laws

$$
\begin{aligned}
d^2 &= 2^2 + 1.5^2 \\
&= 4 + 2.25 = 6.25 \\
d &= 2.50 \\
\cos \theta &= \frac{\text{adjacent}}{\text{hypotenuse}} = \frac{2}{d} = \frac{2}{2.5}
\end{aligned}
$$

Note: you do not need to know the actual value of the angle in degrees.

Using illumination formula and substituting

$$E = \frac{I}{d^2} \cos \theta$$

$$E = \frac{1500}{6.25} \times 0.8 = 192$$

So illuminance at table edge = **192 lx**

Reflection

One method of changing the direction of light is by the process of reflection, which may be of two types as shown in Figure 5.7.

- **Specular reflection** is direct reflection in one direction only. The angle of incidence (*i*) equals the angle of reflection (*r*).
- **Diffuse reflection** is reflection in which the light is scattered in various directions.

Reflectance

Most practical surfaces give a mixture of free and diffuse reflection properties. The amount of reflection at a surface is measured by the *reflection factor* or reflectance.

- **Reflectance** is the ratio of the luminous flux reflected from a surface to the flux incident upon the surface.

Typical values of reflectance are given in Table 5.l. Maximum reflectance has a value near 1, for light, shiny surfaces. Minimum reflectance has a value near 0, for dark, dull surfaces.

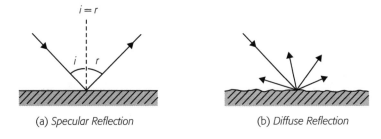

(a) *Specular Reflection*   (b) *Diffuse Reflection*

Figure 5.7 Reflection of light

Table 5.1 Reflectances of building surfaces

Surface	Reflectance
White emulsion paint on plaster	0.8
White emulsion paint on concrete	0.6
Concrete: light grey	0.4
Timber: birch, beech, or similar	0.3
Bricks: red fletton	0.35
Quarry tiles: red	0.1

Reflected components of light are an important factor in illumination. The following is a simple worked example, and the influence of reflectances will be referred to again in the section on lighting design.

Worked Example 5.4

A point source of light with a luminous intensity of 1800 cd is set 3 m above the floor and 1 m below the ceiling which has a reflectance of 0.5. Calculate the direct and reflected components of the illuminance on the floor beneath the light.

(a)   Know $I = 1800$ cd, $d = 3$ m, $E = ?$
   Using illumination formula and substituting

$$E = \frac{I}{d^2}$$

$$E = \frac{1\,800}{3^2} = 200$$

So direct component = **200 lx**

(b)   Know: I = 1800 cd, d = 1 + 1 + 3 = 5 m, E = ?

Using illumination formula and substituting

$$E = \frac{I}{d^2} \times \text{reflectance}$$

$$E = \frac{1\,800}{5^2} \times 0.5 = 36$$

So reflected component = **36 lx**

Luminance

The appearance of an object is affected by the amount of light it emits or reflects compared to the area of its surface. This surface 'brightness' is termed luminance.

- **Luminance** (L) is a measure of the ability of an area of light source, or reflecting surface, to produce the sensation of brightness.

For example, a spot lamp and a fluorescent tube may each have the same luminous intensity. But, because of its larger surface area, the fluorescent tube would have a lower luminance. There are two types of luminance measurement.

Self-luminous sources

Self-luminous sources include light sources and also reflecting surfaces, like the moon, which can be considered as a source of light. The luminance is given by the following formula.

$$L = \frac{\text{Luminous intensity in a given direction}}{\text{Area of source as seen from that direction}}$$

Unit: candela per square metre (cd/m^2)

Reflecting surfaces

For reflecting surfaces only, the luminance can be expressed in terms of the luminous flux emitted per unit area.

$$L = \frac{\text{Luminous flux reflected}}{\text{Area of reflecting surface}}$$

Unit: apostilb (asb) where 1 asb = 1 lumen/m^2

The candela per square metre is the SI unit of luminance. The apostilb is an alternative unit of luminance which is convenient for some calculations. 1 apostilb = $1/\pi$ cd/m^2. This luminance of a surface can be calculated as the product of the illuminance and the reflectance of a surface.

Glare

The eye can detect a wide range of light levels, but vision is affected by the range of brightness visible at any one time.

- **Glare** is the discomfort or impairment of vision caused by an excessive range of brightness in the visual field.

Glare can be caused by lamps, windows and painted surfaces appearing too bright in comparison with the general background. Glare can be further described as disability and discomfort glare.

Disability glare

Disability glare is the glare that lessens the ability to see detail. It does not necessarily cause visual discomfort. For example, excessive reflections from shiny white paper can cause disability glare while reading.

Discomfort glare

Discomfort glare is the glare that causes visual discomfort without necessarily lessening the ability to see detail. An unshielded lamp is a common example of discomfort glare. The amount of discomfort depends on the angle of view and the type of location. If the direction of view is fixed on a particular visual task, then glare caused by lighting conditions will be more noticeable.

Light meters

A *photometer* is an instrument that measures the luminous intensity of a light source by comparing it with a standard source whose intensity is known. The distances between the instrument and the two light sources are adjusted until they each provide the same illuminance at the photometer.

The human eye is a very good judge of this equal illuminance, and photometers generally use some system which allows two screens to be compared. The inverse square law can be used to calculate the unknown intensity. Under conditions of equal illuminance the following formula is true:

$$\frac{I_1}{d_1^{\,2}} = \frac{I_2}{d_2^{\,2}}$$

where
I_1 = intensity of source at distance d_1
I_2 = intensity of source at distance d_2

A photocell light meter is an instrument that directly measures the illuminance on a surface. The electrical resistance of some semiconductors, such as selenium, changes with exposure to light and this property is used in an electrical circuit connected to a galvanometer. This meter may be calibrated in lux.

Directional quality

The appearance of an object is affected by the direction of light as well as by the quantity of light illuminating the object.

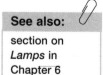

See also:

section on
Lamps in
Chapter 6

- **Illumination vector**: the quantity of light from a specified direction, such as horizontal illumination.
- **Scalar illuminance**: the total illuminance caused by light from all directions, including reflected light.

• **Vector/scalar ratio**: a measure of the directional strength of light at a particular point.

A vector/scalar ratio of 3.0 indicates lighting with very strong directional qualities, such as that provided by spotlights or direct sunlight. The effect is of strong contrasts and dark shadows. A vector/scalar ratio of 0.5 indicates lighting with weak directional qualities, such as that provided by indirect lighting by reflections; this lighting is free of shadows and makes objects look 'flat'.

COLOUR

Colour is a subjective effect that occurs in the brain when the eye is stimulated by various wavelengths of light. It is difficult to specify colour by any method, and especially difficult by means of this black and white printing. But the description and measurement of colour is important in the design of lighting schemes, as well as in photographic films, paints, dyes and inks.

> **See also:**
> section on *Nature of vision* in Resource 3

Spectral energy distribution

The brain can experience the impression of only one colour at a time, and it cannot detect how that colour has been made up. It is possible for different

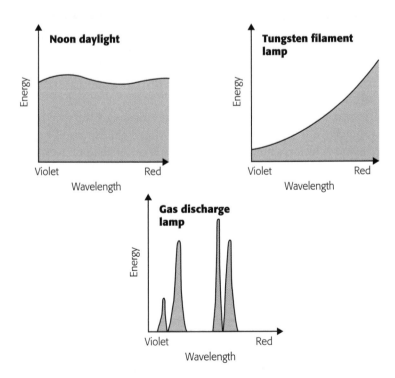

Figure 5.8 Spectral energy distribution

combinations of different wavelengths to produce the same colour impression in the brain.

A *spectrometer* is an instrument that disperses light into its component wavelengths and measures the amount of light energy radiated at each wavelength. This spectral energy distribution reveals whether some light is being emitted at each wavelength (continuous spectrum), or only at certain wavelengths (discontinuous or line spectrum). Some typical spectral curves are shown in Figure 5.8.

Colour systems

CIE colour coordinates

The Commission Internationale de l'Eclairage (CIE) have produced many international lighting standards. The CIE coordinate system describes any colour as a mixture of three monochromatic primary colours. Three coordinates specify how much of each primary colour needs to be mixed to reproduce the colour being described.

Munsell system

The Munsell system is used by some architects and interior designers for classifying the colour of surfaces. It specifies colours in terms of three factors called hue (basic colour), chroma (intensity of colour) and value (greyness).

Colour reproduction

The human vision can distinguish many hundreds of different colours and intensities of colour. Modern systems of colour representation by ink, photography and television can reproduce these many variants to the satisfaction of the eye, yet such systems use just a few basic colours.

It has been shown that white light contains all the colours of the spectrum, which can be recombined to give white light again. White light can also be split into just three colours which recombine to give white light. In addition to white light, any colour can be reproduced by various combinations of three suitable primary colours. Most colour systems use this trichromatic method of reproduction and the eye is believed to send its information to the brain by a similar method of coding.

Newton's discoveries concerning the combination of colours initially seemed to disagree with the experience of mixing paints. This confusion about combining colours can still arise because they can be mixed by two different methods which have different effects: additive mixing and subtractive mixing.

Additive colour

If coloured lights are added together then they will produce other colours. When the three primary additive colours are combined in equal proportions they *add* to produce white light.

Additive primary colours

- Red + Green + Blue = White
- Red + Green = Yellow
- Green + Blue = Cyan
- Red + Blue = Magenta

Cyan is a 'sky-blue' colour
Magenta is a 'purple-red' colour

Applications of additive colour

In everyday life most colour is seen as a result of subtractive colour mixing, as described in the next section. The following applications use additive processes of mixing colour; either by overlapping different colours or placing them so close together that the eye considers them mixed.

- **Stage lighting**: A darkened theatre stage is often lit by strong spotlights of different colours. When the beams from different coloured spotlights overlap on a surface they produce different colours according to the rules of additive colour.
- **Display screens**: Modern television or computer screens use a variety of LCD, LED and plasma technologies to form a matrix of many subpixels which are grouped as trios of red, green and blue. Each subpixel is controlled by a separate electronic signal so that its colour strength can be varied. For example, when a display screen is showing white, a magnified look at the screen will reveal that there is no white there at all, just red, green and blue colours turned on together. The white that you 'see' is actually an effect produced in your brain.
- **Colour printing**: some paper printing processes, such as gravure, produce a mosaic of coloured ink dots on white paper which act in an additive manner when viewed.

Subtractive colour

If colours are subtracted from white light then other colours will be produced. When the three primary subtractive colours are combined in equal proportions they *subtract* (absorb) components of white light to produce darkness.

Subtractive primary colours

Cyan	subtracts	Red	These pairs are termed *complementary* colours
Magenta	subtracts	Green	
Yellow	subtracts	Blue	

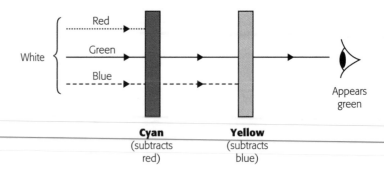

Figure 5.9 Combination of subtractive colours

White light can be considered as a combination of red, green and blue light. Materials that transmit or reflect light absorb selected wavelengths and pass the remaining light to the eye. Combining two or more paints, or coloured layers, has a cumulative effect, as shown in Figure 5.9.

A red surface is defined by the fact that it subtracts green and blue from white light, leaving only red light to reach the eye. A white surface reflects all colours. If the colour content of the light source changes then the appearance of the surface may change. This effect is important when comparing surface colours under different types of light source.

Applications of subtractive colour

Most colour is seen as a result of subtractive processes and some common applications are given below.

- **Paint pigments**: Paint colours are mixed according to subtractive principles, even though, for simplicity, the primary colours are often called red, yellow and blue instead of magenta, yellow and cyan.
- **Colour photographs**: All colour transparencies and colour prints are finally composed of different densities of three basic dyes: cyan, magenta and yellow.

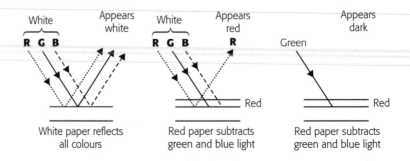

Figure 5.10 Subtractive colour appearance

- **Colour printing**: White paper is over-printed three times with the three basic colours of printing ink which are cyan, magenta and yellow. Black ink is also used to achieve extra density.

EXERCISES

1. A small lamp emits a total luminous flux of 1257 lm in all directions. Calculate the luminous intensity of this light source.

2. A point source of light has an intensity of 410 cd and radiates uniformly in all directions.
   (a) Calculate the quantity of flux flowing into a hemisphere.
   (b) Calculate the average illuminance produced on the inside surface of this hemisphere if it has a radius of 1.5 m.

3. A small lamp has a mean luminous intensity of 80 cd. Calculate the maximum direct illuminance the lamp produces on a surface under the following conditions:
   (a) At a distance of 0.8 m from the lamp.
   (b) At a distance of 3.2 m from the lamp.

4. A street lamp has a uniform intensity of 1200 cd. It is positioned 7 m above the centre line of a road which is 8 m wide.
   (a) Calculate the illuminance on the road surface directly below the lamp.
   (b) Calculate the illuminance at the edge of the roadway.

5. A uniform point source of light emits a total flux of 2500 lm. It is suspended 800 mm above the centre of a square table with sides of length 600 mm. Calculate the minimum and maximum illuminances produced on the table.

6. A photometer is positioned on a direct line between two lamps. When each inside of the photometer receives equal illuminance, the distances from the photometer are 500 mm to lamp A and 650 mm to lamp B. Lamp A is known to have a luminous intensity of 70 cd. Calculate the luminous intensity of lamp B.

7. Use a diagram to predict and demonstrate the appearance of white light after it has passed through a cyan glass and then a magenta glass. Assume that white light is simply composed of red, green and blue.

8. Use a diagram to predict and demonstrate the appearance of a blue surface which is illuminated by yellow light.

Answers are on page 325.

6

Artificial Lighting

CHAPTER OUTLINE

The type of lighting chosen for a building is closely linked to other design decisions for that building, such as the basic plan shape, the type and extent of windows, the type of heating or cooling, and the target energy use. Compared with traditional buildings, where all habitable rooms had windows, many modern buildings have interior spaces that depend totally upon artificial lighting.

The energy used by artificial lighting is around 20 per cent of typical national electricity consumption, and changing the lighting equipment in a building is relatively easy compared with changing other energy properties of a building, such as insulation. The efficacy and technologies of lighting have made rapid advances in recent decades, after changing little for about 100 years after the introduction of the tungsten lamp.

Making changes to lighting design and equipment is therefore a major factor in reducing the energy consumption and carbon emission associated with buildings. The Building Regulations for England and Wales, for example, stipulate the use of low-energy lamps and lighting controls such as automatic switch systems.

The main functions of artificial lighting can be summarised as follows:

▶ **Task**: To provide enough light for people to carry out a particular activity.
▶ **Movement**: To provide enough light for people to move about with ease and safety.
▶ **Display**: To display the features of the building in a manner suitable for its character and purpose.

$\longrightarrow$

To achieve these design aims it is necessary to consider the properties of lamps, of lamp fittings and of the room surfaces that surround lamps. The contents of this chapter will enable you to:

▶ understand the technology of lamps as sources of light
▶ use lamp properties to choose appropriate lamps for different purposes
▶ calculate the total costs of running different lamps
▶ appreciate the trends in modern lamp technology and their applications
▶ understand the purpose and properties of luminaires
▶ understand factors and techniques of lighting design
▶ understand the energy implications of artificial lighting
▶ carry out lighting design calculations.

Where possible, these topics should be considered together with the related principles and practice of natural lighting explained in the next chapter. The previous chapter and Resource 3 of the book also have information that is relevant to the study of natural lighting in this chapter.

LAMPS

The oldest source of artificial light is the flame from fires, from candles and from oil lamps, where light is produced as one of the products of chemical combustion. Modern sources of artificial light convert electrical energy to light energy and are currently of three main types: incandescent sources, gas discharge sources and semiconductor sources.

Incandescent lamps

Incandescent sources produce light by heating substances to a temperature at which they glow and are luminous. This can be achieved by heating with a flame but in an electric lamp, such as the light bulb, a metal wire is heated by an electric current.

See also:

Chapters 5 and 7 on *Properties of lighting* and on *Natural lighting*, and Resource 3

Discharge lamps

Discharge lamps produce light by passing an electric current through a gas or vapour that has become ionised and hence able to conduct electricity. At low gas pressures, a luminous arc or discharge is formed between the electrodes and useful quantities of light are given off. Discharge lamps need special electronic control gear in their circuits and the raw colour quality of the light needs correction, such as in a fluorescent lamp.

Semiconductor sources

Certain semiconductor devices generate light by the process of electroluminescence when the active semiconductor layer is excited by a DC voltage. The colour of the light emitted depends upon the type of semiconductor materials used and the effect of any filter materials, such as with LED lamps.

The characteristics of the electric lamps used in modern lighting are summarised in Table 6.1. Further details of these lamps and their properties are given in the following sections.

Properties of lamps

Luminous efficacy

- **Luminous efficacy** is the ability of a lamp to convert electrical energy to light energy and is measured by the ratio of light output to energy input.

$$\text{Luminous efficacy} = \frac{\text{Luminous flux output}}{\text{Electrical power input}}$$

Unit: lumens/Watt (lm/W)

Lumen: the unit for light energy given out by the lamp

The electrical running costs of a lamp can be calculated by using its luminous efficacy. The efficacy of a lamp varies with its type and its wattage, so exact data should be obtained from the manufacturer.

Because luminous efficacy is a measure of the total luminous flux output in all directions it can be misleading for certain purposes. For example, a globe-shaped lamp gives out light in all directions while a spot lamp focuses its output in limited directions. So a spot lamp might have a lower efficacy rating but deliver more light to a target surface than a globe lamp with a higher efficacy.

See also:

section on *Measurement of lighting* in Chapter 5

Life

The luminous efficacy of a lamp decreases with time, and for a discharge lamp it may fall by as much as 50 per cent before the lamp fails. The nominal life of a lamp is determined by the manufacturer by considering the failure rate of a particular model of lamp combined with its fall in light output. In a large installation it is economically desirable that all the lamps are replaced at the same time on a specified maintenance schedule.

For example, if you had to use specialised labour and hoists to access the lighting in a high space, such as a railway station or airport, it would be unwise to change just the lamps that have obviously failed. If the lamps all started their life at the same time then they all need replacement at the end of their 'service life', even if they appear to be running normally.

Properties of lamps:
light output
energy efficiency
lamp life
colour output
colour rendering

Colour temperature

The qualities of light emitted by heated objects depend upon the temperature of the radiating object, and this fact can be used to describe the colour of light. A theoretically perfect radiator, called a 'black body', is used as the standard for comparison.

Table 6.1 Characteristics of electric lamps

Lamp type	Wattage range	Typical luminous efficacy (lm/W)	Operational life (hours)	Colour temperature (K)	Typical applications
Tungsten filament (GLS)	40–200	12	1000	2700	This traditional 'light bulb' should be replaced by low-energy lamps
Tungsten halogen (T-H)	300–2000	25	2000–4000	2800–3000	Area and display lighting
Compact fluorescent (CFL)	9–20	60	8000+	3000	Homes, offices and public buildings
Tubular fluorescent (MCF)	20–125	80	8000+	3000–6500	Offices and shops
Mercury fluorescent (MCB)	50–2000	60	8000+	4000	Factories and roadways
Mercury halide (MBI)	250–3500	70	8000+	4200	Factories and shops
High-pressure sodium (SON)	70–1000	125	8000+	2100	Factories and roadways
Low-pressure sodium (SOX)	35–180	180	8000+	n/a	Roadways and area lighting

Note: Higher-wattage lamps generally have higher efficiency and longer life; values for the efficacy and life of most discharge lamps are subject to improvements – see manufacturers' data.

- The **correlated colour temperature** (CCT) of a light source is the absolute temperature of a perfect radiator when the colour appearance of the radiator best matches that of the light source.

  Unit: Kelvin (K) – the SI unit of temperature.

Low CT = reddish
High CT = bluish

Specifying colour quality by colour temperature is suitable for light sources that emit a continuous spectrum, such as those giving various types of 'white' light. The lower values of colour temperature indicate light with a higher red content. Some examples of colour temperatures are given below.

Clear sky	12,000–24,000 K
Overcast sky	5000–8000 K
Tubular fluorescent lamps	3000–6500 K
Tungsten filament lamps	2700–3100 K

Colour rendering

The colour appearance of a surface is affected by the quality of light from the source. Colour rendering is the ability of a light source to reveal the colour appearance of surfaces. This ability is measured by comparing the appearance of objects under the light source with their appearance under a reference source such as daylight.

- The **colour rendering index** (R_a) is a number that indicates the accuracy by which a lamp shows surface colours; an ideal index being 100.

An R_a of 100 indicates an ideal lamp. Practical sources of white light have R_a values between 50 and 90. If accurate colour judgments are needed then the R_a should be above 80.

Tungsten filament lamps

Early patents for practical filament lamps
UK: Joseph Swan, 1878
USA: Thomas Edison, 1880

Electric incandescent lamps work by passing an electric current through a filament of metal and raising the temperature to white heat. When the metal is incandescent, at around 2800 K, useful quantities of light are given off. Tungsten is usually used because it has a high melting point and low rate of evaporation.

To prevent oxidation (burning) of the metal the tungsten coil is sealed inside a glass envelope and surrounded by an inert atmosphere of unreactive gases such as argon and nitrogen. During the operation of the lamp, tungsten is evaporated from the filament and deposited on the glass, causing it to blacken. The filament therefore thins and weakens and must eventually break.

General lighting service lamp

The general lighting service (GLS) lamp, or common light bulb, has a coiled filament contained within a envelope (bulb) of glass which may be clear or frosted. Inside the lamp is a fuse, and a typical construction is shown in Figure 6.1. The filament lamp produces a spectral distribution of light which is continuous but deficient in blue, as shown previously in Figure 5.8. This quality of light is seen as 'warm' and is considered generally suitable for social and domestic applications.

The cost of a tungsten filament lamp is low and its installation is simple, but the relatively short life of the lamp can cause the labour costs of replacement to be high. We may not notice the labour of changing a light bulb at home but it can be significant in large buildings and for high installations such as on a high ceiling. The over-riding cost factor against the tungsten filament is that the low luminous efficacy produces high electrical running costs. Only about 5 per cent of the electrical energy is converted to visible light and most of the energy consumed is given off as heat, especially radiant (infrared) heat. The simple light bulb is therefore being phased out in many countries, although some of the more efficient and specialist varieties described below continue to be used.

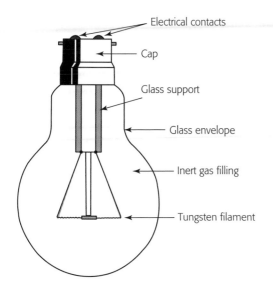

Electrical contacts

Cap

Glass support

Glass envelope

Inert gas filling

Tungsten filament

Figure 6.1 Traditional tungsten filament lamp

Reflecting lamps

The relatively large size of a standard tungsten filament lamp makes it difficult to control the direction of the light. *Spotlamps* (PAR) are filament lamps with the glass bulb silvered inside and shaped to form a parabola with the filament at the focus. This arrangement gives a directional beam of light which is available in different widths of beam. *Sealed beam lamps* use similar techniques. *Crown-silvered lamps* (CSL) are standard filament lamps where the glass bulb is silvered in front. When this lamp is used with a special external reflector it also gives narrow beams of light.

Tungsten-halogen lamps

Tungsten-halogen lamps have filaments that run at higher temperatures than simple tungsten lamps and contain a small quantity of a halogen gas, such as iodine or bromine. When tungsten evaporates from the filament it is deposited on the hot wall of the lamp where it combines with the iodine. This new compound is a vapour which carries the tungsten back onto the lamp and re-deposits it on the hot filament, while the iodine is also recycled.

In order to run at higher temperatures, the envelope of the tungsten-halogen lamp is made of *quartz* instead of plain glass. The heat-resistance of the quartz allows the construction of a very small bulb for applications such as spotlamps, projectors and car headlamps, where directional control of light is important. Tungsten-halogen lamps also have the general advantages over simple tungsten lamps of increased efficiency and longer life.

Low-voltage systems

Tungsten-halogen techniques have allowed the development of low-voltage bulbs where, because a lower resistance is needed, the filaments can be shorter,

Directional light gives sharp edges to items on display

thicker and stronger. A common system uses 12-volt lamps fed from the mains by a transformer or other voltage-reduction device.

The small size of these lamps gives them good directional qualities which make them popular in shops for the display of goods. The relatively low heat output of low-voltage systems is also an important property in stores, where high levels of illumination can cause overheating.

Fluorescent lamps

Fluorescent gas discharge lamps work by passing an electric current through a gas or vapour so that a luminous arc is established within a glass container. The energised gas atoms emit ultraviolet (UV) radiation and some blue-green light. A coating of fluorescent powders on the inside of the glass absorbs the UV radiation and re-radiates this energy in the visible part of the spectrum. The fluorescent coating therefore increases the efficiency of the system and allows the colour quality of the light to be controlled. By using suitable mixtures of the metallic phosphors which make up the fluorescent coating, lamps are available with colour temperatures ranging from 3000 K ('warm light') to 6500 K ('daylight').

2700K = 'warm' light
5000+K = 'cool' light

The common fluorescent lamps discussed in the following sections contain trace amounts of mercury which supply the vapour needed for the gas discharge inside the lamp. Prolonged exposure to mercury vapour is a health hazard but the few milligrams of mercury in a lamp are safely contained by the impervious glass. Lamp manufacturers supply instructions for the safe disposal of expired fluorescent lamps and measures to be taken if the glass of a lamp is broken. Essentially these measures involve capturing the tiny dot of liquid mercury into a sealed container, removing it to a safe place outdoors, and airing the room.

Like all discharge lamps, fluorescent lamps continuously decrease in output and luminous efficacy as they are run, and the lamp should be replaced after a stated number of hours. The lamp will usually run for longer than its stated life but the light output will then fall below the levels specified in the lighting design. The exact life of a discharge lamp also depends on how often the lamp is switched on or off.

Tubular fluorescent lamps

The common tubular fluorescent (MCF) lamp is a form of gas discharge lamp that uses mercury vapour at low pressure. Figure 6.2 illustrates the construction of a typical lamp in a simplified manner. All such lamps need electrical control 'gear' to provide a starting pulse of high voltage, to control the discharge current, and to improve the electrical power factor. Older systems used an arrangement of coil and capacitor but modern systems use relatively small electronic circuits operating at high frequencies which eliminate flicker and allow the use of controls such as dimmers.

The large surface area of this type of lamp produces lighting of a relatively non-directional 'flat' quality and with low glare characteristics that has

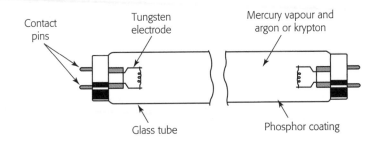

Figure 6.2 Tubular fluorescent lamp essentials

widespread use inside offices and other public buildings. Many years ago it was possible for the cyclic nature of the gas discharge to be annoying and to cause a stroboscopic effect – an apparent change of motion – when viewing moving objects such as machinery. These effects are avoided by the use of modern electronic control gear operating at high frequency, and by the use of new lamps with shielded electrodes and high-efficiency phosphors.

The luminous efficacy of tubular fluorescent lamp system has been continuously improving over the decades. Modern lamps have smaller diameter tubes (26 mm, 16 mm and smaller) compared to the original 38 mm diameter tubes which are now obsolete.

Compact fluorescent lamps

Compact fluorescent lamps (CFLs) are available in sizes comparable to a traditional tungsten filament lamp. They contain multiple miniature fluorescent tubes which are looped or coiled, sometimes within a globe shaped like a traditional light bulb. Most styles of CFL have the electronic control gear incorporated inside the lamp so that they can be installed in a conventional light fitting to directly replace a standard light bulb; other styles have the control gear in the light fitting.

CFLs are available with sockets and styles that match common light bulbs, and their use is an important technique for low-energy lighting in buildings. The colour qualities of most CFLs are designed to match the warm light, about 2800K, of the light bulbs they replace. Not all CFLs will work with a dimmer switch and the lifetime of a CFL will be reduced by frequent on/off cycling.

The cost of a CFL has steadily fallen and is now comparable to the cost of the light bulb that it will replace. Its use will greatly reduce the running costs of the light fitting. Table 6.2 shows how a compact fluorescent lamp that delivers the same amount of light as a tungsten filament light bulb only uses about one fifth the amount of electrical energy or money. In a more commercial situation, Worked Example 6.1 shows how to compare lamp types and calculate the saving of electrical energy and money over several years.

Worked Example 6.1

A certain space needs to be illuminated for 20 hours a day with a total of 18,000 lm. Two types of lamp, detailed below, are considered in the design using a costing period of two years. For the two possible lamp systems:

(a)   Calculate the intial costs and running costs.
(b)   Compare the total costs of the two systems over 2 years.

Given data:

	Lamp A	Lamp B
Lamp type	tungsten filament	tubular fluorescent
Lamp wattage:	100 W	60 W
Lamp efficacy:	12 lm/W	60 lm/W
Lamp life:	1 000 hours	8 000 hours
Installation costs: (luminaire and gear)	£2 each lamp	£10 each lamp
Lamp costs: (parts and labour)	£0.70 each	£3 each
Electricity costs: (same each lamp)	12 pence = £0.08 per kWh	

Working:

	Lamp A	Lamp B
Total output per lamp:	100 × 12 = 1 200 lumens	60 × 60 = 3 600 lumens
so		
Number of installations needed:	$\dfrac{18\,000}{1200}$ = 15 units	$\dfrac{18\,000}{3600}$ = 5 units
so		
Installation costs	£2 × 15 = **£30**	£10 × 5 = **£50**
Total hours: (same each lamp)	2 yrs × 365 days × 20 hrs = 14 600 hrs	
Times each lamp replaced:	$\dfrac{14\,500}{1\,000}$ = 14.6 or 15 units	$\dfrac{14\,500}{8000}$ = 1.83 or 2 units
so		
Replacement costs:	£0.70 × 15 × 15 = £157.50	£3 × 5 × 2= £30
Electrical energy for each lamp:	100/1 000 kW × 14600 h = 1460 kWh	60/1 000 kW × 14600 h = 876 kWh
Total electricity:	1 460 × 15 = 21 900 kWh	876 × 5 = 4 380 kWh
so		
Electricity costs:	£0.12 × 21 900 = £2 628	£0.12 × 4 380 = £525.60
Running costs: (replacement + electricity)	£157.50 + £2 628 = **£2 785.50**	£30 + £525.60 = **£555.60**
Total costs over 2 years: installation + running)	£30 + £2 785.50 = **£2 815.55**	£50 + £555.60 = **£605.50**
	for tungsten filament lamps	for tubular fluorescent lamps

Table 6.2 Energy rating of comparable lamps

Tungsten filament light bulbs:	40 W	60 W	75 W	100 W
Equivalent compact fluorescent lamps:	9 W	11W	15 W	20 W

Note: The lamps in each vertical group give approximately the same light output in lumens; the energy used and cost is directly proportional to the power rating in watts (W).

Discharge lamps

Apart from the well-known tubular fluorescent lamp, gas discharge lamps used to be restricted to outdoor lighting, such as for roadways, where their generally poor colour qualities have not been important. Modern types of discharge lamp have a colour rendering that is good enough for large-scale lighting inside buildings such as factories and warehouses. Continuing technical advances are producing more discharge lamps suitable for interior lighting, and the high luminous efficacy of such lamps can give significant savings in the energy use of buildings.

Mercury discharge lamps

An uncorrected mercury lamp emits sharp spectral peaks of light at certain blue and green wavelengths. A better spectral distribution is obtained by coating the glass envelope with fluorescent powders (MBF lamps).

In the *mercury halide* (MBI) lamp, metallic halides are added to the basic gas discharge in order to produce better colour rendering and to raise the efficacy.

Sodium discharge lamps

Low-pressure sodium (SOX) lamps produce a distinctive yellow light that is virtually monochromatic and gives poor colour rendering. However, the luminous efficacy of the lamps is very high and they have been traditionally used for street lighting.

High-pressure sodium (SON) lamps produce a continuous spectrum without much blue light but with a colour rendering that is more acceptable than the low-pressure sodium lamp. SON lamps are used in modern street lighting and for the economic lighting of large areas such as forecourts and warehouses.

LED lamps

LEDs (light-emitting diodes) produce light when layers of certain semiconductor crystals are electrically excited. Small LED lights have existed since the 1960s and have been used, for example, as tiny red power-on indicators on appliances such as televisions. Groups of LEDs are now combined to produce lamps that can often be used to replace the traditional tungsten filament light bulbs.

Each diode, about 6 mm in diameter, contains an emitter semiconductor mounted on a heat sink and wired to a power supply. The semiconductor is encased in a clear polymer that is shaped to either focus or disperse the light output. An

LED gives off light of one frequency (monochromatic) such as red or blue. White light can be produced by using blue LEDs whose light is then passed through yellow phosphors. The colour quality of the light can be adjusted by the composition of the phosphor. Other colour effects can be obtained by combining LEDs into modules or strips.

The energy efficiencies of commercial LED lamps are comparable to those of fluorescent lamps and are likely to improve as the technology advances. The service life of LED lamps can exceed 50,000 hours and they are mechanically very robust.

LED lamps are available in small powerful sizes that are useful for applications requiring focused light, for example in spotlights and car headlamps. The LEDs can also be arranged in panels which, together with a variety of colour effects, make them effective for purposes such as emergency lighting for paths and display lighting of structures. As the efficiencies of LEDs improve, and manufacturing costs decrease, LED lamps should become an alternative to CFLs for domestic lighting

Lamp developments

The technologies and properties of the types of lamps described in the previous sections highlight the availability of modern lamps that are more energy-efficient than traditional lamps because they give more light per unit of electricity, as measured by their luminous efficacy (lumens/watt). The cost savings achieved by installing these low-energy lamps is backed by calculations, such as Worked Example 6.1, and the purchase price of low-energy lamps is now affordable at a household level.

The use of low-energy lamps in buildings produces significant reductions in national energy consumption and carbon emissions. Therefore many countries are requiring the retirement of old lamp technologies, especially the simple incandescent tungsten filament lamp or light bulb; as well as some earlier generations of discharge lamps. In Europe this retirement of lamps is driven by a 2009 European Union Directive which requires household lamps produced for the EU market to meet minimum energy requirements. This has resulted in the phased withdrawal of simple incandescent light bulbs and the availability of CFL and LED lamps to suit most light fittings. Some specialist incandescent lamps that use tungsten-halogen technology to reduce energy consumption will continue to be available.

In future years we can expect continuing technical improvements to CFL and LED lamps that will bring benefits of higher efficiencies and lower costs. Organic LEDs (OLEDs) use a film of organic compounds as the emissive layer which produces light. Already used as screens for televisions and mobile phones, OLEDs are also being used for lighting spaces in imaginative ways.

Induction lamps are a type of electric lamp with neither filament nor electrodes. An electric induction coil operating at high frequency induces an energy flow which excites a gas, such as mercury vapour, to give off ultra-violet energy. Visible light is then generated when the photons strike a fluorescent coating, as in existing gas discharge lamps. The advantages of the induction lamps are a high quality of white light and an exceptionally long life of up to 60,000 hours.

> A luminaire often redirects light from a lamp

LUMINAIRES

A luminaire is the 'light fitting' that holds or contains a lamp. Luminaires usually absorb and redirect some of the luminous flux emitted by the lamp and, in the design of lighting installations, the choice of lamp must be combined with the choice of luminaire. The specification of luminaires is therefore important, and this section outlines the methods used to classify luminaires according to the effect that they have on light distribution.

Luminaires also serve a number of mechanical and electrical purposes such as positioning the lamp in space, protecting the lamp, and containing electrical control gear. Physical properties may therefore also be relevant in the choice of the luminaire, including its electrical insulation, moisture resistance, appearance and durability.

Polar curves

Polar curves show the directional qualities of light from a lamp or luminaire by a graphical plot on to polar coordinate paper, as shown in Figure 6.3. The luminous intensity of a lamp in any direction can be measured by means of a photometer, the results plotted and joined by a curve which then represents the distribution of light output from the fitting. If the distribution is not symmetrical about the vertical axis, as in a linear fitting, then more than one vertical plane needs to be plotted.

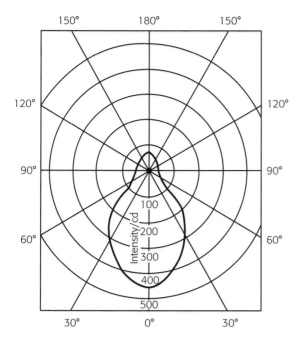

Figure 6.3 Polar curve of a luminaire

Light output ratio

It is convenient to try and describe the distribution of light from a luminaire by a system of numbers. One system is to classify luminaires by the proportion of the

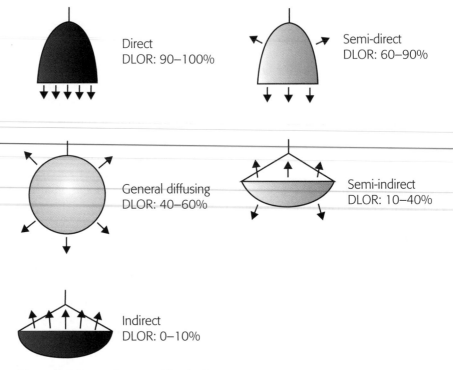

Figure 6.4 General types of luminaire

total light from the luminaire emitted into the upper and lower hemispheres formed by a plane through the middle of the lamp filament. These classifications are as follows.

Light output ratio

$$\mathrm{LOR} \quad = \frac{\text{Total light output of luminaire}}{\text{Light output of its lamp(s)}}$$

Downward light output ratio

$$\mathrm{DLOR} = \frac{\text{Downwards light output of luminaire}}{\text{Light output of its lamp(s)}}$$

Upward light output ratio

$$\mathrm{ULOR} = \frac{\text{Upwards light output of luminaire}}{\text{Light output of its lamp(s)}}$$

So that

$$\mathrm{LOR} = \mathrm{DLOR} + \mathrm{ULOR}$$

Luminaires can be broadly divided into five types, according to the proportion of light emitted upwards or downwards. Figure 6.4 indicates these divisions using luminaires with generalised shapes.

British zonal system

The British Zonal or BZ System classifies luminaires into ten standard light distributions with BZ numbers from 1 to 10. BZ1 indicates a distribution of light that is mainly downwards and BZ10 a distribution that is mainly upwards. The BZ number is obtained from a combination of the properties of the luminaire and the room surfaces, which are measured by the following terms.

- **Direct ratio** is the proportion of the total downward flux from the luminaires which falls directly on the working plane.
- **Room index** is a number which takes account of the proportions of the room by the following formula:

$$RI = \frac{L \times W}{H_m(L + W)}$$

where L = length of the room

W = width of the room

H_m = mounted height of the luminaire above the working plane.

A plot of direct ratio against room index for a particular luminaire gives a curve which can be described by the BZ number of the standard curve to which it is the closest fit. Figure 6.5 illustrates the standard curves of some BZ numbers. The actual curves of most luminaires cross from one BZ zone to another as the room index varies.

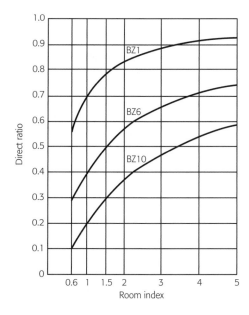

Figure 6.5 BZ curves for luminaires

LIGHTING DESIGN

The lighting we arrange in our homes probably evolves through a feeling of providing 'enough light' to be comfortable and through personal taste in decoration. However, the lighting arrangements for commercial buildings need to designed and specified in advance of installation using specifications, calculations and drawings that provide authoritative technical evidence of the requirements listed below.

- sufficient 'quantities' of light for people to carry out particular tasks, to move about with ease and safety, and to display features of a building and its contents
- appropriate 'qualities' of light in terms of colour rendering, directional effects, surface brightness and glare
- lamps and fittings that are cost effective to install, and simple and safe to maintain
- lamps and lighting circuits with high energy efficiency
- controls that are convenient and energy saving
- avoidance of effects that annoy others or pollute the night sky
- minimum impact on the environment from equipment and by light pollution.

The previous sections have explained the technical principles and properties of the sources of light called lamps and the light fittings called luminaires that are major factors in lighting design. Other factors are described in the sections that follow. Lighting designs may also need to use natural daylight and this topic is described in the next chapter.

Illuminance levels

1 lux = 1 lumen/ (metre)^2

The quantity of light on a certain surface is usually the primary consideration in the design of a lighting system. This quantity is specified by the density of luminous flux, or illuminance, and measured in lux. The illuminance level changes with time and varies across the working plane, so an average figure is used.

- **Service illuminance** is the mean illuminance achieved during the maintenance cycle of a lighting system, and averaged over the area being considered.

The illuminance needed for a particular task depends upon the visual difficulty of the task, the average standard of eyesight involved and the type of performance expected. It can be shown that the speed and accuracy of various types of work depend upon having the level of illuminance high enough.

Table 6.3 gives some example values of standard service illuminance levels recommended for a variety of interiors and tasks. The values represent current good practice, which takes into account visual needs, practical experience and

Table 6.3 Typical values of service illuminance

Location	Illuminance (lux)
Public entrance halls, foyers	200
Public passageways, stairs	100
Restaurant: bars, tables	150
Retailing: sales areas, displays	500–750
Office: general	500
Office: workstations	300–500
Office: drawing boards, fine work	750
Workplace: fine work	1 000
Workplace: medium work	300
Workplace: casual work	200
Education: general classrooms	300
Education: display boards	500
Home: kitchens, study areas	300
Home: halls, landings	150

See also:

section on *Measurement of lighting* in Chapter 5

Notes:
Values typical of UK, EU and AUS/NZ standards and lighting codes.
The illuminance quoted is for the appropriate 'working plane' of the task such as a desk.

efficient use of energy relating to the situation. Some design techniques modify the target illuminance level by a correction factor which takes into account the age of the work force, the reflectance of surroundings and the relative importance of speed and accuracy of perception

The source of the recommendations varies with country but authorities used include international standards, European Directives, national standards, health and safety laws, building codes, and codes of lighting practice such as those issued in the United Kingdom by the Society of Light and Lighting.

Calculation of illuminance

Chapter 5 explained how illuminance on a surface depends upon the distance from the light source and the angles between the illuminated surface and the light source, and also on the reflectances of the surrounding surfaces. Worked Example 5.3 shows the use of the general formula ($E = I \cos \theta / d^2$) for calculating the illuminance provided by a single point source of light.

In an interior where there is more than one light source, the formula needs to be repeatedly applied, together with calculations for the effect of reflecting surfaces. Such multiple calculations are tedious for humans but easy for computer programmes. Practical lighting design therefore makes good use of software packages for lighting calculations, although the task is not impossible using a spreadsheet and the basic formulas already given. Fortunately, some simplified techniques such as the *lumen method* below have been developed and are found to give satisfactory results.

Table 6.4 Utilisation factors for some luminaires

Description of fitting/ typical outline LOR	Basic down-ward LOR (%)		Reflectances								
		Ceiling	0.7			0.5			0.3		
		Walls	0.5	0.3	0.1	0.5	0.3	0.1	0.5	0.3	0.1
		Room index									
Aluminium industrial reflector, aluminium or enamel high-bay reflector	70	0.6	0.39	0.36	0.33	0.39	0.36	0.33	0.39	0.35	0.33
		0.8	0.48	0.43	0.40	0.46	0.43	0.40	0.46	0.43	0.40
		1.0	0.52	0.49	0.45	0.52	0.48	0.45	0.52	0.48	0.45
		1.25	0.56	0.53	0.50	0.56	0.53	0.49	0.56	0.52	0.42
		1.5	0.60	0.57	0.54	0.59	0.57	0.53	0.59	0.57	0.53
		2.0	0.65	0.62	0.59	0.63	0.60	0.58	0.63	0.59	0.57
		2.5	0.67	0.64	0.62	0.65	0.62	0.61	0.65	0.62	0.60
		3.0	0.69	0.66	0.64	0.67	0.64	0.63	0.67	0.64	0.62
		4.0	0.71	0.68	0.67	0.69	0.67	0.65	0.69	0.66	0.64
		5.0	0.72	0.70	0.69	0.71	0.69	0.67	0.71	0.67	0.66
Near-spherical diffuser, open beneath	50	0.6	0.28	0.22	0.18	0.25	0.20	0.17	0.22	0.18	0.16
		0.8	0.39	0.30	0.26	0.33	0.28	0.23	0.27	0.25	0.22
		1.0	0.43	0.36	0.32	0.38	0.34	0.29	0.31	0.29	0.26
		1.25	0.48	0.41	0.37	0.42	0.38	0.33	0.34	0.32	0.29
		1.5	0.52	0.46	0.41	0.46	0.41	0.37	0.37	0.35	0.32
		2.0	0.58	0.52	0.47	0.50	0.48	0.43	0.42	0.39	0.36
		2.5	0.62	0.56	0.52	0.54	0.50	0.47	0.45	0.42	0.40
		3.0	0.65	0.60	0.56	0.57	0.53	0.50	0.48	0.45	0.43
		4.0	0.68	0.64	0.61	0.60	0.56	0.54	0.51	0.48	0.46
		5.0	0.71	0.60	0.65	0.62	0.59	0.57	0.53	0.50	0.48
Recessed louvre trough with optically designed reflecting surfaces	50	0.6	0.28	0.25	0.23	0.28	0.25	0.23	0.28	0.25	0.23
		0.8	0.34	0.31	0.28	0.33	0.30	0.28	0.33	0.30	0.28
		1.0	0.37	0.36	0.32	0.37	0.34	0.32	0.37	0.34	0.32
		1.25	0.40	0.38	0.35	0.40	0.37	0.35	0.40	0.37	0.35
		1.5	0.43	0.41	0.38	0.42	0.40	0.38	0.42	0.39	0.38
		2.0	0.46	0.44	0.42	0.45	0.43	0.41	0.44	0.42	0.41
		2.5	0.48	0.46	0.44	0.47	0.45	0.43	0.46	0.44	0.43
		3.0	0.49	0.47	0.46	0.48	0.46	0.45	0.47	0.45	0.44
		4.0	0.50	0.49	0.48	0.49	0.48	0.47	0.48	0.47	0.46
		5.0	0.51	0.50	0.49	0.50	0.49	0.48	0.49	0.48	0.47

Lumen method

The lumen method is a simplified method of lighting design which is valid if the luminaires are mounted overhead in a regular pattern. The luminous flux output (lumens) of each lamp needs to be known, as do details of the luminaires and the room surfaces. Usually the illuminance level is already specified, the designer chooses suitable lamps and luminaires, and then wishes to know how many fittings are required to meet the specification. The number of lamps is given by the following formula.

$$N = \frac{E \times A}{F \times UF \times LLF}$$

where

N = number of lamp fittings required

E = illuminance level required (lux)

A = area at working plane height (m²)

F = initial luminous flux output of each lamp (lm)

UF = utilisation factor (an allowance for the distribution effects of the luminaire and the room surfaces, as described below)

LLF = light loss factor (an allowance for a reduction in light output caused by lamp deterioration and dirt; details are given below).

Utilisation factor

- The **utilisation factor** is the ratio of the total flux reaching the working plane compared to the total flux output of the lamps.

The utilisation factor can be calculated directly or, more usually, obtained from tables which combine the distribution properties of the luminaire with the room index and with the reflectances of the room surface. Table 6.4 lists the utilisation factors for some representative types of luminaire.

Light loss factor

- The **light loss factor** is the ratio of the illuminance provided at some given time compared to the initial illuminance.

The light loss factor is calculated as the product of the three other factors.

$$LLF = \frac{\text{lamp maintenance}}{\text{factor}} \times \frac{\text{luminaire}}{\text{maintenance factor}} \times \frac{\text{roof surface}}{\text{maintenance factor}}$$

where:

the *lamp maintenance factor* is an estimate of the decline in output of the lamp source over a set time

the *luminaire maintenance factor* is an estimate of the reduction in light output caused by dirt on the luminaire over a set time

the *room surface maintenance factor* is an estimate of the effect of dirt deposited on the room surfaces over a set time.

A range of values for light loss factors is given in Table 6.5. The determination of

Table 6.5 Typical light loss factors

12-month LLF	Direct lighting	Indirect lighting
Air-conditioned building	0.95	0.9
Dirty industrial area	0.7	0.35

Notes:

Room is of average cleanliness, like an office.

Lamps run for average 8 hours per working day.

particular values depends on the type of lamps in use, the cleanliness of air in the district and the building, and whether dirt can easily collect on fittings.

The *maintenance factor*, previously used instead of the light loss factor, took account of lighting loss caused by dirt but did not include deterioration of lamps, for which an average output was assumed.

Layout

The number of lamps needed, as calculated by the lumen formula, usually needs to be rounded up to a convenient figure and the layout of the luminaires decided upon. In order that the illuminance provided does not fall below a minimum value, the fittings must be placed in a regular grid pattern and their spacing must not exceed certain distances. This maximum spacing depends on the type of luminaire and the height at which they are set. Typical values are as follows.

For fluorescent tubes in diffusing luminaires

$$S_{max} = 1.5 \times H_m$$

For filament lamps in direct luminaires

$$S_{max} = 1.0 \times H_m$$

where

S_{max} = maximum horizontal spacing between fittings

H_m = mounted height of fitting above the working plane, as shown in Figure 6.6. Take working plane as 0.85 m above floor level unless otherwise specified.

Energy use of lighting

A useful check on the overall energy effectiveness of a lighting design is the amount of electrical power the installation uses per unit of floor area.

- **Power density** of a lighting installation is the total power rating of the installation divided by the total floor area of the installation.

    Unit: watts/m².

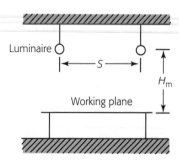

Luminaire

Working plane

H_m

S

Figure 6.6 Spacing of luminaires

Table 6.6 Typical power density values

Task illuminance (lux)	Power density range (W/m²)
300	10–12
500	10–15
1000	15–30

Note: Assumes workplace conditions.

The power density will increase for higher levels of illuminance and decrease for more efficient lamps which have a higher efficacy (lumens per watt). Ranges of typical values are shown in Table 6.6.

Another method of assessing the energy efficiency of a lighting system is to evaluate the annual energy consumption per square metre. (kWh/m²/year).

Worked Example 6.2

A factory space measuring 40 m by 12 m by 4 m in height requires a service illuminance of 500 lux on the workbenches which are set 1 m above the floor. The 65 W tubular fluorescent lamps chosen have a luminous efficacy of 80 lm/W. They are to be mounted on the ceiling in luminaires which have a DLOR of 50 per cent. The room reflectances are 0.5 for the ceiling and 0.3 for the walls; the initial light loss factor is 0.7.

(a)   Use the lumen method of design to calculate the number of lamps required.
(b)   Suggest a suitable layout for the lamp fittings.

Know

$$E = 500 \text{ lx}, F = 65 \times 80 = 5\,200 \text{ lm}$$
$$L = 40 \text{ m } W = 12 \text{ m}, H_m = 4 - 1 = 3 \text{ m}$$
$$A = 40 \times 12 = 480 \text{ m}^2, LLF = 0.7$$

Using the room index formula and substituting

$$RI = \frac{L \times W}{H_m(L + W)}$$

$$= \frac{40 \times 12}{3(40 + 12)} = 3$$

Reflectances: ceiling 0.5, walls 0.3
            UF = 0.46 (from Table 6.4 using above data)

Using the lumen method formula and substituting

$$N = \frac{E \times A}{F \times UF \times LLF}$$

$$= \frac{500 \times 480}{5200 \times 0.46 \times 0.7} = 143.3$$

So number of lamps required = **144 lamps**

Suggested layout: 9 rows of 16 luminaires

Check spacing using $S_{max} = 1.5 \times H_m$
$$= 1.5 \times 3 = 4.5 \text{ m}.$$

So the suggested layout is satisfactory provided that the distance between lamps is not greater than 4.5 m.

Lighting control

Switching off unwanted lights is an obvious method of saving energy and money. Building regulations for energy usually require that buildings achieve certain levels of energy saving by control of their lighting installations in order to make maximum use of daylight and to reduce energy when a space is unoccupied. Lighting control can be achieved by the following methods.

- **Timer control**: where timers are set to switch off lighting for periods of known inactivity, such as at the end of the working day.
- **Daylight control**: where lights are switched on or off, or dimmed, according to the level of daylight detected in a room.
- **Occupation control**: for example by sensors which detect noise or movement in an area. The sensors turn lighting on when there is someone in the area, and off again after a time delay if there is nobody in the room.
- **Local switching**: where it is possible to switch on lights only in the part of the room that is being occupied.

Glare rating

Glare was defined in the previous chapter as the discomfort or impairment of vision caused by an excessive range of brightness in the visual field. The usual causes of glare in buildings are bright skies seen through windows and direct views of bright lamps. The glare is most likely to be discomfort glare which, over a period of time, can cause annoyance and affect efficiency.

- **Glare rating** is an index or numerical measure of discomfort glare which enables glare to be assessed and acceptable limits recommended.

A glare index, such as the unified glare rating (UGR) is calculated from the following factors: the positions of the source and the viewpoint; the luminances of the source and the surroundings; and the size of the source. The calculated index may be compared with the maximum index recommended for the particular environment.

The UGR index ranks discomfort glare on a scale from 13 to 28, where the higher the index the higher the level of glare.

Lighting criteria

Many factors may be relevant to the design of a particular lighting system. The following list summarises the main factors that usually need to be considered; the priorities will depend upon the type of situation.

- **Light quantity**: depends upon the nature of the task and the light output of lamp and luminaire. It is usually specified by illuminance level in lux.
- **Natural light**: may be used as a complete source of light or to supplement artificial light sources. Daylighting is the topic of the next chapter.
- **Colour quality**: depends upon the requirements of the task and the colour rendering properties of the source. Methods of specifying colour quality include spectral distribution, colour temperature and the colour rendering index.
- **Glare**: depends upon the brightness and contrast of light sources and surfaces, and the viewing angles. It is can be specified by a glare index or rating.
- **Directional quality**: depends upon the three-dimensional effect required and the nature of the lamp and luminaire. It can be specified by vector and scalar luminance.
- **Energy use**: depends upon the electrical efficiency of the lamps and the use of switches. All lamps give off heat and are a source of heat gain in a building. Windows providing natural light can also be a significant source of heat loss and solar gain.
- **Costs**: depend upon the initial cost of the fittings, the cost of replacing the lamps (including labour), and the electricity consumption of the lamps.
- **Physical properties**: include size, safety, appearance and durability of fittings.

Typical lighting operating costs:
93 per cent electrical energy
4 per cent replacement lamps
3 per cent labour

EXERCISES

1. A certain space needs to be illuminated with a total luminous flux of 18,000 lm. The compact fluorescent lamps used are rated at 11 W each and have a luminous efficacy of 60 lm/W.
   (a) Calculate the number of such lamps required (assuming an even distribution)
   (b) Calculate the number of kilowatt hours of electrical energy used by the lamps in a 12-hour period (1 kWh = 1 kW × 1 hour).

2. The level of illumination described in Exercise 1 is required 12 hours a day for two years. Compare the differences in initial costs, replacement costs and the energy costs using the following types of lamps:
   (a) CFL lamps
   (b) LED lamps.
   Use manufacturers' data or your own estimates for costs of lamps, fittings and electrical energy. Table 6.1 and Worked Example 6.2 may also be useful.

3. An area measuring 18 m by 8 m is to have a service illuminance of 300 lx. The tubular fluorescent lamps each have a luminous flux output of 2820 lm and the luminaires give a utilisation factor of 0.4. The light loss factor assumed is 0.8. Calculate the number of lamps required and suggest a layout for them.

4. A classroom with an area 10 m by 6 m is illuminated by 18 tubular fluorescent lamps, where each lamp is 60 W with a luminous efficacy of 80 lm/W. The utilisation factor is 0.46 and the light loss factor used is 0.8. Calculate the average illuminance in the classroom.

5. A workshop is 12 m by 6 m by 4 m high and has workbenches 1 m high. Discharge lamps, each with an output of 3700 lm, are to be fitted in aluminium industrial reflectors at ceiling level. The surfaces have reflectances of 0.7 for the ceiling and 0.5 for the walls. The light loss factor is 0.7. The illuminance required on the workbenches is 400 lx.
   (a) Find the utilisation factor for the room (use Table 6.4).
   (b) Calculate the number of lamps required.

Answers are on page 326.

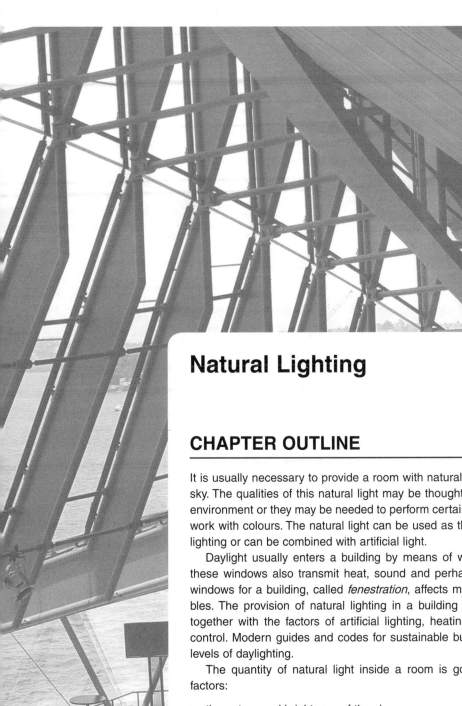

7

Natural Lighting

CHAPTER OUTLINE

It is usually necessary to provide a room with natural light from the sun or the sky. The qualities of this natural light may be thought desirable for a pleasant environment or they may be needed to perform certain tasks, such as exacting work with colours. The natural light can be used as the sole source of interior lighting or can be combined with artificial light.

Daylight usually enters a building by means of windows or skylights; but these windows also transmit heat, sound and perhaps air. So the design of windows for a building, called *fenestration*, affects many environmental variables. The provision of natural lighting in a building needs to be considered together with the factors of artificial lighting, heating, ventilation and sound control. Modern guides and codes for sustainable buildings require minimum levels of daylighting.

The quantity of natural light inside a room is governed by the following factors:

▶ the nature and brightness of the sky
▶ the size, shape and position of the windows
▶ reflections from surfaces inside the room
▶ reflections and obstructions from objects outside the room.

By analysing these factors it is possible to describe daylight numerically and to predict its effects in a room. The contents of this chapter will enable you to:

▶ understand the sources and nature of natural light
▶ know the properties of different standard skies

▶ use the daylight factor measurement
▶ understand the significance of different daylight factor values
▶ appreciate the methods available for predicting daylight factors
▶ calculate average daylight factors for rooms
▶ understand the role of daylight in well-being and sustainability
▶ appreciate the techniques of combined lighting design.

These topics should be considered together with the related principles and practice of artificial lighting explained in Chapter 6. The Resource sections of the book also contain supporting information that can be used for both revision of principles and further investigation of topics.

NATURAL LIGHT SOURCES

See also:

Chapters 5 and 6 on *Properties of lighting* and on *Artificial lighting*, and Resource 3

All natural daylight comes from the Sun, by way of the sky. But, while the Sun can be considered as a reasonably constant source of light, the light from the sky varies with the time of the day, with the season of the year and with the local weather.

In parts of the world with predominantly dry, sunny weather much of the natural light inside buildings is direct sunlight which has been reflected. In Britain, and similar countries where sunshine is unreliable, the overcast sky is considered as the main source of natural light. Because the sky continually varies it is necessary to define certain 'standard skies', with constant properties, for use in design work.

See also:

section on *Sun controls* in Chapter 3

Direct sunlight

The levels of illuminance on the ground provided by light direct from the Sun may be as high as 100,000 lx in the summer. Direct sunlight should generally be avoided inside working buildings because it can easily cause unacceptable glare. However, in domestic buildings a certain amount of sun penetration is considered desirable by most people who live in temperate climate zones such as North-West Europe. One guideline for Britain is that sides of all residential buildings which face east, south and west, should have at least one hour of sunshine on March 1st.

Uniform standard sky

See also:

section on *Measurement of lighting* in Chapter 5

The uniform sky is a standard overcast sky which is taken to have the same luminance in every direction of view. Viewed from an unobstructed point on the ground, the sky effectively makes a hemispherical surface around that point. This surface emits light and the uniform sky assumes that every part of the surface is equally bright.

The illuminance at a point on the ground provided by an unobstructed overcast sky continually varies, but a constant 5000 lx is taken as one standard for daylight calculations in Britain. This represents the light from a heavily overcast sky and is a conservative design assumption: the 5000 lx is actually exceeded for 85 per cent of normal working hours throughout the year.

Although the overcast sky of countries like Britain is not strictly uniform in luminance the model is useful for some purposes. A sky of uniform luminance is also a reasonable description of the clear skies in sunny climates.

> *Illuminance:* lighting effect
> *Luminance:* surface brightness

CIE standard sky

The CIE sky is a standard overcast sky in which the luminance steadily increases above the horizon. This sky was defined by the Commission Internationale de l'Eclairage as being described by the following formula and illustrated in Figure 7.1:

> *Daylight or skylight* = diffuse light from the whole sky
> *sunlight* = the direct solar beam

$$L_\theta = L_z \, 1/3 \, (1 + 2 \sin \theta)$$

where

L_θ = luminance of the sky at an altitude θ degrees above the horizon
L_z = luminance of the sky at the zenith.

The luminance of the CIE sky at the zenith is *three* times brighter than at the horizon. The CIE model is found to be a good representation of conditions in many parts of the world, especially in regions like North-West Europe.

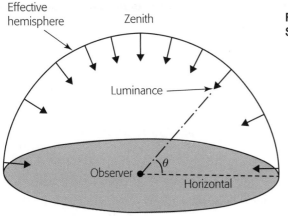

Figure 7.1
Sky hemisphere

DAYLIGHT FACTORS

The natural light that provides illumination inside a room is usually only a small fraction of the total light available from a complete sky. The level of illuminance provided by the sky varies as the brightness of the sky varies, so it is not possible to specify daylight by a fixed illuminance level in lux.

The amount of daylight inside a room can be measured by comparing it with the total daylight available outside the room. This ratio, or daylight factor, remains constant for a particular situation because the two parts of the ratio vary in the same manner as the sky changes.

- The **daylight factor** is the ratio between the actual illuminance at a point inside a room and the illuminance possible from an unobstructed hemisphere of the same sky.

    Unit: percentage

Direct sunlight is excluded from both values of illuminance, and the daylight factor can be expressed by the following formula:

$$DF = \frac{E_i}{E_o} \times 100$$

where

DF = daylight factor at a chosen reference point in the room (per cent)
E_i = illuminance at the reference point (lx)
E_o = illuminance at that point *if* the sky was unobstructed (lx).

The definition is a theoretical one as it is not possible to instantly measure both types of illuminance at the same place. For purposes of specification and design, a standard sky is assumed to give a minimum level of illuminance on the ground, and 5000 lx is a commonly used value.

Worked Example 7.1

A minimum daylight factor of 4 per cent is required at a certain point inside a room. Calculate the natural illuminance that this represents, assuming that an unobstructed standard sky gives an illuminance of 5000 lx.

Know $DF = 4\%$, $E_o = 5\,000$ lx, $E_i = ?$

Using formula for daylight factor and substituting

$$DF = \frac{E_i}{E_o} \times 100$$

$$4 = \frac{E_i}{5\,000} \times 100$$

$$E_i = \frac{4 \times 5\,000}{100} = 200$$

So illuminance = **200 lx**

Table 7.1 Typical daylight levels

Location	Average daylight factor	Minimum daylight factor	Surface
General office	5	2	desks
Classroom	5	2	desks
Entrance hall	2	0.6	working plane
Library	5	1.5	tables
Drawing office	5	2.5	boards
Sports hall	5	3.5	working plane

Typical daylight levels

Daylight factors can be used to specify recommended levels of daylight for various interiors and tasks. Table 7.1 lists a selection of recommendations for interiors where daylight from side windows is a major source of light. Daylight factors vary for different points within a room, so it is usual to quote average values or minimum values.

A room in daytime with an average daylight factor of less than 2 per cent will seem gloomy and occupants will probably need the electric lighting turned on. A room with an average daylight factor above 5 per cent will seem strongly lit by daylight, and the windows producing this effect will be relatively large and therefore liable to give high heat losses or gains. An acceptable range of daylight factor is between 2 per cent and 5 per cent, with supplementary electric lighting available when needed.

The UK *Code for Sustainable Homes* gives an indication of daylighting levels in homes that both improve the quality of life and reduce the energy used for lighting.

Effect of DF values:
5 per cent and above – seems bright
2 per cent and below – seems gloomy

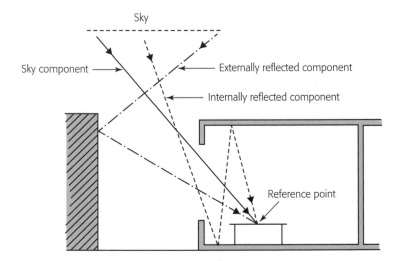

Figure 7.2 Components of daylight factor

Daylighting is placed under the category of Health and Well-being, and includes the following requirements:

- Kitchens: minimum average daylight factor of 2 per cent.
- Living rooms, dining rooms and studies: minimum average daylight factor of 1.5 per cent.
- The working plane in each kitchen, living room, dining room and study: 80 per cent of the working plane must receive direct light from the sky.

Daylight factor components

The daylight reaching a particular point inside a room is made up of three principal components. The three components arrive at the same point by different types of path, as indicated in Figure 7.2.

- **Sky component** (SC): the light received directly from the sky.
- **Externally reflected component** (ERC): the light received directly by reflection from buildings and landscape outside the room.
- **Internally reflected component** (IRC): the light received from surfaces inside the room.

The use of daylight factor components is a step towards predicting the daylight factors at a point. The three components can be analysed and calculated separately. The final daylight factor is the sum of the three separate components, so that:

$$DF = SC + ERC + IRC$$

where

DF = Daylight factor
SC = Sky component
ERC = Externally reflected component
IRC = Internally reflected component.

Daylight factor contours

Daylight factor contours represent the distribution of daylight inside a room by means of lines which join points of equal daylight factor. The contours are commonly shown on a plan of the room at working-plane height, as shown in Figure 7.3. Contours around windows have characteristic lobe shapes which converge at the edge of the windows. Tall windows provide greater penetration of contours and multiple windows cause contours to join.

Areas with no direct view of sky are likely to have low DF

Assessment of daylight factor

Once a building exists, the values and the distribution of daylight inside the building can be measured directly. A *daylight factor meter* is a specially calibrated light meter (photometer) which gives direct readings of the daylight factor at any point.

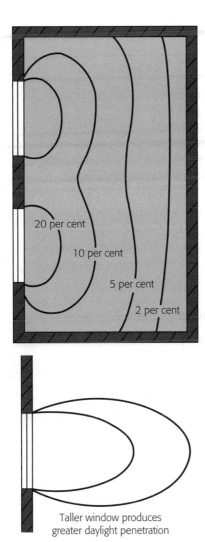

20 per cent

10 per cent

5 per cent

2 per cent

Taller window produces
greater daylight penetration

Figure 7.3 Daylight factor contours

The prediction of daylight factors at the design stage requires a knowledge of the proposed building and its surroundings. It is possible to calculate the three components of the daylight factor by using information about the size of the windows and room, the size of any external obstructions and the proposed reflectances of the surfaces. The sky component is the major contributor to a daylight factor and can be considered as the percentage of an unobstructed sky that is visible from the reference point.

No-sky line:
marks the start of
areas of the room
that cannot
receive light
directly from the
sky

The following methods can be used for predicting the daylight factor in a building:

- calculation of average daylight factors
- tables of window and room dimensions

- grids of the sky such as the Waldram diagram
- computer programs
- physical models measured in an artificial sky room.

As with other practical methods of lighting design, there are commercial software packages which predict and show the effects of daylighting in buildings. Special tables, diagrams and protractors can also be used to help bypass the many repetitive calculations needed to predict the daylight factor at each point within a room. The *Waldram diagram*, for example, is a specially scaled grid representing half the hemisphere of sky. Using scaled plans of the room, the area of sky visible through the window from the reference point is plotted onto the grid. The area of grid covered by this plot is proportional to the sky component at the reference point. An other, older manual method of predicting daylight factors at the design stage was to use scale drawings and measure the sight lines between the window opening and the view point with the help of specially calibrated protractors developed by the Building Research Establishment.

Computer simulations

Computers are easily programmed to repeatedly make the tedious calculations needed for the prediction of daylight factor. Modern software packages for the prediction of daylight ask you to enter the details of the room, the windows and reflecting surfaces, and to specify a grid pattern within the room. After the daylight factor expected at each grid point have been calculated, the results may be shown on screen as daylight factor contours between the points and also given as a print-out. Similar calculation techniques can be combined with CAD programs to give an onscreen 3D visualisation of the building interior that show the daylight effects from different viewpoints as you move through the CAD model.

Models and artificial skies

An artificial sky is a large room fitted with an arrangement of lamps that can be controlled to simulate sky conditions at any time of the day in any part of the world. The daylight factors in a perfectly scaled model will agree with those in a full-scale building A scale model of a building is therefore placed in the centre of the artificial sky and the daylight effects observed and measured in the model by light meters. For the process to be valid, all aspects of the expected conditions need to be modelled or allowed for, including obstructions outside the building, the transmittance of the glass and the reflectances of the room surfaces.

Average daylight calculation

The average daylight factor can be predicted at the design stage using knowledge of the glazing area, the floor area, the average 'angle of the sky' at the window, the type of glass and the overall reflectance of the surfaces. The formula below

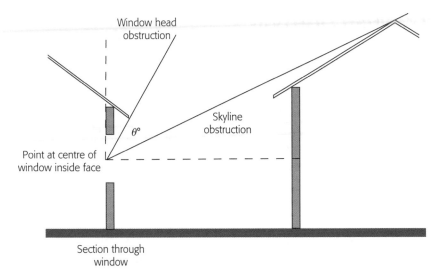

Figure 7.4 Angle of visible sky

will give the average daylight factor, or it can be transposed to give the area of glazing required to give a certain daylight factor.

$$DF = \frac{A_g \, \theta \, T}{A \, (1 - R)}$$

where

DF = average daylight factor (per cent)

A_g = glazed area of windows (excluding frames or obstructions) in m²

θ = angle of visible sky, measured in section from the centre of the window opening, as in Figure 7.4

T = transmittance of glazing to diffuse light, including the effect of dirt

A = total area of enclosing room surfaces: ceiling + floor + walls (including windows) in m²

R = reflectance of surrounding room surfaces.

Approximate rule:
Glazed area at least 1/25 floor area gives daylight factor at least 2 per cent

Typical transmittance values	Approximate reflectance values
Clear single glazing: $T = 0.8$	Normal office or living room: $R = 0.3$
Clear double glazing: $T = 0.7$	White ceiling and light-coloured walls: $R = 0.5$
Reduce by at least 5% to allow for dirt buildup	

COMBINED LIGHTING

People prefer to work by natural light but it is difficult to provide adequate levels of natural illumination to all parts of an interior. Even areas that are quite close to windows may require extra illumination for certain purposes, and artificial lighting then needs to be combined with the natural lighting. This system of combined lighting requires careful design so as to preserve the effect of daylight as much as possible, and to make the best use of energy.

- **Principle for combined lighting**: to light some areas of the interior for the whole time by artificial light, which is designed to balance and blend with the daylight and give the feeling of natural lighting

See also:

section on
Energy regulations in Chapter 3

Good combined lighting design can achieve most of the psychological benefits of full daylighting but allows the use of deeper room plans, which can save energy because of lower heat losses or may be necessary because of building shape. The guiding principle of this type of lighting design is to provide illumination that appears to be of good daylight character, even though most of the working illumination might be supplied by unobtrusive artificial light sources. This type of lighting design has also been termed PSALI (permanent supplementary artificial lighting of interiors).

To achieve an appropriate combined lighting design the factors discussed below should be considered.

Distribution of light

See also:

section on
Lighting criteria in Chapter 6, p159

The total illuminance should gradually increase towards the windows. The illumination to be provided can be determined by choosing a final illumination curve, as shown in Figure 7.5, and then subtracting average daylight values. The illuminance over the main working areas should not vary by more than about

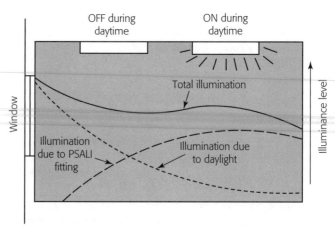

Figure 7.5 Example of combined lighting

3 to 1. Sudden changes in luminance between room surfaces should be avoided. In general this requires neutral colour schemes which have the same appearance under both natural and artificial light.

Choice of lamps

The lamps chosen for the artificial lighting should match the natural light in colour appearance. Daylight is variable in colour quality, with a spectral output different from any lamp, so a compromise must be made. Tubular fluorescent lamps with colour temperatures in the range 4000–6500 K are usually employed. The fittings should be unobtrusive and are often recessed in the ceiling.

Switching

A combined lighting system requires lighting controls which can provide combined lighting during daylight hours and complete artificial lighting after dark. Figure 7.5 indicates a simple system where extra lights would be switched on at night or during dull days. Various systems of switching controls can control the artificial lighting with photoelectric cells which sense changing daylight levels and switch lights on and off as necessary. Such switching systems are also required by building regulations in order to reduce energy loads.

EXERCISES

1. Calculate the luminance of a CIE standard sky at an altitude of 20° if the luminance at the zenith is taken to be 2200 cd/m².

2. The natural illuminance at a point inside a room is 430 lx and the illuminance given by an unobstructed sky is assumed to be 5380 lx. Calculate the daylight factor at that point.

3. Draw a plan of a convenient classroom or office and sketch the shapes of the daylight factors that would be expected from the windows. If possible, investigate the actual distribution of daylight with a light meter and/or computer program.

4. Measure the floor area and glazed area of a convenient existing room and use the formula for average daylight factors to calculate an average daylight factor for this room. Compare this value with your subjective assessment of the room and compare it with the descriptions in the text.

Answers are on page 326.

8

Aspects of Sound

CHAPTER OUTLINE

The ability to speak and to hear are important factors in our personal lives and in our environment. Activities such as making music, sound recording and reproduction, telephony, architectural acoustics and noise control also have strong associations with our sensation of hearing, and require an understanding of the principles and effects of sound.

Unwanted sound in the environment is perceived as a nuisance and can cause emotional effects such as annoyance, irritation and sleep disturbance. The main sources of environmental noise issues are transportation noise, industrial noise, construction noise, and noise from leisure and entertainment. The measurement of noise exposure is an important step towards protecting people from hearing damage and in creating satisfactory environments for our living.

Good practice in the design of buildings and the construction of buildings therefore involves a consideration of sound in the environment. Common topics of concern are the reduction of noise at source, the exclusion of external noise, the reduction of sound passing between rooms and the quality of sound inside rooms.

Before these topics are studied in other chapters, this chapter describes certain properties of sound, its measurement and its effects on hearing. The contents of this chapter will allow you to:

- understand how the energy content of sound interacts with human hearing
- make use of the terms and units associated with sound and its measurement

⟶

- calculate sound levels and combinations of sound levels
- appreciate how sound travels and changes as it moves
- understand the effect of sound on human hearing
- appreciate the causes and mechanisms of hearing damage and hearing loss
- appreciate the technical concept of loudness.

The two chapters following this one describe noise, sound insulation and acoustic quality. The Resource 3 section of this book contains supporting information about the basic science and principles of sound.

MEASUREMENT OF SOUND LEVELS

The scientific nature of sound and the way it travels is described in the Resource 3 section of this book. Human hearing detects rapid variations in air pressure that reach the ear after travelling as a wave motion. The wave motions means that the particles of air, or other material, carrying the sound vibrate about a fixed position but do not travel themselves. Wave motions however do convey energy from one place to another and the strength or 'loudness' of a sound depends upon the energy content of the sound wave. The amplitude of the sound wave, the maximum displacement of each air particle, is greater for stronger sounds as shown in Figure 8.1. Notice that the frequency, or rate of vibration, of the sound remains unaltered and therefore the pitch of a note, in terms of lowness (bass) or highness (treble) that we hear, does not change with loudness.

See also:

Chapters 9 and 10 on *Noise and sound insulation* and on *Room acoustics*, and Resource 3

Velocity of sound wave = 344 m/s (770 mph) in air at 20°C

Measurement

To specify the strength of a sound, it is usually easiest to measure or describe some aspect of its energy or its pressure. Even so, sound does not involve large amounts of energy and its effect depends upon the high sensitivity of our hearing.

Sound power

- **Sound power** (P) is the rate at which sound energy is produced at the source.

    Unit: watt (W)

Sound power is a fundamental property of a sound *source* but is difficult to measure directly. The maximum energy output of a voice is about 1 mW, which explains why talking does not usually exhaust us! In terms of sound power a typical jet engine produces only several kilowatts.

Sound intensity

- **Sound intensity** (I) is the sound power distributed over unit area.

    Unit: watts per square metre (W/m²)

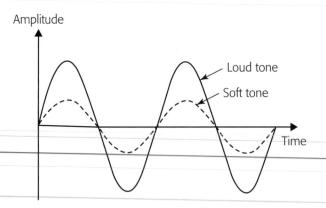

Figure 8.1 Waveforms of soft and loud sounds

Sound intensity is a measure of the rate at which energy is received at a given surface. If, for example, a source is radiating sound in all directions then the sound spreads out in the shape of a sphere and the intensity is found from the following formula:

$$I = \frac{P}{4\,\pi r^2}$$

where I = intensity at distance (W/m²)
P = sound power of the source (W)
r = distance from the source (m)

Sound pressure

- **Sound pressure** (p) is the average variation in atmospheric pressure caused by the sound.

    Unit: pascal (Pa)

The pressure continuously varies between positive and negative values of amplitude, so it is measured by its *root mean square* (RMS) value, which is a type of average having positive values only.

> RMS values are also used to describe the nature of AC electricity

- **RMS pressure** = 0.707 maximum pressure

Intensity–pressure relationship

The intensity of a sound is proportional to the square of its pressure and is expressed by the following formula

$$I = \frac{p^2}{\rho v}$$

where I = intensity of the sound (W/m²)
   p = pressure of the sound (Pa)
   ρ = density of the material (kg/m³)
   v = velocity of sound (m/s).

Thresholds

The weakest sound that the average human ear can detect is remarkably low and occurs when the membrane in the ear is deflected by a distance less than the diameter of a single atom.

Threshold of hearing

- The **threshold of hearing** is the weakest sound that the average human ear can detect.

The value of the threshold varies slightly from person to person, but for reference purposes it is defined to have the following values at 1000 Hz:

I_0 = 1 × 10⁻¹² W/m² when measured as intensity
p_0 = 20 × 10⁻⁶ Pa when measured as pressure

Threshold of pain

- The **threshold of pain** is the strongest sound that the human ear can tolerate.

Very strong sounds become painful to the ear. Excessive sound energy will damage the ear mechanism and very large pressure waves will have other harmful physical effects, such as those experienced in an explosion. The threshold of pain has the following approximate values.

I = 100 W/m² or p = 200 Pa

Decibels

Although it is quite accurate to specify the strength of a sound by an absolute intensity or pressure, it is usually inconvenient to do so. For instance, the range of values between the threshold of hearing and the threshold of pain is a large one and involves awkward numbers. It is also found that, for the same change in intensity or pressure, the ear hears different effects when listening at high intensities and at low intensities. The ear judges differences in sound by ratios so that, for example, the difference between 1 and 2 Pa pressure is perceived to be the same difference as between 5 and 10 Pa.

For practical measurements of sound strength it is convenient to use a decibel scale based on constant ratios, a scale which is also used in some electrical measurements.

Electrical measurements may also use decibels in a related but different way

- The **decibel** (dB) is a logarithmic ratio of two quantities.

The decibel is calculated by the following formulas using either values of sound intensity or sound pressure. Velocity is proportional to the pressure squared.

$$N = 10\log_{10}\left\{\frac{I_2}{I_1}\right\} = 10\log_{10}\left\{\frac{p_2}{p_1}\right\}^2$$

where
N = number of decibels
I_1 and I_2 are the two intensities being compared
or p_1 and p_2 are the two pressures being compared.

Sound levels

In the measurement of sound levels the decibel ratio is always made with reference to the standard value for the threshold of hearing. This produces a scale of numbers that is convenient and gives a reasonable correspondence to the way that the ear compares sounds.

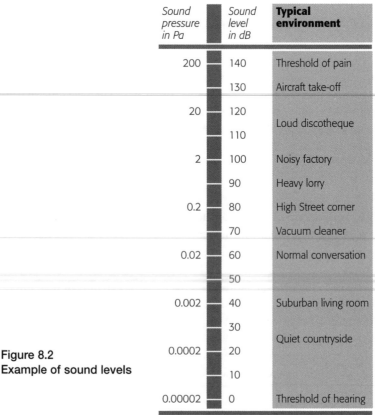

Figure 8.2
Example of sound levels

Sound pressure in Pa	Sound level in dB	Typical environment
200	140	Threshold of pain
	130	Aircraft take-off
20	120	Loud discotheque
	110	
2	100	Noisy factory
	90	Heavy lorry
0.2	80	High Street corner
	70	Vacuum cleaner
0.02	60	Normal conversation
	50	
0.002	40	Suburban living room
	30	Quiet countryside
0.0002	20	
	10	
0.00002	0	Threshold of hearing

Table 8.1 Changes in sound levels

Sound level change	Effect on hearing
+/− 1 dB	negligible
+/− 3 dB	just noticeable
+ 10 dB	twice as loud
− 10 dB	half as loud
+ 20 dB	four times as loud
− 20 dB	one-quarter as loud

Figure 8.2 shows the total range of sound levels in decibels between the two thresholds of hearing and gives typical decibel values of some common sounds. Precise values would depend upon the frequencies contained in the sounds and the distances from the source.

The smallest change in sound level that the human ear can detect is 1 dB, although a 3 dB change is considered the smallest difference that is generally significant. A 10 dB increase or decrease makes a sound seem approximately twice as loud or half as loud, as shown in Table 8.1.

Calculation of sound levels

Values of sound intensity or sound pressure are converted to decibels by comparing them with the standard value of the threshold of hearing. The word 'level' indicates that this reference has been used.

Sound intensity level

If the sound strength is considered in terms of intensity then a sound intensity level (SIL) is given by the formula:

$$SIL = 10\log_{10}\left(\frac{I}{I_0}\right)$$

where

I = the intensity of the sound being measured (W/m^2)

I_0 = the intensity of the threshold of hearing taken as 1×10^{-12} W/m^2

Sound pressure level

Most practical instruments measure sound by responding to the sound pressure. The sound pressure level (SPL) is then given by the formula:

$$SPL = 20\log_{10}\left(\frac{p}{p_0}\right)$$

where

p = the RMS pressure of the sound being measured (Pa)

p_0 = the RMS pressure of the threshold of sound taken as 20 μPa

Worked Example 8.1

A sound has a pressure of 4.5×10^{-2} Pa when measured under certain conditions. Calculate the sound pressure level of this sound. Threshold of hearing pressure = 20×10^{-6} Pa.

Know

$$p = 4.5 \times 10^{-2} \text{ Pa}, \qquad p_0 = 20 \times 10^{-6} \text{ Pa}, \qquad \text{SPL} =?$$

Using formula for SPL and substituting

$$\text{SPL} = 20\log(p/p_0)$$

$$= 20\log\left\{\frac{4.5 \times 10^{-2}}{20 \times 10^{-6}}\right\} = 20\log(2250)$$

$$= 20 \times 3.3522 = 67.04$$

So SPL = **67 dB**

> 'log' on a calculator button indicates $\log_{10}$

Worked Example 8.2

Calculate the change in sound level when the intensity of a sound is doubled.

Let I = initial intensity, so $2I$ = final intensity.

Let L_1 = initial sound level and L_2 = final sound level.

> This calculation shows that doubling the sound energy gives a 3 dB increase in sound level

Using the formula for SIL and taking the difference between the two values:

$$L_2 - L_1 = 10\log\left\{\frac{2I}{I_0}\right\} - 10\log\left\{\frac{I}{I_0}\right\}$$

$$= 10\log\left\{\frac{2I}{I_0} \div \frac{I_0}{I}\right\} \text{ (by the rules of logarithms)}$$

$$= 10 \times 0.3010 = 3.010$$

So change in SIL = **3 dB**

For most practical purposes, the SIL and the SPL give the same value in decibels for the same sound.

Combination of sound levels

If two different sounds arrive at the same time then the ear is subject to two pressure waves. Because the decibel scale is logarithmic in origin the simple addition of sound levels in decibels does *not* give the sound level of the combined sounds. Two jet aircraft, for example, each with an SPL of 105 dB do not, fortunately, combine to give a total effect of 210 dB, which is well above the threshold of pain.

> Decibel values *cannot* be added directly

Although decibels cannot be directly added, intensities can be added, or the squares of pressures can be added, using the following formulas:

$$I = I_1 + I2$$

or $P = \sqrt{(p_1{}^2 + p_2{}^2)}$

When interpreting the results of combined sound levels, you should keep in mind the following guidelines.

- **3 dB** increase in sound level is caused by doubling the sound energy.
- **10 dB** increase in sound level seems approximately twice as loud.

Worked Example 8.3

Calculate the total sound level caused by the combination of sound levels of 95 dB and 90 dB. Threshold of hearing intensity $= 1 \times 10^{-12}$ W/m².

Let
$I_1 =$ intensity of 95 dB,
$I_2 =$ intensity of 90 dB, and
$I_3 =$ intensity of the combined sounds.

Using formula for sound level and substituting both sets of values

$$SIL = 10 \log(I / I_0)$$

$$95 = 10\log\left\{\frac{I_1}{I_0}\right\} \qquad \text{and} \quad 90 = 10\log\left\{\frac{I_2}{I_0}\right\}$$

Using the definition of logarithms each expression can be converted to exponent form and

$$\log\left\{\frac{I_1}{I_0}\right\} = \frac{95}{10} = 9.5 \qquad \log\left\{\frac{I_2}{I_0}\right\} = \frac{90}{10} = 9$$

$$\frac{I_1}{I_0} = 10^{9.5} \text{ (or inv log9.5)} \qquad \frac{I_2}{I_0} = 10^9 \text{ (or inv log9)}$$

$$\frac{I_1}{1 \times 10^{-12}} = 1 \times 10^{9.5} \qquad \frac{I_2}{1 \times 10^{-12}} = 1 \times 10^9$$

$$I_1 = 1 \times 10^{-12} \times 10^{9.5} \qquad I_2 = 1 \times 10^{-12} \times 10^9$$

$$I_1 = 3.16 \times 10^{-3} \text{ W/m}^2 \qquad I_2 = 1 \times 10^{-3} \text{ W/m}^2$$

Combined intensity $I_3 = I_1 + I_2$

$$= (3.16 \times 10^{-3}) + (1 \times 10^{-3}) = 4.16 \times 10^{-3}$$

$$\text{Combined SIL} = 10\log\left\{\frac{I_3}{I_0}\right\} = 10\log\left\{\frac{4.16 \times 10^{-3}}{1 \times 10^{-12}}\right\}$$

$$= 10\log (4.16 \times 10^{-9})$$

$$= 10 \times 9.619$$

So total sound level = **96 dB**

Approximate combination of sound levels

The addition of decibel values is made easier with the aid of a scale such as that shown in Figure 8.3. The fact that the ear cannot detect differences less than 1 dB makes the inaccuracies of the table acceptable. From the scale for the addition of sound levels it can be seen that if the difference between two sound levels is greater than 15 dB then the addition of the lower level will produce a negligible effect on the higher sound level.

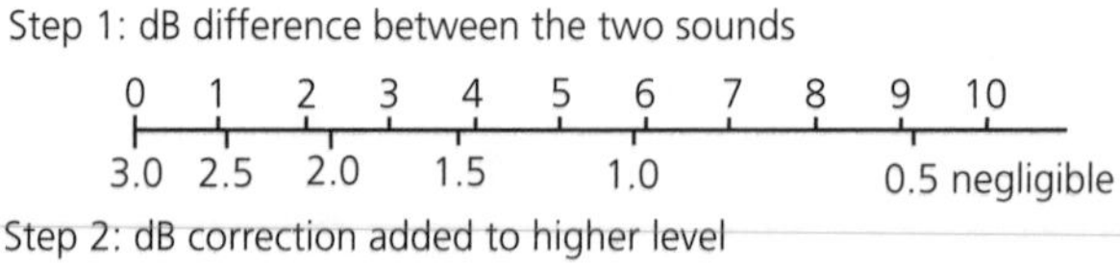

Step 1: dB difference between the two sounds

```
      0   1    2    3   4   5   6   7   8   9   10
      ├───┼────┼────┼───┼───┼───┼───┼───┼───┼───┤
      3.0  2.5  2.0  1.5     1.0              0.5 negligible
```

Step 2: dB correction added to higher level

Figure 8.3 Addition of sound levels

In practical situations:
70 dB + 70 dB
    = 73 dB
70 dB + 64 dB
    =71 dB
70 dB + 60 dB
    = 70 dB

These results mean that a significant sound, such as one of 80 dB, will not be heard above a similar type of sound at 95 dB. People can therefore be run over by site vehicles because the sound of the vehicle is masked by nearby machinery noise. For the lower sound level to be noticed it should have a significantly different frequency content, such as that of a bell, siren or telephone warbler.

ATTENUATION OF SOUND

As sound waves spread out from a source they attenuate: that is, their amplitude decreases and the sound level drops. Except for some absorption by the air, the total energy of the wave front remains constant but the area of the wave front constantly increases. The energy therefore spreads over larger areas and the density of this energy (intensity) or the sound pressure measured at any point must decrease.

The manner in which the sound wave spreads and attenuates is affected by any directional effects of the source – a jet engine, for example, emits more noise to the rear than to the front. The propagation of the sound is also affected by any reflecting or blocking objects in the path.

A *free field* is one in which the sound waves encounter no objects. If there is an object in the sound path then some of the sound will be reflected, some will be absorbed and some will be transmitted through the object. The exact effects depend on the nature of the object and the wavelength of the sound. In general, the size of the object must be greater than one wavelength of the sound wave in order to significantly affect the wave. A wavelength of one metre is typical of sound produced by voices.

Typical wavelengths:
bass note: 5 m
voices: 1 m
whistle: 0.1 m

For the initial prediction of the sound in the open air it is necessary to assume free field conditions and to consider the behaviour of sound being emitted from a point or a line type of source. These results may then be modified to take account of the reflective conditions encountered in most practical situations.

Point source of sound

In a free field the sound wave from a point source spreads out uniformly in all directions in the shape of a sphere, as shown in Figure 8.4. The surface area of a sphere increases in proportion to the square of its radius and so the intensity of sound energy varies in an inverse manner.

Inverse square law

- The **sound intensity** from a point source of sound decreases in inverse proportion to the square of the distance from the source.

This relationship can also be expressed in the following equation:

$$I \propto \frac{1}{d^2}$$

where I is the sound intensity measured at distance d from the source.

The ratio of any two intensities is given by the formula:

$$\frac{I_1}{I_2} = \frac{d_2^{\,2}}{d_1^{\,2}}$$

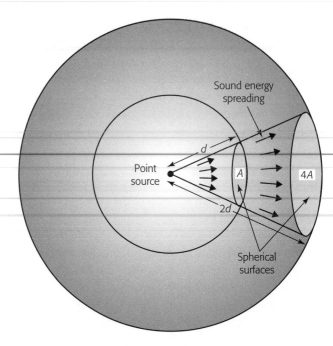

Figure 8.4 Attenuation of sound from a point source

where
I_1 is the sound intensity measured at distance d_1 from the source and
I_2 is the sound intensity measured at distance d_2 from the source.

Note the 'cross-over' within the formula because of the inverse relationship.

Worked Example 8.4 illustrates the following general effect of attenuation from a point source of sound.

- The SPL decreases by 6 dB each time the distance is doubled from a point source of sound in free space.

If the source is on a perfectly flat and reflecting surface, then all the sound radiates into a hemisphere. The SPL then decreases by 3 dB (half of 6 dB) when the distance is doubled. A single motor car or other machine on a paved surface approximates to this condition, and the sound attenuates by only 3 dB for each doubling of distance.

In practical situations: doubling the distance from an object reduces sound by about 3 dB

Line source of sound

The sound wave from a line source spreads out in the shape of a cylinder, as shown in Figure 8.5. The surface area of a cylinder increases in simple proportion to its radius. Sound intensity from a line source therefore is in simple inverse proportion to the distance from the source.

Worked Example 8.4

A microphone measures sound at a position in a free field 5 m from a point source. Calculate the change in SPL if the microphone is moved to a position 10 m from the source.

Let

L_1 = SPL at distance d_1 = 5 m
L_2 = SPL at distance d_2 = 10 m

Using formulas

$$L_1 - L_2 = 10\log(I_1/I_2) \text{ and } I_1/I_2 = (d_2^2/d_1^2)$$

gives

$$L_1 - L_2 = 10\log\left\{\frac{d_2^2}{d_1^2}\right\} = 10\log\left\{\frac{10^2}{5^2}\right\}$$

$$= 10\log\left\{\frac{100}{25}\right\} = 10\log 4$$

$$= 10 \times 0.6021 = 6.021$$

So change in SPL = **6 dB** decrease.

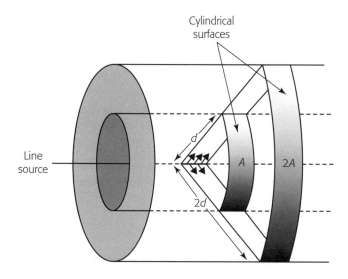

Surface area increases in proportion to the distance.

Figure 8.5 Attenuation of sound from a line source

The attenuation from a line source in a free field can be shown by the general relationship:

$$I \propto \frac{1}{d}$$

where I is the sound intensity measured at distance d from the source.

The ratio of any two intensities is given by the following formula:

$$\frac{I_1}{I_2} = \frac{d_2}{d_1}$$

where

I_1 is the sound intensity measured at distance d_1 from the source

I_2 is the sound intensity measured at distance d_2 from the source.

For a perfect line source the SPL decreases by 3 dB each time that the distance is doubled. For a line source radiating all its energy into half a cylinder, the attenuation is therefore 1.5 dB (half of 3 dB) for each doubling of the distance. A line of cars on a busy road tends to approach this condition.

In practical situations near buildings, the decrease in sound level caused by doubling the distance is between 1.5 and 3 dB. Such changes in sound level of less than 3 dB are heard as a negligible change in loudness. Moving a building further away from the road therefore produces little effect within the area of most building plots; other methods, such as sound reflection and sound insulation, are more effective for the control of road noise.

> *In practical situations*: doubling the distance from a road reduces sound by about 2 dB

Attenuation by air

The attenuation of sound caused by its spreading out from a point source or line source is an effect of energy distribution. The factors discussed below may also affect the passage of sound through the air.

Air absorption

> *In practical situations*: sound is likely to be heard further on a foggy night

Some of the energy of a sound wave is spent in alternately compressing and expanding air. The effect is negligible at low frequencies and at 2000 Hz causes a reduction of about 0.01 dB/m of travel. The attenuation is increased at low humidities.

Temperature gradients

The velocity of sound is greater in warm air than in cold air. Open air has layers at different temperatures, and sound waves crossing these layers are deflected by the process of *refraction*. One result of this refraction is that, in general, sound travels along the ground better at night than during the day. This is the effect of relative changes in the temperature of the air lying next to the ground. Sound is refracted upwards during the day and at night it is refracted downwards.

Wind effects

Sound waves will be affected by any wind blowing between the source and the receiver. The velocity of wind increases with height above the ground and this gradient deflects the sound waves upwards or downwards.

Ground attenuation

It is possible for some sound energy to be absorbed by passing over the surface of the ground. This effect is quite local and only applies within 6 m of the ground, which must be free of obstructions. Hard surfaces such as paving provide little attenuation, but surfaces such as grassland can provide a reduction of overall noise level of up to 5 dB.

NATURE OF HEARING

Sound waves can be detected and measured by instruments without the aid of human senses, but the aspect of sound that interests us most is the human perception of sound waves. The sense of hearing involves the ear and the brain, and the effect of sound can therefore vary from person to person. However, most people share the basic characteristics of hearing outlined in this section.

The ear

Most of the mechanism of the ear is situated inside the head, and the structure of the ear can be divided into three main parts, as shown diagrammatically in Figure 8.6.

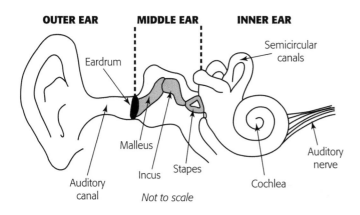

Figure 8.6 The ear

Outer ear

The outer ear is the part of the ear that can be seen. It collects the sound waves and funnels them to the eardrum, a membrane which vibrates when sound waves fall upon it. Sound waves between 2 and 6 kHz can resonate within the ear canal, causing higher pressures and therefore making these frequencies the ones to which we are most sensitive.

Middle ear

The middle ear is an air-filled cavity, connected to the throat, which passes the vibrations of the eardrum to the inner ear. This transfer is achieved by means of the three small bone levers. The mechanical link between the bones amplifies the vibrations to adjust for the difference between the air of the middle ear and the fluid of the inner ear.

The middle ear also has an *aural reflex* mechanism which reacts to loud sounds and tightens the eardrum, thereby lessening the force which is transmitted into the inner ear. This defence mechanism reacts too slowly to protect against sudden loud sounds such as gunshots or some impulsive effects from recorded music.

Inner ear

The inner ear converts the mechanical vibrations of sound into electrical impulses which are transmitted to the brain. The *cochlea* is a hollow coil of bone, filled with liquid, in which the sound waves vibrate. Dividing the cochlea along its length is the *basilar membrane*, which contains approximately 25,000 nerve endings. The fine hairs attached to these nerves detect the sound vibrations in the fluid, and the information is transmitted to the brain by the auditory nerve. The inner ear is situated near the three semicircular canals which contain fluid and are associated with the sense of balance.

Hearing loss

Sounds of certain intensity and duration will damage the hearing system and cause a temporary or permanent hearing loss. This hearing loss can range from being barely noticeable to being a severe impairment that can have a major impact on personal communication and quality of life, particularly for people who have not been brought up in a positive culture of deafness.

See also:

section on *Noise control* in Chapter 9

Loss of hearing not only affects sensitivity to the level of sounds but also affects the frequencies at which sounds can be detected. The various causes and categories of hearing loss are described below. The hearing damage caused by over-exposure to noise is cumulative over a lifetime and cannot be reversed. Measurement of noise and strategies for its control are discussed in Chapter 9.

Conductive hearing loss

Conductive hearing loss involves a failure of sound to be carried or conducted through the outer ear or the middle ear, or both. The failure might be caused by a blockage in the ear canal, a broken ear, or a break or stiffening in the system of connecting bones. The resulting hearing loss affects the transmission of low tones rather than high tones and can usually be treated.

Sensory hearing loss

Sensory hearing loss is caused by damage to the nerve endings of the cochlea in the inner ear and, more unusually, by a failure in the nerve system that carries information to the brain. This type of hearing loss can be caused by infections, by head injuries and by exposure to high levels of noise, in addition to being a condition at birth.

Exposure to noise

Hearing loss caused by exposure to noise can give the following two effects:

- **Temporary threshold shift** (TTS) is a temporary loss of hearing, which usually recovers in one or two days after the exposure to noise.
- **Permanent threshold shift** (PTS) is a permanent loss of hearing caused by longer exposure to noise.

> TTS recovers between noise exposures
> PTS accumulates with noise exposures

The first effect of *noise-induced hearing loss* (NIHL) is loss of hearing in the region of high tones, around frequencies between 3000 and 4000 Hz. This may not affect speech reception at first, and the hearing loss usually remains unnoticed until it begins to affect ordinary conversation. Because the receptor cells in the inner ear are damaged, this sensory hearing loss is irreversible and cannot be helped by amplification.

Acoustic trauma

Mechanical damage occurs when the ear is exposed to sudden or high sound levels, such as those from an explosion. This damage can occur throughout the ear, from the eardrum to the cochlea.

Tinnitus

Tinnitus is experienced as ringing, whistling, buzzing or other sounds, usually of medium to high frequencies, that are heard inside both ears or heard generally within the head. Tinnitus may be a warning of hearing loss occurring and it can become a permanent effect, only masked by louder noise.

Presbyacusis

Presbyacusis is a gradual loss of hearing sensitivity due to age and is experienced by everybody. The higher frequencies are affected first but the effect is not usually noticeable until around the age of 65 years or above. However, loss of hearing caused by age adds to any loss of hearing due to noise exposure earlier in life and the combination of the two effects is usually troublesome.

Audiometry

Audiometry is the measurement of human hearing in terms of sensitivity at various frequencies across the noise spectrum. The threshold of hearing is measured at each frequency for each ear and the results can be presented graphically as an *audiogram*.

The results shown on an individual's audiogram can be compared with previous results for the same person and with the average sensitivities for a person of the same age. Such readings allow noise-induced hearing loss caused by a person's job to be distinguished from the normal presbyacusis that occurs with age.

Loudness

As the intensity of a sound increases it is heard to be 'louder'. This sensation of loudness is a function of the ear and the brain, and it depends upon the frequency as well as the amplitude of the sound wave. Human hearing is not equally sensitive at all frequencies and tones of different pitch will be judged to be of different loudness, even when their SPL is the same. For example, a 50 Hz tone must be boosted by 15 dB if it is to sound as loud as a 1000 Hz tone at 70 dB.

The results of many measurements of human hearing response can be presented in the form of standard contours, as in Figure 8.7. The contours show how the SPL in dB of pure tones needs to change to create the same sensation of loudness when at different frequencies. It can be seen that the ear is most sensitive in the frequency range between 2 kHz and 5 kHz, and least sensitive at low or extremely high frequencies. This effect is more pronounced at low SPLs than at high SPLs, and Figure 8.7 has a 'family' of curves for different SPLs.

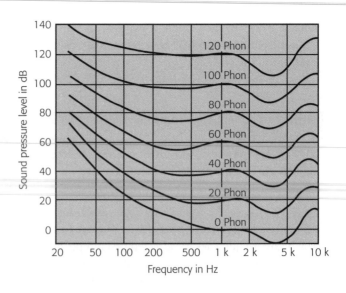

Figure 8.7 Equal loudness contours

Phon scale

If two different tones seem equally loud to the ear, then it can be useful to have a scale which gives them the same value, even though the two tones have different SPLs. The *phon* scale of loudness level is obtained from the family of equal loudness curves shown in Figure 8.7.

- **Loudness level** (L_N) of any sound is numerically equal to that SPL, in decibels, of a 1000 Hz pure tone which an average listener judges to be equally loud.

    Unit: phon

For example, it can be seen from Figure 8.7 that
    60 phons = 60 dB at 1000 Hz
    but
    60 phons = 78 dB at 50 Hz.

This difference between the phon and the decibel values reflects the fact that human hearing is less sensitive at low frequencies.

Sone scale

The *sone* scale of loudness is a re-numbering of the phon scale so that the sone values are directly proportional to the magnitude of the loudness. For example, a value of 2 sones is twice as loud as 1 sone. One sone is equivalent to 40 phons. Loudness, in sones, is doubled each time the loudness level is increased by 10 phons.

EXERCISES

1. A certain sound has an intensity of 3.16×10^{-4} W/m². Calculate the sound intensity level in decibels if the threshold of hearing reference intensity is 1×10^{-12} W/m².

2. Calculate the actual pressure of a sound which has an SPL of 72 dB. The threshold of hearing reference pressure is 20×10^{-6} Pa.

3. Calculate the total sound pressure level caused by the combination of sounds with the following SPLs:
   (a) 85 dB and 87 dB.
   (b) 90 dB and 90 dB and 90 dB.

4. The intensity of a point source of sound, measured at a distance of 6 m from the source, is 3.4×10^{-6} W/m².

(a) Calculate the intensity of the sound at a distance 20 m from the source.

(b) Calculate the decibel change between the two positions.

5. Calculate the difference between SPLs measured at 10 m and at 63 m from a perfect line source of sound. State the factors that might modify this result in a practical situation.

Answers are on page 326.

9

Noise and Sound Insulation

CHAPTER OUTLINE

Noise is unwanted sound. The results of national surveys of typical exposure to noise show that over 50 per cent of the population are exposed to day-time noise levels that exceed the World Health Organisation (WHO) ratings for significant community annoyance. Other surveys report that around 50 per cent of people find their home in some way unsatisfactory because of noise intrusion. These results include people living in new homes built to modern building codes which require certain standards of noise control.

A satisfactory environment of sound quality therefore needs to be a major consideration when we create our built environment. This chapter deals with the technical and practical aspects of noise measurement and sound insulation so that you can:

▶ appreciate the environmental factors that affect human tolerance of noise
▶ understand the technical aspects of measuring and assessing noise
▶ be familiar with the common indices of noise assessment
▶ understand how noise exposure is assessed in the workplace, and in recreation
▶ appreciate the levels of noise exposure that causes hearing loss
▶ understand how noise annoyance is assessed for contexts such as road noise and construction site noise
▶ know the principles of noise control and their practical applications
▶ appreciate how sound is transmitted within buildings
▶ understand the difference between airborne sound and impact sound

$\longrightarrow$

▶ know the principles of sound absorption and sound insulation
▶ appreciate how typical walls and floors and other building elements perform in providing sound insulation
▶ be familiar with the common measurements of sound insulation
▶ understand the typical requirements for sound insulation of building regulations and codes.

Your knowledge will be enhanced by also referring to the related chapters on *Aspects of Sound* and on *Room Acoustics*. The Resource sections at the back of the book contain information about the science of sound and references that are useful for both revision of principles and for extended investigation of topics.

The environmental and social definitions of noise take account of the effect of the sound rather than its technical nature. So even if a sound consists of the finest music it can be considered as noise if it occurs in the middle of the night! The following are some of the detrimental effects that noise can have on people and their environment:

- **Hearing loss**: excessive exposure to noise causes loss of hearing, as discussed in the previous chapter.
- **Quality of life**: noisy environments, such as areas near busy roads or airports, are considered unpleasant and undesirable.
- **Interference**: interference with significant sounds such as speech or music can be annoying and, in some situations, dangerous.
- **Distraction**: distraction from a particular task can cause inefficiency and inattention, which could be dangerous.
- **Expense**: the measures needed to control excessive noise are expensive. Businesses may also suffer loss of revenue in a noisy environment.

> Noise annoyance is 'a feeling of displeasure evoked by noise' (the World Health Organization)

MEASUREMENT OF NOISE

> **See also:**
>
> Chapters 8 and 10 on *Aspects of sound* and on *Room acoustics*, and Resource 3

The acceptance of noise by people obviously depends upon personal hearing sensitivity and personal preferences, so individuals vary widely in their response to the same level of noise, even when it arises from the same source. The acceptance of noise is also affected by the external factors outlined below.

- **Type of environment**: acceptable levels of surrounding noise are affected by the type of activity. A library, for example, has different requirements from those for a workshop.
- **Frequency structure**: different noises contain different frequencies, and some frequencies are found to be more annoying or more harmful than others. For example, the high, whining frequencies of certain machinery or jet engines are more annoying than lower frequency rumbles.
- **Duration**: a short period of high-level noise is less likely to annoy than a long period. Such short exposure also causes less damage to hearing.

The measurement of noise must take these factors into account, and the index used to assess them should be appropriate to the type of situation. Although individuals vary in their response to a certain level of noise, when a large number of people are exposed to the same sources of noise the average or community response is relatively stable. It is therefore possible to establish the community average degree of annoyance associated with long-term average noise exposure.

The concept of annoyance is well established for identifying the noise impacts from roads, railways, airports and building sites. A number of different indices and scales have evolved to assess situations causing risk and annoyance from noise, and this section describes some of those in common use around the world.

Sound level meter

A sound level meter is an instrument designed to give constant and objective measurements of sound level. Figure 9.1 shows the main sections of a typical sound level meter. The meter converts the variations in air pressure to variations of voltage which are amplified and displayed on an electrical meter calibrated in decibels. Sound level meters can be small enough to be hand-held and are supplied in several grades of accuracy.

A typical sound level meter takes an RMS (root mean square) value of the signal, which is a type of average that is found to be more relevant than peak values. The meter may have 'fast' and 'slow' response settings but may not be fast enough to accurately record an impulse sound, such as a gunshot.

Frequency components

Most practical noise is sound that contains a spectrum of different frequencies which can be detected by the microphone of a sound level meter. The interpretation of these different frequencies needs consideration, especially because human hearing judges some frequencies to be more important than others.

If a noise contains *pure tones* of single frequencies, then it is usually more annoying than broadband noise with a spectrum of frequencies. Pure tones are

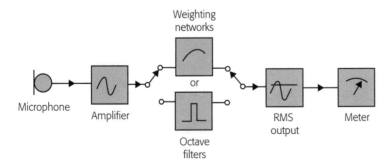

Figure 9.1 Construction of a sound level meter

often present in the noise given off by industrial equipment such as high-speed fans or other machinery, and by electrical generators and transformers.

Weighting networks

One method of dealing with the frequency content of sound is for the sound level meter to emphasise or give 'weight' to the same frequencies that human hearing emphasises. The weighting networks in a meter are electronic circuits whose response to frequency is similar to that of the ear. The response to low frequencies and to very high frequencies is reduced in a specified manner, and four different weightings have been standardised internationally as the A, B, C and D scales.

> dB(A) or dBA indicates a reading using the A-weighting

The *A scale* has been found to be the most useful weighting network. Many measurements of noise incorporate decibels measured on the A scale and the symbol dB(A) indicates a standard treatment of frequency content.

Frequency bands

A sound level measurement that combines all frequencies into a single weighted reading is not suitable when it is necessary to measure particular frequencies. Some sound level meters allow the use of filters, which pass selected frequencies only. The sound pressure level (SPL) in decibels is then measured over a series of frequency bands, usually octaves or one-third octaves.

The *noise spectra* produced by such a series of measurements can be presented as a dB-frequency diagram, like that shown in Figure 9.2. Narrower frequency

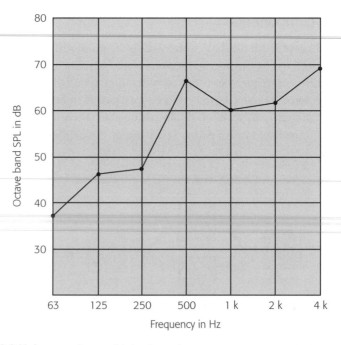

Figure 9.2 Noise spectrum of telephone buzzer

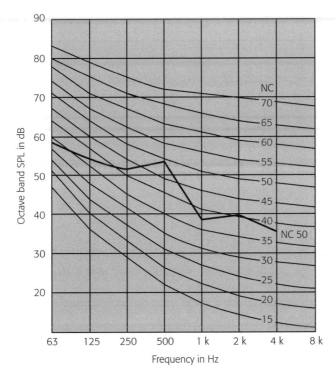

Figure 9.3 NC (noise criterion) curves

bands give more information about the frequency content of the sound than wide frequency bands.

Noise limiting curves

One method of specifying acceptable levels of sound at different frequencies is to use standard curves of noise. These rating curves, which are plots of SPL against set frequencies, are based on the sensitivity of the human ear. The noise being analysed is measured at octave intervals and these results are plotted on top of the standard curves, as shown in Figure 9.3.

The aim of most systems is to produce an index or criterion which is a single-figure rating for the noise. This figure is obtained by comparing the plotted curve against the standard curves and determining the closest fit according to published rules. Commonly-used systems are as follows.

- **Noise criterion** (NC): obtained from NC curves.
- **Preferred noise criterion** (PNC): obtained from PNC curves.
- **Noise rating** (NR): obtained from NR curves.

The NC curves have been widely used for assessing the noise made by heating and ventilating equipment, and PNC curves are a development of the NC curves. NR curves are commonly used for other industrial measurements of noise. Acceptable noise limits for different purposes can be specified by numbers read

Table 9.1 Acceptable levels of background noise from services installations

Environment	NC/NR/PNC approx. index
General offices	40
Libraries	35
Homes, hospitals	30
Theatres, cinemas	25
Concert halls, studios	20

from the standard curves, and Table 9.1 indicates some typical limits for levels of background noise caused by services installations.

Time components

Human response to noise greatly depends upon the total time of the noise and upon the variation in sound levels during that time. There is more tolerance, for example, of high but steady levels of background sound than of a lower level background level with frequent noise intrusions.

Statistical measurement

During a chosen period of time, such as 12 hours, it is possible to record many instantaneous readings of sound level in dB(A). The variations in these readings can then be combined into one single number or 'index' by a statistical process using a percentile level, such as 10 per cent.

- L_{A10} is the noise level, measured as A-weighted sound level, exceeded for 10 per cent of a given measurement time.

> Index = single value made up from multiple values

The traffic noise index, described below, uses the 10 per cent level for estimating maximum noise levels. Another percentile level, of 90 per cent, is used to estimate background noise.

These statistical measurements of noise are best made by a noise level analyser attached to the sound level meter which records the percentage of time the noise

Nomenclature

The exact method of writing and printing a sound index varies slightly from document to document, especially as the use of subscripts can get changed while printing. The general form is usually as follows, illustrated with an example of $L_{Aeq,8hr}$

- L indicates a sound level index
- A indicates that the sound level in dB is measured using the A-scale
- eq is shorthand for the particular type of index, in this case the equivalent continuous sound level
- 8hr indicates the time period used.

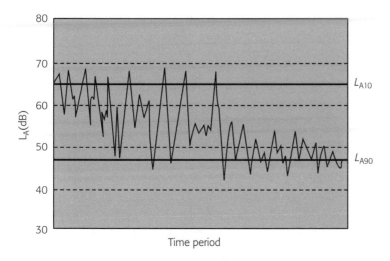

Figure 9.4 Time distribution of noise level

level has spent in each portion of the decibel scale. The analyser produces a single figure in dB(A) as the percentile level and some systems can also produce a graph of results, such as that shown in Figure 9.4.

Traffic noise index L_{A10}

Noise caused by road traffic usually changes in level during the day, and the way it varies has a considerable effect on the nuisance it causes. The traffic noise index L_{A10} takes these variations into account.

- **Traffic noise index L_{A10}** is an average of the 18 hourly L10 values taken between 0600 and 2400 hours on a normal weekday.

Measurements of L_{A10} must be taken at certain standard distances from the roadway with corrections made for any wind and reflections. A typical measurement at 10 m from the edge of a motorway is $L_{A10,18hr}$ of 74 to 78 dB(A). The traffic noise index has been found to give reasonable agreement with surveys of the dissatisfaction caused by road traffic noise. A typical basis for UK Government compensation to households affected by new or improved roads is an $L_{A10,18hr}$ of 68 dB.

> Any result expressed as a sound level index, L, combines a range of information into a single figure

Equivalent continuous sound level L$_{Aeq}$

Another method of assessing noise that varies in sound level over time is to use an average value related to total sound energy. The equivalent continuous sound level compares a varying sound level to a theoretical constant sound which gives an equivalent amount of sound energy.

Although human hearing does not judge loudness in terms of energy, the L$_{Aeq}$ measurement of accumulated sound energy is found to correlate well to the annoyance caused by noise and also to damage to hearing.

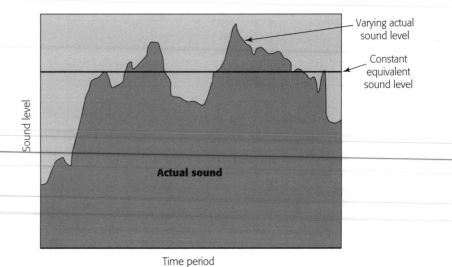

Figure 9.5 Equivalent continuous sound level

- **Equivalent continuous sound level**, $L_{Aeq,T}$, is that constant sound level which, over the same period of time T, provides the same total sound energy as the varying sound.

  Unit: dB(A)

In a simple example, a doubling of sound energy increases the SPL by 3 dB, so that the following combinations of sound levels and time periods could give the same $L_{Aeq,8hr}$

	90 dB(A) for 8 hours	All give the same value of $L_{Aeq,8hr}$ = 90 dB
or	93 dB(A) for 4 hours	
or	99 dB(A) for 1 hour	

In most practical situations the noise levels tend to vary continuously; integrating sound level meters are employed to sample the noise regularly and to produce a L_{Aeq} figure as the 'noise history' over a particular period of time. Values greater than L_{Aeq} = 55 dB are considered by the World Health Organisation (WHO) to mark significant community annoyance during daytime hours.

Railway noise

An equivalent continuous sound level $L_{Aeq,18hr}$ has been used to assess the noise near a railway line. It is difficult to directly compare road noise and railway noise because of differences in their nature, especially in duration and regularity of the noise. In terms of annoyance, an approximate conversion relationship is:

- $L_{Aeq,T}$ approximately equals $L_{A10,T} - 3$

Hearing risk

Because the risk of damage to hearing is largely dependent on the total energy reaching the ear in a given period, $L_{A,eq}$ is the basis of safe exposure to noise. Recommended levels are usually in the range $L_{Aeq,8hr} = 80$ to 90 dB, as described below.

More sound energy = more hearing damage

Occupational noise index $L_{EP,d}$

The $L_{EP,d}$ index is a measure of *daily personal noise exposure*. It measures the cumulative noise energy received by a person during a standard eight-hour working day. For working days and weeks with different pattern of hours, the index can be made equivalent to an eight-hour day by using published formulas. In UK legislation, which is typical of that required for Europe, the duties of employers and employees are linked to the values of $L_{EP,d}$ described below.

- **Lower exposure action level**: $L_{EP,d} = 80$ dB(A-weighted), and a peak sound pressure of 135 dB(C-weighted).
- **Upper exposure action level**: $L_{EP,d} = 85$ dB(A-weighted), and a peak sound pressure of 137 dB(C-weighted).
- **Exposure limit value**: $L_{EP,d} = 87$ dB(A-weighted), and a peak sound pressure of 140 dB(C-weighted).

When the exposure exceeds the action levels, the regulations prescribe actions to be taken involving various degrees of staff information, assessments of noise, assessment of health and the use of hearing protectors. Personal noise exposure must never exceed 87 dB(A) after taking into account the effect of hearing protection.

Remember: A sound level increase of just 3 dB may not be noticed but:
- the sound energy received doubles
- the hearing damage doubles

Noise dose

The noise dose index makes 100 per cent dose equal to a fixed noise exposure such as 85 or 90 db(A) for eight hours. Individual noise exposure can be assessed by wearing a personal dosemeter.

Peak

The peak is the highest pressure produced by an explosive sound, such as that from a cartridge tool or a gunshot. Although this peak only lasts for a brief instant, its pressure is high and damaging to hearing. The peak value is measured by a sound level meter which can hold and display the information.

Estimates for the UK: over 2 million workers with noise exposures greater than 80 dB(A)

A peak of 200 Pa (equivalent to 140 dB) is used as an industrial limit. At this level of sound, an eight-hour L_{eq} exposure of 90 db(A) is reached in about a fifth of a second.

Construction site noise

The activities on a construction site often generate noise which can:

Recreation and noise

You don't have to work in 'heavy industry' to suffer hearing damage. Some leisure activities, such as those described below, can provide noise doses that exceed industrial limits and hence introduce a high risk of hearing damage.

- Informal surveys of some popular restaurants have recorded average noise levels of over 88 dB(A). Exposure to this level of sound for eight hours will exceed the upper limit for personal noise exposure of 85 dB(A).
- Many well-known music performers, including old Rockers, have acknowledged the hearing loss they have accumulated from prolonged exposure to amplified music. The risk of hearing loss applies to performers, sound and lighting engineers, venue staff – and the audience.
- Listening to music on a personal audio device with earphones can cause the same hearing risks as loud restaurants or rock concerts. If the volume needs to be increased to counteract a noisy environment such as a train, then it is easy to be doubling or trebling the noise dose – and also the hearing damage.
- Other leisure activities that can expose you to hazardous noise doses include: sports shooting, motor racing, DIY tools and indoor swimming pools during busy periods.

People exposed daily to high sound levels, such as bar and restaurant staff, will often be suffering from temporary or permanent threshold shift, as described in the previous chapter. Therefore to get the same personal 'effect' they often increase the volume of the amplified music; which unfortunately helps make their own hearing loss permanent and endangers other people.

Young people are as vulnerable to hearing loss as anyone else – and have more to lose.

- cause hearing damage to people working on the site
- cause annoyance to people nearby.

The hearing risk to construction workers can be controlled by appropriate assessment and working methods, some of which are outlined in the later section on noise control. The rights of people to be protected from unreasonable noise annoyance can be controlled by legislation that acts on measurements of sound level in a specified manner, such as the 12-hour continuous equivalent sound level described below.

Occupational noise index $L_{A,EP,d}$ for site workers

The $L_{A,EP,d}$ index measures the personal noise exposure received by a person working on a building site. Employers have a duty to assess the likely risk to site and

Table 9.2 Noise exposure of construction activities

Activity		Likely noise exposure	
		$L_{EP,d}$ average	Range
Blasting		100+	
Bricklayer		83	81–85
Carpenter		92	86–96
Concrete	chipping/drilling	85+	
	floor finishing	85	
	grinding	85+	
Concrete worker		89	
Crushing	mill worker	85+	
Driver	crawler tractor	85+	
	dumper	85+	
	excavator	<85	
	loader	<85	
	roller	85+	
	wheeled loader	89	
	wheeled tractor	<85	
Engineer	supervising pour	96	
	surveying	<80	
Foreman	supervising workers	80	
Formwork setter		92	89–93
Ganger	concrete pour	93	92–93
	general work	94	
Guniting		85+	
Labourer	concrete pour	97	95–98
	digging/scabbling	100	
	general work	84	
	shovelling hardcore	94	
	shuttering	91	
M&E installer	general	89	82–96
	small work	84	78–89
Piling operator		85+	
Piling worker		100+	
Reinforcement worker	building site	86	82–89
	bending yard	84	77–87
Sandblasting		85+	

Source: Health and Safety Executive for the UK.

workers, and Table 9.2 gives some likely noise exposure figures for typical site activities. When personal noise exposure levels exceed 80 dB(A) then action must be taken.

Twelve-hour L_{Aeq} index for site noise

Periods of high site noise levels must be 'balanced' by less noisy periods

An $L_{Aeq,12hr}$ continuous equivalent sound level index is used in the United Kingdom to assess the noise associated with operations on a building site. The annoyance caused to the environment can be equated to the total sound energy received at the boundary of the site during the course of the day. An $L_{Aeq,12hr}$ value of 75 dB is a limit that has been used, above which site operations can be stopped by legal action.

As with other cumulative measurements, different patterns of noise during the 12 hours can give the same value. For example, it is possible to run plant with sound levels over 100 dB(A) for several hours and still remain within the $L_{Aeq,12hr}$ of 75 dB. The prediction of such results can be made using published tables. Site measurements can be made at the edge of a site using the integrating sound level meters and techniques described in the sections above.

Other noise measurements

Sound exposure level (SEL)

The sound exposure level is an index of transient noise levels, such as those produced by passing road vehicles or aircraft. The SEL is a measure of acoustic energy, which is a factor in the annoyance caused by a noise but is different from the maximum sound level in dB(A).

- **Sound exposure level**, $L_{A,SEL}$, is that constant sound level in dB(A) which, during one second, provides the same total sound energy as the measured noise.

Perceived noise level (L_{PN})

L_{PN} is an index of aircraft noise that takes account of those higher frequencies in aircraft engine noise which are known to cause annoyance. The noisiness is rated in *noys* and converted by calculation to perceived noise decibels PNdB.

An approximate value for perceived noise level L_{PN} is obtained by adding 13 dB to a noise level measured in dB(A).

Noise and Number Index (NNI)

NNI is an index of airport noise that includes the average perceived noise level and the number of flyovers heard in a given period. This index has been used to predict and to measure annoyance resulting from noise near airports.

Sixteen-hour L_{Aeq} index for airport noise

$L_{Aeq,16hr}$ = 57 dB is one limit for noise near an airport (this has to measured over three months)

An $L_{Aeq,16hr}$ continuous equivalent sound level index can be used to assess the noise in the environment of an airport. The noise from aircraft movements are expressed as the $L_{Aeq,16hr}$ produced by readings taken over a six-hour day, 0700 to 1300 hours local time, during mid-June to mid-September. Sets of readings are taken at set places around an airport and used to prepare maps which show contour lines joining up points of the same $L_{Aeq,16hr}$ value.

Speech Interference Level (SIL)

SIL is a measure of the level of background noise at which the noise will interfere with speech in a particular situation. The type of voices and distances involved are taken into account.

NOISE CONTROL

The many types of noise that cause concern in buildings can be grouped into the following three action areas.

- **Source**: Sources of sound may be outside the building, such as from a road, or may be within the building, such as noise from occupants.
- **Path**: The sound path may be through the air from the source to the building, or the path may be within the building.
- **Receiver**: The receiver of the sound may be the building itself, it may be a particular room, or it may be the person hearing the noise.

Sound reductions in all three areas of source, path and receiver are relevant to the design of quieter buildings, which is the main focus of this chapter.

Noise levels in and around buildings are also affected by wider issues such as the design of quieter vehicles and machinery, the location of industry and transport, and the type of construction techniques in use. Sometimes these matters need to be enforced by regulations, so that legislation also becomes a tool for noise control.

Noise control actions

The following actions are useful to control noise from many industrial sources, including factories and construction sites. The actions help prevent damage to the hearing of workers and to reduce noise pollution of the environment.

Design and elimination

Noise control is most effective when it is considered at the design stage. Using alternatives in the design of a building and its construction methods can eliminate potentially noisy equipment and techniques.

Choice of equipment

Some types of equipment are quieter than other types and there are options at the planning stage to use quieter machinery and tools. Techniques of elimination and substitution include:

- using a later model with improved noise insulation
- using electric motors instead of petrol/diesel motors
- fitting anti-vibration mountings between engines and other structures

Frequency of sound: The performance of many solutions to noise control depends upon the frequency of the sound

- using hydraulic systems instead of pneumatic systems
- use of tools with hush kits.

Work planning

Some methods of working are inherently quieter than others and can be incorporated at the design and work planning stage. Examples include:

- hydraulic piling rather than hammer piling
- avoiding oversized elements and therefore minimising cutting, scabbling and similar noisy operations.

Work practices

The actions of people carrying out operations have an important effect on noise control. For example:

- Correctly using equipment: for instance staff should wear their own hearing protectors, and implement techniques such as closing doors.
- Where it is necessary to use equipment which produce high levels of sound then a skilled operator will use it for shorter periods of time. Good training and education is therefore effective.

Distance and location

For a source of sound which acts as an approximate point, such as a single machine, the noise level falls by between 3 and 6 dBA every time the separation distance is doubled. Positioning the source of noise in an isolated position, such as on a building site, therefore provides an easy opportunity for noise control. To be effective, this technique requires reasonable distances so it is not usually appropriate for controlling noise within buildings.

Acoustic screens and barriers

Devices to deflect and absorb sound waves can be effective if they are placed close to the source of sound, or close to the recipient of the noise.

Personal protection

If other methods fail to reduce the noise energy reaching the ear, then individual hearing needs to be protected by wearing hearing defenders over the ears. The two broad types of ear protectors available are ear plugs inserted in the ear canal and ear muffs which cover the entire ear. When choosing and using hearing defenders the following properties should be considered:

- sufficient level of protection
- sufficient comfort to wear for as long as required
- durability for as long as required
- hygiene in practical operation.

Active Noise Reduction (ANR)

A sound wave can be neutralised by an identical wave that, if it is half a wavelength out of phase, provides equal and opposite compressions and rarefactions. Equipment is available that rapidly analyses a sound and uses a speaker to produce appropriate 'anti-noise' which lowers the sound level. ANR techniques work best for relatively constant sounds at lower frequencies. Current applications include ear muffs, personal headsets, confined machinery spaces, industrial chimneys and aircraft cabins.

> ANR 'cancels' noise with sound waves of opposite phase

NOISE TRANSFER

Noise is transferred into buildings and between different parts of buildings by means of several different mechanisms. It is necessary to identify the types of sound involved as being of two main types:

- airborne sound
- impact sound.

> Impact sound may also be known as structure-borne sound

Airborne sound

- **Airborne sound** is sound which travels through the air *before* reaching a partition.

Notice that this definition is not as simple as merely saying the 'sound travels through the air'. The vibrations in the partition under consideration must be started by sound that has travelled through the air. Sound transfer by airborne sound is shown in Figure 9.6; typical sources of airborne sound include voices, radios, musical instruments, traffic and aircraft noise.

> *Airborne sound sources*: voices, music, traffic, aircraft

Impact sound

- **Impact sound** is sound which is generated on a partition.

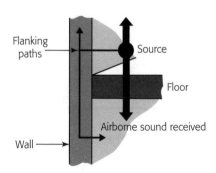

Figure 9.6 Airborne sound transmission

Impact sound sources: footsteps, doors, switches, boilers

The transfer of impact sound is shown in Figure 9.7; typical sources of impact sound include footsteps, slammed doors and windows, noisy pipes and vibrating machinery. A continuous vibration can be considered as a series of impacts, and impact sound is also termed structure-borne sound. Notice that such impact sound will normally travel through the air to reach your ear; but it is not the same as airborne sound.

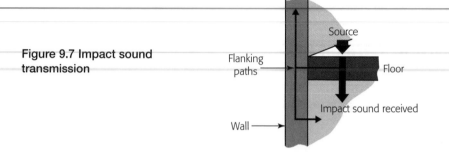

Figure 9.7 Impact sound transmission

Practical sound sources

It is important to distinguish between airborne and impact sound because the best methods of controlling them can differ. A single source of noise may also generate both types of sound, so the definition of airborne and impact sound must be applied to the sound which is being heard in the receiving room. For example, footsteps on a floor would be heard mainly as impact sound in the room below but heard as airborne sound in the room above.

Sound can also pass into a receiving room by flanking transmission, as shown in Figures 9.6 and 9.7. These indirect sound paths can be numerous and complex. The effect of flanking transmissions increases at high levels of sound insulation and often limits the overall noise reduction that is possible.

Absorption and insulation

The techniques used to control sound are described by some terms that may appear to be interchangeable but are, in fact, very different in their effect. Poor understanding of these terms leads to incorrect and wasted efforts in the control of sound.

Sound insulation

- **Sound insulation** is the reduction of sound energy transmitted into an adjoining air space.

Insulation is the most useful method for controlling noise in buildings and is discussed in the next section.

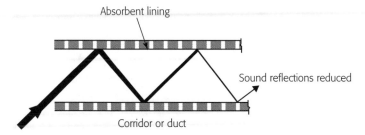

Figure 9.8 Noise control by absorption

Sound absorption

- **Sound absorption** is a reduction in the sound energy reflected by the surfaces of a room.

See also:

sections on *Acoustic principles* in Chapter 10

Absorption usually has little effect on noise control but has an important effect on sound quality, which is discussed in Chapter 10 on acoustics.

It can be shown by calculation that if the amount of absorption in a room is doubled then the sound energy in the room is halved, but the sound level drops by only 3 dB. Such a change in absorption may, however, make a difference to the subjective or apparent sound level in the room because this is also influenced by the acoustic quality.

Sound absorption can also be useful in the control of noise that spreads by reflections from the ceiling of offices and factories, or by multiple reflections along corridors or ducts, as shown in Figure 9.8. The sound insulation of insulating panels discussed below, such as a wall with multiple leafs, will also be improved by absorbing reflections inside cavities.

SOUND INSULATION

Insulation is the principal method of controlling both airborne sound and impact sound in buildings. The overall sound insulation of a structure depends upon its performance in reducing the airborne and impact sound transferred by all sound paths, direct and indirect. The assessment of sound insulation initially considers one type of sound transfer at a time.

Sound Reduction Index (*R*)

The difference in sound levels on either side of a partition, as shown in Figure 9.8, can be used as an index of airborne sound insulation.

- **Sound reduction index** (*R* or *SRI*) is a measure of the insulation against the direct transmission of airborne sound.

The measurement of an SRI can be made in a special laboratory where no flanking sound paths are possible around the partition under test. The sound levels in the

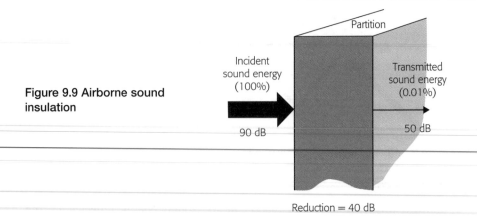

Figure 9.9 Airborne sound insulation

rooms on each side of the partition are measured and a normalised SRI obtained by adjusting for the area of the partition and for the absorption in the receiving room.

Because insulation varies with frequency, the sound reduction index needs to be measured and considered for different frequency bands, such as specified octave intervals. The results of measurements can be plotted against frequency on a graph, as shown in Figure 9.10, which then gives a visual presentation of performance over the range of frequencies. Values of sound insulation for typical constructions in a building are given in Table 9.3. While it is sometimes convenient to consider only a single-figure value, such as the average of readings or the single reading at 500 Hz, that single figure may not reveal deficiencies in performance that occur at certain frequencies.

Sound at 500Hz = high-pitched voice

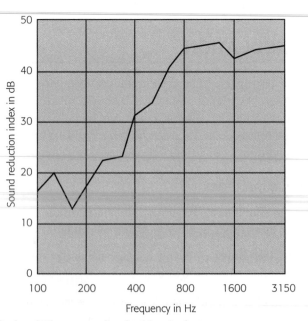

Figure 9.10 An insulation curve for double glazing

Table 9.3 Values of sound insulation

Building elements	Sound reduction index (R) Octave intervals (centre frequencies in Hz)				
	125	250	500	1k	2k
Walls					
brick, cavity, block	41	45	45	54	58
timber frame, lightweight cladding	24	34	40	45	49
Windows, well-sealed					
4 mm single glazing, sealed	20	22	28	32	33
6 mm single glazing, sealed	20	24	31	35	27
4 mm double glazing, 12 mm air gap	24	20	25	34	37
6 mm double window, 150 mm air gap	29	35	45	56	52
Door, well-sealed					
Solid core, 25 kg/m3	20	25	27	28	32
Roofs					
Tiles/slates, 12 mm plasterboard	21	26	33	33	35
Tiles/slates, sound absorbing layer, 25 mm plasterboard	27	37	43	48	52

A working value for the SRI of a partition, at a frequency of 500 Hz, can be obtained by using the formula

$$R = 14.5 \log_{10} M + 10$$

where M = mass per unit area of the partition (kg/ m²).

Insulation principles

Good sound insulation depends upon the following general principles:

- heaviness
- completeness
- flexibility
- isolation.

The effectiveness of each technique of insulation can differ with the type of sound, but in most constructions all the principles of insulation are relevant. Details of the principles are described in the following sections.

Heaviness

Heavyweight structures with high mass transmit less sound energy than light-weight structures. The high density of heavyweight materials restricts the size of the sound vibrations inside the material so that the final face of the structure, such as the inside wall of a room, vibrates with less movement than for a lightweight material.

Because the vibrations of this 'loudspeaker' effect are restricted, the amplitude of the sound waves re-radiated into the air is also restricted. Although a reduction in the amplitude of sound waves affects the 'strength' or 'loudness' of a sound, it does not affect the frequency (pitch) of that sound.

Mass Law

- The **Mass Law** states that the sound insulation of a single-leaf partition has a linear relationship with the surface density (mass per unit area) of the partition, and increases with the frequency of the sound.

Single-leaf construction includes composite construction such as plastered brickwork, as long as the layers are bonded together. Theory predicts an insulation increase of 6 dB for each doubling of mass, but for practical constructions the following working rule is more suitable.

- Sound insulation increases by 5 dB whenever the mass is doubled.

For example, the average SRI of a brick wall increases from 45 dB to 50 dB when the thickness is increased from 102.5 mm to 215 mm. This doubling of mass does not have to be achieved by a doubling of thickness as the mass of a wall for sound insulation purposes is specified by its surface density measured in kilograms per square metre (rather than per cubic metre). Concrete blocks of different densities can produce the same surface density by varying the thicknesses of the blocks.

The effectiveness of sound insulation depends upon frequency and the Mass Law also predicts the following effect on frequency.

- Sound insulation increases by about 5 dB whenever the frequency is doubled.

Frequency examples:
100 Hz = bass note
400 Hz = voice

Any doubling of frequency is a change of one octave. For example, a brick wall provides about 10 dB more insulation against 400 Hz sounds than against 100 Hz sounds. This change, from 100 to 200 Hz and then 200 to 400 Hz, is a rise of two octaves.

The illustrations of construction for good sound insulation, shown later in this chapter, illustrate the Mass Law in operation. Techniques for increasing sound insulation often involve increasing the thickness of masonry, plaster and glass. Where a construction does not obey the Mass Law it is because other factors such as airtightness, stiffness and isolation have an effect.

Completeness

Areas of reduced insulation or small gaps in the construction of a wall have a far greater effect on overall insulation than is usually appreciated. The completeness of a structure depends upon airtightness and uniformity.

Airtightness

As insulation against airborne sound is increased, the presence of gaps becomes more significant. For example, if a brick wall contains a hole or crack which in size represents only 0.1 per cent of the total area of the wall, the average SRI of that wall is reduced from 50 dB to 30 dB.

Air gaps often exist because of poorly constructed seals around partitions, particularly at the joins with floors, ceilings, windows, doors, service pipes and ducts. Some materials may be also porous enough to pass sound through the small holes in their structure; brick and blockwork should therefore be plastered or sealed. Doors and openable windows should be airtight when closed, and the type of sealing used to increase thermal insulation is also effective for sound insulation. In general, 'sound leaks' should be considered as carefully as leaks of water.

> *Common air gaps*:
> **Wall–floor gaps**
> **Gaps around doors**
> **Poor window seals**
> **Unsealed pipe**
>   **runs**
> **Unsealed cable**
>   **runs**
> **Porous**
>   **blockwork**

Uniformity

The overall sound insulation of a construction is greatly reduced by small areas of poor insulation. For example, an unsealed door occupying 25 per cent of the area of a half-brick wall reduces the average SRI of that wall from around 45 dB to 23 dB. The final sound insulation is influenced by relative areas but is always closer to the insulation of the poorer component than to the better component.

Windows and doors are necessary parts of a building but a knowledge of the uniformity principle can prevent effort being wasted on the insulation of the wrong areas. To improve the insulation of a composite structure the component with the lowest insulation should be improved first of all. Walls facing noisy roads should contain the minimum of windows and doors, and they should be well insulated.

Flexibility

Stiffness is a physical property of a partition and depends upon factors such as the elasticity of the materials and the fixing of the partition. High stiffness can cause loss of insulation at certain frequencies where there are resonances and coincidence effects. These effects upset the predictions of the Mass Law, as indicated in Figure 9.11.

Resonance

Loss of insulation by resonance occurs if the incident sound waves have the same frequency as the natural frequency of the partition. The increased vibrations that occur in the structure are passed on to the air and so the insulation is lowered. Resonant frequencies are usually low and most likely to cause trouble in the air spaces of cavity construction.

Coincidence

Loss of insulation by coincidence is caused by the bending flexural vibrations, which can occur along the length of a partition. When sound waves reach a partition at angles other than 90°, their transmission can be amplified by the

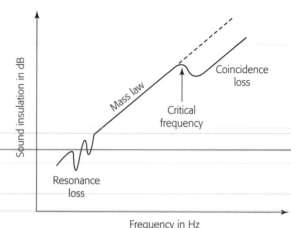

Figure 9.11 Resonance and coincidence loss

flexing inwards and outwards of the partitions. The sound-wave frequency and the bending-wave frequency coincide at the critical frequency. For several octaves above this critical frequency the sound insulation tends to remain constant and less than that predicted by the Mass Law. Coincidence loss is greatest in double-leaf constructions, such as cavity walls or hollow blocks.

Flexible (non-stiff) materials, combined with a high mass, are best for high sound insulation. Unfortunately flexibility is not usually a desirable structural property in a wall or a floor.

Isolation

Discontinuous construction can be effective in reducing the transmission of sound through a structure. As the sound is converted to different wave motions at the junction of different materials, energy is lost and a useful amount of insulation is gained. This is the principle behind the effectiveness of air cavities in windows, of floating floors, of carpets and of resilient mountings for vibrating machines. Some broadcasting and concert buildings achieve very high insulation by using the completely discontinuous construction of a double structure separated by resilient mountings.

Sound isolation is easily ruined by strong flanking transmissions through rigid links, even by a single nail. Cavity constructions must be sufficiently wide for the air to be flexible, otherwise resonance and coincidence effects can cause the insulation to be reduced at certain frequencies.

Sound insulation regulations

The building codes of many countries include standards for sound insulation in buildings, especially where walls and floors are shared between dwellings, such as those between flats and joined houses. These codes have usually evolved from a desire to protect people from annoyance or potential ill health caused by noise from traffic, industry or neighbours.

The following parts of buildings are those where regulations are typically used to ensure suitable standards of sound insulation.

- **Building envelope**: such as the roof, external walls and windows which separate a dwelling from external sources such as noise from industry, road traffic, trains and aircraft.
- **Separating walls**: between dwellings and within the same dwelling.
- **Separating floors**: between dwellings.

A dwelling can be taken to mean a place where a family lives, by day and night. Regulations for sound insulation sometimes ignore noise annoyance between parts of the same dwelling, assuming perhaps that it is a family matter! However, all dwellings benefit from noise control and there are some buildings, such as hotels, student hostels, nursing accommodation and rest homes, where sound insulation between rooms and corridors must be considered and regulations may apply.

Building codes for sound insulation have been updated over recent years in response to surveys showing that around a quarter of households are bothered by noise. One aspect of general improvements in living standards has been the use of home entertainment systems with increased power, especially at bass or lower frequencies. Accompanying this change in living habits is a change in attitude as people are now generally less tolerant of noise disturbance.

Check:
which building codes apply to your country and which version is current.
  Requirements shown here are good examples but they may not be complete or current.

Requirements of regulations

Building regulations, such as those for England and Wales, require certain minimum standards of sound insulation for the building envelope and those internal walls and floors which protect the rooms in dwellings where people live. The distinction between airborne and impact sound transfer, as explained earlier in the chapter, is important because the need for insulation varies with the type of sound transfer. Figure 9.12 indicates typical sound insulation requirements for walls and floors separating dwellings in a simplified block of flats. In general, walls require insulation against airborne sound while floors require insulation against both airborne sound and impact sound.

The sound insulation performance of a building element such as a wall or floor can be expressed using the following types of measurement:

Airborne sound travels through the air before a partition
Impact sound is generated on the partition

- **Level (*L*)** of sound level in a space.
- **Difference (*D*)** between the sound levels in the spaces on either side of a partition such as a wall.

The type of measurement used depends upon the type of sound transfer. For example, effectiveness against airborne sound is usually measured by a difference in the sound levels between the rooms on either side of a partition. For impact sound, the effectiveness of insulation needs to be measured by recording sound level that is heard in the room when it receives sound made by an agreed standard of impact on the partition.

Earlier sections of this book have described the general measurement of sound

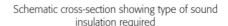

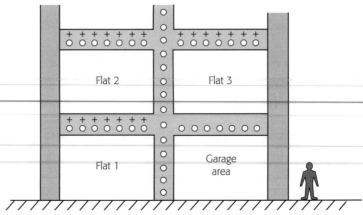

Note
This diagram does not show requirements for sound insulation of the external envelope

Key
o   Airborne sound insulation
+   Impact sound insulation

Figure 9.12 Internal walls and floors needing sound insulation

pressure levels (SPL) and sound insulation levels (SIL or R). These measurements are developed when the technical requirements for sound measurements are defined in published standards such as British Standards (BS), Euro Standards (EN) and International Standards (ISO). Conditions affecting the insulation measurement are taken account of and noted in the nomenclature or shorthand describing the sound insulation index. For example:

- Frequency spectrum is noted by C, C_{tr} and C_I.
- Reverberation of a receiving room is noted by T.
- Normalisation or standardisation between different room properties is noted by n.
- Weighting or averaging of frequency values is noted by w.

Sound insulation of building envelope

The levels of noise outside and inside a building naturally vary throughout the day and throughout the night. This noise is therefore measured by the equivalent continuous sound level L_{Aeq}, which assesses varying sound levels over specified time periods such as 16 hours for daytime and eight hours for night-time. This index is used to express a comfortable target for noise level inside the building, and Table 9.4 shows typical modern target levels for internal noise.

The targets shown in Table 9.4 are a reasonable basis for the sound insulation of building envelopes, but there may be some types of noise events which,

Table 9.4 Typical target sound levels for internal noise

Location	Maximum noise level
All rooms during daytime (0700ñ2300 hours)	40 dB L$_{Aeq,16h}$
All bedrooms during night-time (2300ñ0700 hours)	30 dB L$_{Aeq,8h}$

although falling within the target levels, cause disturbance, particularly at night. Some sounds may also have characteristics, such as high or low frequencies, which make them more noticeable and annoying than is revealed by a statistical measurement like the equivalent continuous sound level.

The external noise level on a new housing site needs to be estimated using measurements of current noise, and future developments likely to cause noise must be predicted. There are recommended methods for predicting road traffic noise, railway noise and aircraft noise. The sound reduction index R needed in order to provide insulation for each habitable room can be calculated by comparing the external noise level and the target internal noise level, and standardising the result. Table 9.5 shows some typical constructions for building envelopes which satisfy the requirements of the target internal noise levels.

Table 9.5 Sound insulation standards for separating walls and floors – houses and flats

Location	Airborne sound insulation Minimum value of difference D$_{nT,w}$ + C$_{tr}$	Impact sound insulation Maximum value of level L$_{nT,w}$ + C$_l$
Purpose-built dwellings		
Walls	45 dB	n/a
Floors and stairs	45 dB	62 dB
Dwellings formed by conversions		
Walls	43 dB	na
Floors and stairs	43 dB	62 dB

Notes:

$D_{nT,w}$ + Ctr is a single-number quantity used to characterise the airborne sound insulation between rooms using a particular noise spectrum (defined in standards from BS, EN and ISO).

$L_{nT,w}$ + C$_l$ is a single-number quantity used to characterise the impact sound insulation of floors using a particular frequency spectrum (defined in standards from BS, EN and ISO).

These requirements are from typical building regulations for England & Wales. The values cannot be directly compared with previous regulations as the measurements and units are different.

Regulations apply to stairs when they form part of the separation between dwellings.

Requirements for rooms for residential purpose within buildings such as hostels are the same as the above except for the purpose-built walls, which have a value of 43 dB.

Other requirements

This chapter has dealt mainly with wall and floor constructions that provide sound insulation from outside and between dwellings. Listed below are other situations in buildings where sound insulation can be provided and may be required by regulations:

- internal walls and floors within a dwelling
- change of use, such as buildings being converted into flats
- reverberation and sound transmission down corridors
- between a habitable room and a room containing a WC
- walls separating refuse chutes from habitable rooms and kitchens
- noise transmission and acoustic quality in schools
- sound insulation properties of ventilators.

Meeting sound insulation standards

There has been a trend to improve the standards of sound insulation in buildings, and modern regulations require better sound insulation compared with past practice. For example, some everyday sounds that can currently just be heard in existing buildings should not be heard at all with modern standards of sound insulation. As ever, this will only be true when the construction is carried out according to the specification.

Robust Design Details are published to help meet building regulations requirements for England and Wales

A common method of meeting the requirements of sound-insulation regulations is to adopt standard approved constructions. Some standard constructions are summarised in Tables 9.6 and 9.7 and illustrated in Figures 9.13 to 9.16. Attention must be paid to details of the construction, especially with seals and at junctions.

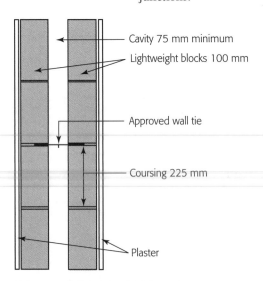

Figure 9.13 Block cavity wall for sound insulation

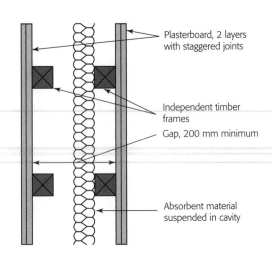

Figure 9.14 Timber frame wall for sound insulation

Table 9.6 Construction for sound insulation of building envelopes

For external noise levels not exceeding 55 dB $L_{Aeq,16h}$ or 45 $L_{Aeq,8h}$ (a low-noise site)	
Any construction which complies with other aspects of building regulations, such as thermal insulation, is likely to meet requirements	
For external noise levels not exceeding 60 dB $L_{Aeq,16h}$ or 50 $L_{Aeq,8h}$	
Wall	solid brickwork, or brick/block cavity or timber frame with brick cladding or timber frame with lightweight cladding
Window	any practical window construction, well-sealed when closed
Roof	tile/slate roof, 9 kg/m² plasterboard ceiling
Ventilator	trickle ventilators
For external noise levels not exceeding 65 dB $L_{Aeq,16h}$ or 60 $L_{Aeq,8h}$	
Wall	solid brickwork, or brick/block cavity or timber frame with brick cladding or timber frame with lightweight cladding
Window	double glazing, 10/6 mm glass, 12 mm gap
Roof	tile/slate roof, 9 kg/m² plasterboard ceiling, 100 mm sound-absorbing layer above ceiling
Ventilator	bedrooms: mechanical ventilator in bedrooms living rooms: acoustic trickle ventilator
For external noise levels not exceeding 75 dB $L_{Aeq,16h}$ or 65 $L_{Aeq,8h}$	
Wall	solid brickwork, or brick/block cavity or timber frame with brick cladding
Window	double window with 100 mm gap, well-sealed when closed window area limited to 2.5 m²
Roof	tile/slate roof, 20 kg/m² plasterboard ceiling, 100 mm sound-absorbing layer above ceiling, timber boarding on top of ceiling joists
Ventilator	all ventilation mechanical

Notes:

These forms of construction have satisfied the requirements of building regulations for England and Wales.

The constructions listed are examples, and other forms and variations are available.

When defining the construction of elements it is also necessary to define the properties of materials and products contained in the construction. The following properties and features may need to be specified:

- density of materials used in kg/m³. For bricks and blocks use the density measurement at 3 per cent moisture content
- mass per unit area of wall in kg/m² of wall
- wall ties of a specified type and stiffness
- cavity width as a minimum value.

See also:

later section about *Calculating mass per unit area*

Construction for sound insulation

Figures 9.13, 9.14, 9.15 and 9.16 illustrate the construction of some forms of wall and floor that are considered to provide adequate sound insulation. Variations of

Table 9.7 Construction for sound insulation of separating walls and floors

Construction type/construction details	Notes
Separating walls	
Concrete cast in situ, plaster on both faces • plaster 13 mm, minimum area density 10 kg/m² • concrete 190 mm, minimum density 2200 kg/m³ • plaster 13 mm, minimum area density 10 kg/m²	• All masonry joints must be filled • All joints must be sealed with mortar • Sockets on either side of wall must be staggered
Solid brickwork, plaster on both faces • plaster 13 mm, minimum area density 10 kg/m² • brick 215 mm, minimum density 1~610 kg/m³ • plaster 13 mm, minimum area density 10 kg/m²	• Bricks laid with frogs up and with 75 mm coursing • All masonry joints must be filled • All joints must be sealed with mortar
Two leaves of lightweight block, 75mm cavity, plaster on both faces • plaster 13 mm, minimum area density 10 kg/m² • block 100 mm, minimum density 1~375 kg/m³ • cavity 75 mm minimum • block 100 mm, minimum density 1~375 kg/m³ • plaster 13 mm, minimum area density 10 kg/m²	• Blocks with 225 mm coursing • Do not insert mortar or concrete into the cavity
Dense concrete block, independent panels on both faces • panel with 2 sheets plasterboard, staggered joints • gap 25 mm • block core 140 mm, minimum density 2~200 kg/m³ • gap 25 mm	• Each plasterboard panel: 18 kg/m³ minimum overall density • Blocks with 110 mm coursing • Gap size: 25 mm from core. If panels on frame then at least 5 mm between frame and core
Double leaf frames, absorbent material • panel with 2 sheets plasterboard, staggered joints • independent frame • gap 200 mm minimum gap between linings • absorbent material, minimum density 10 kg/m³ • independent frame • panel with 2 sheets plasterboard, staggered joints	• Each lining: two layers of plasterboard, minimum area density 9 kg/m² • Absorbent material thickness: 25 mm if suspended, 50 mm if fixed to one frame
Separating floor	
Concrete base, ceiling, soft floor covering • floor covering of resilient material, 4.5 mm uncompressed thickness • solid concrete slab, minimum area density 365 kg/m²	• Concrete slab cast in situ or with permanent shuttering • Do not use non-resilient floor finishes, such as ceramic floor tiles and wood-block floors • Floor base must not bridge a cavity in a cavity masonry wall • Ceiling fixed by any normal fixing method (e.g. timber battens)

Table 9.7 continued

Construction type/construction details	Notes
Concrete base, ceiling, screed floating floor • sand cement screed floating layer, 65 mm, minimum area density 80 kg/m² • mineral wool, 25 mm thickness, density 36 kg/m³ • solid concrete slab, minimum area density 300 kg/m² • single layer of plasterboard, minimum area density 9 kg/m²	• Floating layer should be laid loose on the resilient layer • Mineral wool layer with paper faced on the upper side to prevent the screed entering the resilient layer. Concrete slab cast in situ or with permanent shuttering • Ceiling fixed by any normal fixing method, such as with timber battens
Timber frame structural floor base with deck, ceiling, platform floor • wood-based board, tongued and grooved edges and glued joints 18 mm, spot bonded to a substrate of 19 mm plasterboard with joints staggered, minimum total mass per unit area 25 kg/m² • mineral wool 25 mm, density 60ñ100 kg/m³ • timber joists with a deck • absorbent layer of at least 100 mm of mineral wool (10 kg/m³) laid In cavity above the ceiling • ceiling of at least 2 layers of plasterboard staggered joint, minimum area density 20 kg/m²	• Deck can be of any suitable material with a minimum mass per unit area of 20 kg/m²

Notes:
These forms of construction have satisfied the requirements of building regulations for England and Wales.
The constructions listed are examples, and other forms and variations are available.

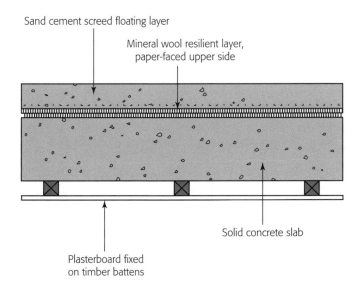

Figure 9.15 Concrete floor for sound insulation

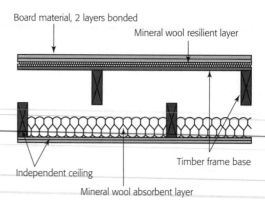

Board material, 2 layers bonded

Mineral wool resilient layer

Independent ceiling

Timber frame base

Mineral wool absorbent layer

Figure 9.16 Timber floor for sound insulation

these constructions can also provide sufficient sound insulation and some are described in Tables 9.6 and 9.7 of standard constructions. The junctions between elements such as floors and walls require careful detailing and constructing. To make the link between theory and practice it is worthwhile to evaluate the purpose of the special construction features outlined in the diagrams and in Tables 9.6 and 9.7 in terms of the principles of sound insulation described in the previous sections.

> *Sound insulation principles:*
> Heaviness
> Completeness
> Flexibility
> Isolation
> (see details earlier in chapter)

Walls

Heavyweight walls generally provide high levels of airborne sound insulation because of the effect of the Mass Law. Cavities can increase sound insulation by the principle of isolation but practical cavity walls with wire ties behave little better in practice than a double skin of bricks.

Lightweight walls can provide adequate levels of airborne sound insulation if attention is paid to the principles of isolation and airtightness. Each side of the wall is built on separate timber frames and an absorbent blanket of mineral fibre is used to provide acoustical isolation between the sides of the wall. The mass provided by multiple layers of dense plasterboard contributes significantly to the insulation of the wall.

> *Common problem areas:*
> Wall–floor gaps
> Gaps around doors
> Poor window seals
> Unsealed pipe runs
> Unsealed cable runs
> Porous blockwork

Floors

The natural mass of concrete floors provides insulation against airborne sound but concrete can transfer impact sound; therefore a resilient layer is also needed to provide insulation against impact sound. Older methods of construction increased the mass of timber floors by the use of sand *pugging* within the joist space. More recent methods of mass enhancement include the use of thick layers of composite floorboard or multiple layers of plasterboard.

The insulation of a floor must be maintained at all junctions with the surrounding walls in order to prevent flanking transmission. The separation of the two parts of a floating floor must continue around all the edges by the use of resilient materials and airtight techniques.

Windows

Glass has relatively high density so increasing the thickness of glass provides improved sound insulation because of the Mass Law. Air cavities can provide worthwhile increased sound insulation if their width is greater than 150 mm. The cavity should be lined with absorbent material to minimise resonance and the two frames must be isolated from one another by some construction technique such as a resilient layer. Such efforts for increased sound insulation are a waste of time unless the windows also shut with a good seal in order to provide airtightness.

> Detailed installation diagrams are available from manufacturers

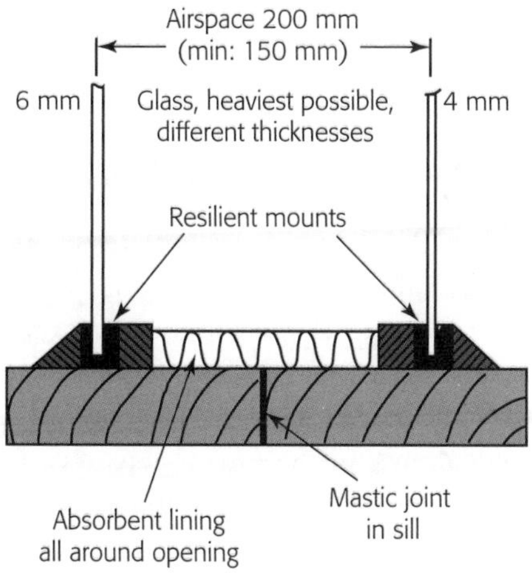

Figure 9.17 Double window for sound insulation

Measurement of sound insulation

A convenient way to satisfy the requirements of building regulations for sound insulation is to use standard construction types which have been approved, such as those described in Tables 9.6 and 9.7. Alternative methods of satisfying building regulations involve testing the sound insulation of buildings or parts of buildings by measurements made in buildings and in sound laboratories.

The principles for measuring sound and noise are described earlier in this chapter, but making practical measurements of sound levels on site and in laboratories requires careful use of approved equipment and the following of established procedures. These requirements are outside the scope of this chapter but are specified by common national and international standards such as the BS, EN and ISO series.

Measurements made in finished buildings are also important to establish whether performance targets are actually being met. Studies have revealed that up to a quarter of finished buildings do not meet their design targets for sound insulation. A typical field test and report needs to include the following information:

- details of organisation and person conducting test
- details of equipment and test procedures
- sketch showing dimensions of rooms and their relationships
- descriptions of separating walls, partitions and floors, their materials and finishes
- mass per unit area of walls, partitions and floors
- dimensions of any step and stagger between rooms
- dimensions of and positions of windows and doors
- results of tests shown in tabular and graphic form for third octave bands
- single-number quantity and data used to calculate the quantity.

Calculations for sound insulation

Calculating mass per unit area

One of the important principles of good sound insulation is the effectiveness of mass. To specify the amount of mass in a building element such as a wall or floor, it is common to use the concept of mass per unit area expressed in kilograms per square metre (kg/m^2). This value, in kilograms per square metre of wall construction, is the same as you would get if you could cut out one square metre of wall and weigh it. Mass per unit area is used to specify some features of wall and floor constructions that meet the sound insulation requirements of building regulations. This form of target allows the designer to have a choice of building materials, and their combinations, providing that the total mass per unit area meets the specification.

Although individual bricks or blocks have a reasonably constant density and mass they must be laid with mortar, which has a different density and may vary in thickness. Therefore the final brick or blockwork needs to be described in terms of mass per unit area. Manufacturers provide values of mass per unit area for their materials in use. Otherwise the practical formula given below provides a sufficiently accurate result.

$$\text{mass per unit area in kg/m}^2 = \frac{M_B + \rho_m + [Td(L + H - d) + V]}{LH}$$

where

M_B is mass of the block (kg), at an appropriate moisture content
ρ_m is density of the mortar (kg/m^3)
T is block thickness (m), unplastered
d is mortar thickness
L is coordinating length
H is coordinating height
V is volume of any frog or void filled with mortar (m^3)

Notes:

Coordinating length is length of a block plus one vertical mortar joint.

Coordinating height is height of a block plus one horizontal mortar joint.

For cavity walls: calculate each leaf separately and add together.

With surface finishes: add mass per unit area of surface finish to mass per unit area of wall.

Calculating sound reduction index R

The sound reduction index used to measure the airborne sound insulation of a partition depends upon the amount of sound energy transmitted across the partition, as shown in Figure 9.9.

The proportion of energy transmitted through the partition is measured by the transmission coefficient, T, where

$$T = \frac{\text{transmitted sound energy}}{\text{incident sound energy}}$$

The *sound reduction index, R,* is then defined by the following formula:

$$R = 10\log_{10}\left[\frac{1}{T}\right]$$

Unit: decibel (dB)

Worked Example 9.1

A wall transmits 1 per cent of the sound energy incident upon the wall at a given frequency. Calculate the sound reduction index R of the wall at this frequency.

Let   Incident sound energy $= 100$
           Transmitted sound energy $= 1$

So T $=$ (transmitted sound/incident sound) $= 1/100 = 0.01$

Using reduction index formula and substituting

R $= 10 \log (1/T)$
R $= 10 \log (1/0.01) = 10 \log (100)$
      $= 10 \times 2$

So R $= \textbf{20 dB}$

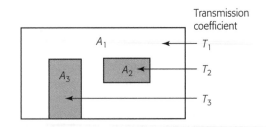

Figure 9.18 Transmission coefficients of a composite partition

Composite partitions

A window placed in a well-insulated wall can greatly reduce the overall sound insulation of the wall, as was discussed in the section on completeness. Because values of sound reduction index have been calculated using a logarithmic formula, they cannot be simply averaged according to area. But the overall transmission coefficient can be calculated using the transmission coefficients and areas of the individual components, such as shown in Figure 9.18.

$$T_0 = \frac{(T_1 \times A_1) + (T_2 \times A_2) + (T_3 \times A_3)}{A_1 + A_2 + A_3}$$

where
T_0 = overall transmission coefficient
T_1 = transmission coefficient of one component
A_1 = area of that component etc.

The overall sound reduction index for the complete partition can then be calculated using the overall transmission coefficient.

Worked Example 9.2

A wall of area 10 m² contains a window of area 2 m². The values of sound reduction index are 50 dB for the brickwork and 18 dB for the window. Calculate the overall sound reduction index for the wall.

For brickwork let: T_1 = ?, A_1 = 10 − 2 = 8 m²,   SRI = 50 dB.
For window let: T_2 = ?, A_2 = 2 m²,   SRI = 18 dB.

Using reduction index formula and substituting

$R = 10 \log (1/T)$
$50 = 10 \log (1/T_1)$ and $18 = 10 \log (1/T_2)$

$\log \dfrac{1}{T_1} = 5$ $\log \dfrac{1}{T_2} = 1.8$

$$\frac{1}{T_1} = 10^5 \qquad\qquad \frac{1}{T_2} = 63.10$$

$$\frac{1}{T_1} = 10^5 \qquad\qquad T_2 = 1.585 \times 10^{-2}$$

Using averaging formula and substituting

$$T_0 = \frac{(T_1 \times A_1) + (T_2 \times A_2)}{A_1 + A_2}$$

$$= \frac{(10^5 \times 8) + (1.585 \times 10^{-2})}{8 + 2} = \frac{3.18 \times 10^{-2}}{10}$$

$$= 3.18 \times 10^{-3}$$

Using reduction index formula and substituting

$$R = 10\log(1/T_0)$$

$$= 10\log \frac{1}{3.18 \times 10^{-3}} = 10\log 314.5 = 25$$

So the overall SRI = **25 dB**

EXERCISES

1. A doubling of sound energy increases the sound pressure level by 3 dB. Use this information and a table to show the increases in sound energy reaching the ear when the SPL level of a personal stereo system is adjusted from 70 db to 85 dB.

2. When the SRIs are measured for a certain double-glazing unit the results obtained are those shown in Figure 9.10. Explain the likely reason for the dips in the insulation curve and outline some techniques that would help to reduce these effects.

3. Use the information given in Tables 9.6 and 9.7, or other sources, to draw an annotated section of any wall that is considered to give sufficient sound insulation to meet regulations in your country or region. Briefly explain how each feature of the wall works to provide the sound insulation.

4. 800 units of sound energy are incident upon a wall and ten of these units are transmitted through the wall.
   (a) Calculate the sound reduction index of this wall.
   (b) If a window has a sound reduction index of 33 dB, then calculate the transmission coefficient of this window.

5. An external brick cavity wall is to be 4 m long and 2.5 m high. The wall is to contain one window 1.2 m by 800 mm and one door 750 mm by 2m. The relevant sound reduction indexes are: brickwork 53 dB; window 25 dB; door 20 dB. Calculate the overall sound reduction index of the completed partition.

Answers are on page 326.

Room Acoustics

CHAPTER OUTLINE

The term 'acoustics' can be used to describe the study of sound in general, but the subject of room acoustics is concerned with the control of sound within an enclosed space. The general aim is to provide the best conditions for the production and the reception of sound such as music or speech.

The sound quality of a large auditorium such as a concert hall can be difficult to get right, and acoustics has sometimes been described as an art rather than a science. But, as with thermal and visual comfort, there are technical properties that do affect our perception and these make the best starting point for designing or improving the environment. The contents of this chapter will allow you to:

▶ appreciate the nature of and the requirements for good acoustics
▶ understand the purposes and acoustic qualities of different types of auditorium
▶ understand the mechanisms and effects of sound reflections
▶ understand the nature of echoes and their effects
▶ appreciate the influence of shape in the design and acoustic effect of an auditorium
▶ understand the mechanisms and effects of sound absorption
▶ understand the principles and uses of different types of absorber
▶ appreciate the nature of reverberation and its effects on acoustic performance
▶ know the principles of acoustic assessment and reverberation time
▶ calculate reverberation times for practical contexts.

$\longrightarrow$

Noise control is treated separately in Chapter 9 but the exclusion of unwanted noise is an important element of room acoustics. Similarly, the acoustic quality of sound in a room can affect the way that people judge noise levels. The Resource sections at the back of the book also contain supporting information that can be used for both revision of principles and extended investigation of topics.

ACOUSTIC PRINCIPLES

General requirements

See also:

sections on *Noise control* in Chapter 9, and Resource 3

The detailed acoustic requirements for a particular room depend upon the nature and the purpose of the space, and the exact nature of a 'good' sound is partly a matter of personal preference. The general requirements for good acoustics are summarised as follows:

- adequate levels of sound
- even distribution to all listeners in the room
- rate of decay (reverberation time) suitable for the type of room
- background noise and external noise reduced to acceptable levels
- absence of echoes and similar acoustic defects.

Types of auditorium

An auditorium is a room, usually large, designed to be occupied by an audience. The acoustic design of auditoria is particularly important, and detailed acoustic requirements vary with the purpose of the space, as outlined below.

Speech

The overall requirement for the good reception of speech is that the speech is intelligible. This quality will depend upon the power and the clarity of the sounds. Examples of auditoria that are especially used for speech are conference halls, law courts, theatres and lecture rooms.

Some notable concert halls and opera houses:
Festival Hall, London
Symphony Hall, Birmingham
Concertgebouw, Amsterdam
Musikverein, Vienna
Metropolitan Opera, New York
Opera House, Sydney

Music

There are many more acoustic requirements for music than for speech. Music consists of a wide range of sound levels and frequencies which all need to be heard. In addition, some desirable qualities of music depend upon the individual listener's judgment and taste. These qualities are difficult to define but terms in common use include 'fullness' of tone, 'definition' of sounds, 'blend' of sounds and 'balance' of sounds. When music is loud, it is usually perceived as being more lively and intimate.

Examples of auditoria designed exclusively for music are concert halls, opera houses, recording studios and practice rooms.

Multi-purpose auditoria

The previous sections indicate that there are some conflicts between the ideal acoustic conditions for music and for speech. Compromises have to be made in the design of auditoria intended for more than one purpose, and the relative importance of each activity needs to be decided upon. Churches, town halls, conference centres, school halls and some theatres are examples of multi-purpose auditoria.

Sound paths in rooms

A *sound path* or *sound ray* is the directional track made by the wave vibrations as they travel through a material such as air. The scaled geometrical drawing of sound rays is a useful technique for predicting acoustic effects. The behaviour of sound paths inside an enclosed space can be affected by the mechanisms of reflection, absorption, transmission and diffraction, as shown in Figure 10.1. As with other wave forms, such as light, *diffraction* is an effect which occurs at the edges of objects and is one reason why it is possible to hear sounds around corners.

    Reflection and *absorption* play the largest roles in room acoustics, the final result depending upon the particular size and shape of the enclosure and the nature of the materials used for the surfaces.

See also:

Nature of sound in Resource 3

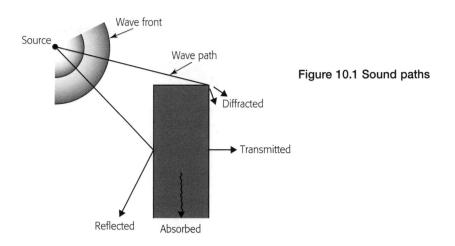

Figure 10.1 Sound paths

REFLECTION

Sound is reflected in the same way as light, provided that the reflecting object is larger than the wavelength of the sound concerned. The angle of reflection equals the angle of incidence of the wave, as shown in Figure 5.7 for light rays. Remember that this angle is measured from a line drawn at right angles to the surface (the normal), and not measured from the surface itself. Using the rule of reflection, the straight lines representing sound rays can be drawn on plans and used to predict some of the effects of reflection. The special case of rapid reflections or reverberation is treated in a later section.

Types of reflector

Reflecting surfaces in a room are used to help the even distribution of sound and to increase the overall sound levels by reinforcement of the sound waves. The following sections describing various reflection effects also indicate that there can be unwanted reflections (echoes):

- Reflections near the source of sound can be useful.
- Reflections at a distance from the source may be troublesome.

Plane reflectors

The effect of a plane or flat reflector is shown in Figure 10.2. Adjustable plane reflectors are often suspended above a stage to give an early reflection into the audience. Reflectors should be wide enough to reflect sound across the full width of the audience and the reflected sound must not be significantly delayed.

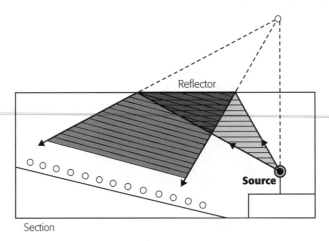

Figure 10.2 Plane reflector of sound

Curved reflectors

At curved surfaces, the angle of reflection still equals the angle of incidence but the geometry gives varying effects. Figure 10.3 illustrates the main type of reflection and shows the two types of curved surface.

Concave: inwards curve
Convex: outwards curve

- **Concave** surfaces tend to focus sound.
- **Convex** surfaces tend to disperse sound.

Convex surfaces are useful for distributing sound over larger areas. When the curves are of tight radius the sound will be scattered at random, which is a useful

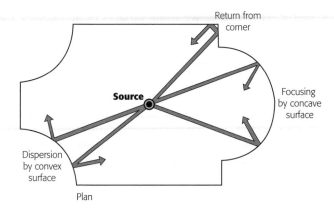

Return from corner

Focusing by concave surface

Source

Dispersion by convex surface

Plan

Figure 10.3 Reflection from room surfaces

effect produced by the ornaments and elaborate plasterwork found in early theatres and cinemas.

Concave surfaces tend to concentrate sound energy in particular areas and this is often acoustically risky. The domed ceilings of some public buildings, such as the Royal Albert Hall in London, have often contributed to unsatisfactory acoustics and have required remedies. A shallow concave surface may be satisfactory if the focal point is well outside the enclosure, and the tight curves found in decoration will contribute to scattering the sound.

Echoes

An echo is a delayed reflection. Initially a reflected sound reinforces the direct sound, but if the reflection is delayed and is strong then this echo causes blurring and confusion of the original sound. The perception of echoes depends upon the power and the frequency of sounds.

There is a risk of a distinct echo if a strong reflection is received later than 1/20th second (50 ms) after the reception of the direct sound. At a velocity of 340 m/s this time difference corresponds to a path difference of 17 m. This difference in length between direct sound paths and reflected paths can be checked by geometry and is most likely to affect seats near the front of a large auditorium. Late reflections can be minimised by the use of absorbers on those surfaces that cause the echoes.

Flutter echoes are rapid reflections which cause a 'buzzing' quality as sound decays. They are caused by repeated reflections between smooth parallel surfaces, especially in smaller rooms. The flutter can be avoided by using dispersion and absorption at the surfaces.

See also:

section on Reverberation in this chapter

Standing waves

Each frequency of a sound has a wavelength. Sometimes the distance between parallel walls in a room may equal the length of half a wave, or a multiple of a half wavelength. Repeated reflections between the surfaces cause standing waves or *room resonances*, which are detected as large variations in sound level at different positions. Standing wave effects are most noticeable for low-frequency sounds in smaller rooms and, in general, parallel reflecting surfaces should be avoided.

Hall shapes

Acoustic requirements are not the only factors deciding the internal shape of an auditorium. Everyone needs to see the stage, for example, so good sight-lines from seats are a major requirement. Fortunately, this requirement of being able to see the stage also helps to provide a strong component of *direct sound*, which is

'The ideal concert hall is obviously that into which you make a not very pleasant sound and the audience receives something that is quite beautiful.'
Conductor Sir Adrian Boult

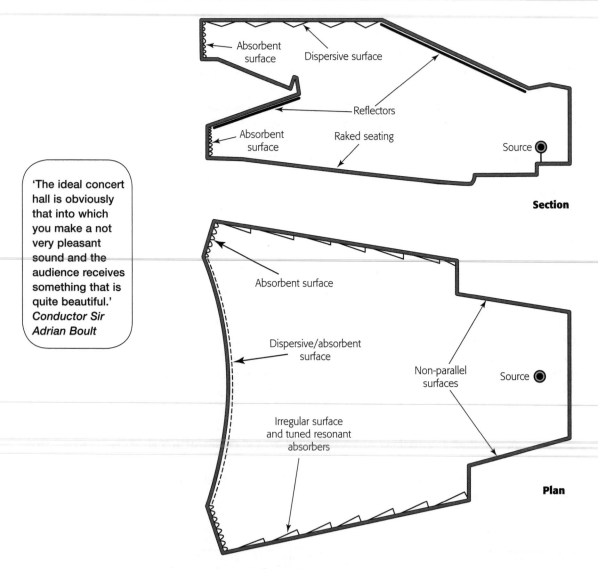

Figure 10.4 Acoustic features of a concert hall

an important feature of good acoustics. The aspects of shape described below are likely to affect the acoustics of an auditorium. Figure 10.4 shows one possible plan and section for a concert hall which includes many of the features necessary for good acoustics.

Rectangular

A rectangular plan or 'shoebox' is the traditional shape of many successful older concert halls. The sound waves tend to establish themselves along the length of the hall and all listeners receive a strong component of direct sound and also receive beneficial side reflections. Reflectors can be used to direct sound to the rear of the hall and absorbers used to prevent unwanted reflections. Traditional ratios of dimensions for height, width and length are about 2 : 3 : 5.

Wide fan

A fan shape or short-wide hall allows more people in the audience to be near the source of sound and gives them better views of the stage. Wide curves were used for the seating of ancient outdoor theatres and amphitheatres and can give agreeable acoustics, although the acoustics are not always optimised for music.

Horseshoe

A horseshoe shape is common in traditional opera houses where the tall concave shape at the back of the hall is broken up by tiers of seats and boxes. The audience and furnishings in these tiers also act as absorbers.

Terracing

A design for halls that has evolved over the last half century involves placing the audience in subdivided areas or terraces at different levels. The sides of the terraces give beneficial reflections that provide good acoustical qualities, as well as good sightlines. Halls using this technique include St David's Hall, Cardiff and The Waterfront Hall in Belfast.

Raked seats

Most theatres have their rows of seats rising towards the rear. This raking provides a good view of the stage and prevents absorption of direct sound paths. Raked seating is not essential for music quality and the floor of a traditional opera house is relatively flat.

ABSORPTION

Sound absorption is a reduction in the sound energy reflected from a surface. In Chapter 9 sound absorption was distinguished from sound insulation because the two concepts have different effects and applications. Sound absorption is a major factor in producing good room acoustics, especially when controlling reverberation.

Absorption coefficient

The absorption coefficient is a measure of the amount of sound absorption provided by a particular type of surface. The amount of sound energy not reflected is compared with the amount of sound energy arriving at the surface in the following formula:

$$\text{Absorption coefficient (a)} = \frac{\text{Absorbed sound energy}}{\text{Incident sound energy}}$$

Unit: none – its value is expressed as a ratio.

A surface that absorbs 40 per cent of incident sound energy has an absorption coefficient of 0.4. Note that the coefficient of 'absorption' is a surface consideration and is not affected by what actually happens to the sound energy that is not reflected. A strange example of a perfect absorber, for example, is an open window.

Different materials and constructions have different absorption coefficients, and the coefficient for any one material varies with the frequency of the incident sound. Table 10.1 lists the average absorption coefficients of some common materials at the standard frequencies used in acoustic studies.

Total absorption

The effective absorption of a particular surface depends on both the absorption coefficient of the surface material and the area of that particular surface exposed to the sound. A measure of this total absorption is obtained by multiplying the two factors together.

$$\text{Absorption of surface} = \text{Area of surface} \times \frac{\text{Absorption coefficient}}{\text{of that surface}}$$

Units: m² sabins or 'absorption units'

The total absorption of a room is the sum of the absorptions provided by each surface in the room. This total is the sum of the products of all areas and their respective absorption coefficients as expressed in the following formula:

$$\text{Total absorption} = \Sigma(\text{Area} \times \text{Absorption coefficient})$$

This calculation is usually best set out in a table, as shown later in the chapter, or on a computer spreadsheet. People and soft furnishings absorb sound, and air also absorbs sound at higher frequencies. Absorption factors for these items are given in Table 10.1.

Absorption coefficients:
Maximum for perfect absorber
a = 1
↓
a = 0
Minimum for poor absorber (or perfect reflector)

Table 10.1 Absorption coefficients

Common building materials		Absorption coefficient		
		125 Hz	500 Hz	2000 Hz
Brickwork	plain	0.02	0.03	0.04
Clinker blocks	plain	0.02	0.06	0.05
Concrete	plain	0.02	0.02	0.05
Cork	tiles 19 mm, solid backing	0.02	0.05	0.10
Carpet	thick pile	0.10	0.50	0.60
Curtains	medium weight, folded	0.10	0.40	0.50
	medium weight, straight	0.05	0.10	0.20
Fibreboard	13 mm, solid backing	0.05	0.15	0.30
	13 mm, 25 mm airspace	0.30	0.35	0.30
Glass	4 mm, in window	0.30	0.10	0.07
	tiles, solid backing	0.01	0.01	0.02
Glass fibre	25 mm slab	0.10	0.50	0.70
Hardboard	on battens, 25 mm airspace	0.20	0.15	0.10
Plaster	lime or plaster, solid backing	0.02	0.02	0.04
	on laths/studs, airspace	0.30	0.10	0.04
Plaster tiles	unperforated, airspace	0.45	0.80	0.65
Polystyrene tiles	unperforated, airspace	0.05	0.40	0.20
Water	swimming pool	0.01	0.01	0.01
Wood blocks	solid floor	0.02	0.05	0.10
Wood boards	on joists/battens	0.15	0.10	0.10
Wood wool	25 mm slab, solid backing	0.10	0.40	0.60
	25 mm slab, airspace	0.10	0.60	0.60

Special items		Absorption coefficient		
Air	per m^3			0.007
Audience	per person	0.21	0.46	0.51
Seats	empty fabric, per seat	0.12	0.28	0.28
	empty metal, canvas, per seat	0.07	0.15	0.18

Types of absorber

The materials and the devices used especially for the purpose of absorbing sound can be classified into the following three main types of absorber, which have maximum effect at different frequencies, as indicated in Figure 10.5:

- porous absorbers for high frequencies
- panel absorbers for lower frequencies
- cavity absorbers for specific lower frequencies.

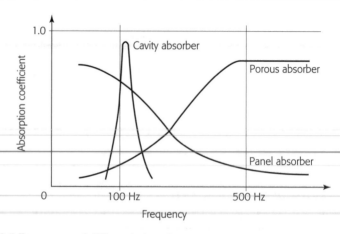

Figure 10.5 Response of different absorbers

Porous absorbers

Porous absorbers consist of cellular materials such as fibreglass and mineral wool. The air in the cells provides a viscous resistance to the sound waves which then lose energy as frictional heat. The cells should interconnect with one another, and the closed cell structure of some foamed plastics is not always the most effective form for sound absorption.

Porous materials used for sound absorption include acoustic tiles, acoustic blankets and special coatings such as acoustic plaster. The absorption of porous materials is most effective at frequencies above 1 kHz; the low-frequency absorption can be improved slightly by using increased thickness of material.

> Window glass is flexible and acts as a panel absorber

Panel absorbers

Panel or *membrane* absorbers are constructed from fixed sheets of continuous materials with a space behind them; the space may be of air or may contain porous absorbent. The panels absorb the energy of sound waves by converting them to mechanical vibrations in the panel which, in turn, lose their energy as friction in the clamping system of the panel. The panels may be made of materials such as plywood or they may already exist, for example in the form of windows or suspended ceilings.

The amount of absorption depends on the degree of damping in the system. The *resonant frequency* of the system, at which maximum absorption occurs, depends on the mass of the panel and the depth of the airspace beyond. The resonant frequency is given by the formula:

$$f = \frac{60}{\sqrt{md}}$$

where m and d are the measurements shown in Figure 10.6.

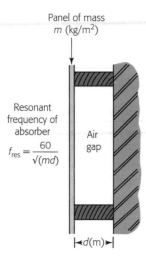

Figure 10.6 Panel absorber

A panel absorber is most effective for low frequencies in the range 40 to 400 Hz. A typical response curve is shown in Figure 10.5.

Cavity absorbers

Cavity absorbers or *Helmholtz resonators* are enclosures of air with one narrow opening. The opening acts as an absorber when air in the opening is forced to vibrate and the viscous drag of the air removes energy from the sound waves. In practice, the cavity may contain material other than air and be part of a continuous structure, such as a perforated acoustic tile.

A cavity can provide a high absorption coefficient over a very narrow band of frequencies. The maximum absorption occurs at the resonant frequency of the cavity which is estimated by the formula

$$f = 55 \sqrt{\frac{a}{dV}}$$

where A, v and d are the measurements shown in Figure 10.7.

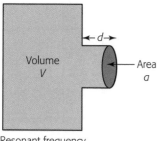

Figure 10.7 Cavity absorber

The ability to tune a cavity absorber to specific frequencies is useful for controlling certain sounds inside rooms; a typical response curve is shown in Figure 10.5.

Practical absorbers

The acoustic tiles used on walls and ceilings often absorb sounds by a combination of several different methods, depending upon the frequency content of the sound. The basic material of the tile, such as fibreboard, is porous and acts as an absorbent for higher frequencies. The tile material may also be drilled with holes which then act as cavity absorbers.

Some tiles have a perforated covering, the holes in which form effective resonators. The tiles may also act as panel absorbers if they are mounted with an airspace behind them, as in a suspended ceiling. In general: panel absorbers are used for low frequencies; perforated panels are used for frequencies in the range 200 to 2000 Hz, and porous absorbers are used for high frequencies.

> Acoustic materials also need ~~resistance~~ to:
> heat
> fire
> chemical attack
> insect attack
> UV decay

REVERBERATION

If the main source of sound in a room suddenly stops, it is unlikely that the sound in the room will also stop suddenly. A single hand-clap demonstrates this effect. There is a continuing presence of sound, known as reverberation, which is particularly noticeable in a large reflective interior such as a church.

* **Reverberation** is a continuation and enhancement of a sound caused by rapid multiple reflections between the surfaces of a room.

> Reverberation depends upon the rate of decay of sound

Reverberation is *not* the same as an echo, as the reflections reach the listener too rapidly for them to be heard as separate sounds. Instead, the reverberative reflections are heard as an extension of the original sound.

When a pulse of sound is generated in an enclosed space the listener first receives sound in a direct path from the source. This direct sound is quickly supplemented by the sound reflected from the surfaces of the room. Some

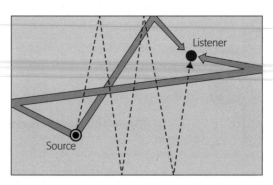

Figure 10.8 Multiple reflections of reverberation

simplified rebeverant sound paths are shown in Figure 10.8. The sound loses some energy at each reflection, depending on the nature of the surface, and absorption usually limits the number of reflections causing reverberation.

   If the original source of sound is continuous, then the reverberant sound combines with the direct sound to produce a continuing *reverberant field*. A room totally without reverberation is termed *anechoic*, and is achieved by using special absorption techniques at each surface.

> Small spaces and soft furnishings give low reverberation

Reverberation time

If the source of sound stops then the reverberant sound level dies away with time, as shown in Figure 10.9. The rate at which the sound decays is a useful indication of the reverberation quality and is measured by a reverberation time with a standard definition.

- **Reverberation time** is the time taken for a sound, when stopped, to decay by 60 dB.

     Unit: seconds (s)

A decrease in sound level of 60 dB is the same as a drop to one millionth of the original sound power and represents the decay of a moderately loud sound to inaudibility. The time taken for this decay in a room depends upon the following factors:

- areas of exposed surfaces
- sound absorption at the surfaces
- distances between the surfaces
- frequency of the sound.

> *Reverberation times*:
> Outdoors
>   0.0 s
> Living room
>   0.4 s
> Opera House,
>   Sydney
>   2.0 s
> Symphony Hall,
>   Birmingham
>   2.4 s
> St Paul's
>   Cathedral
>   13 s

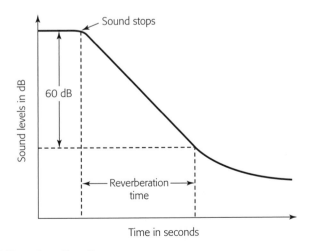

Figure 10.9 Reverberation time

Reverberation time is an important index for describing the acoustical quality of an enclosure. The reverberation time of an existing auditorium can be determined by generating pulse sounds, such as a gunshot from a starter's pistol, and measuring the decay time. The reverberation time of a planned auditorium can be calculated in advance from a knowledge of the above factors that affect reverberation time.

Ideal reverberation time

Typical reverberation times vary from a fraction of a second in small rooms to 5 seconds or more in very large enclosures like a cathedral. Different activities require reverberation times in the following ranges:

- **Speech**: 0.5 to 1 seconds reverberation time.
- **Music**: 1 to 2 seconds reverberation time.

> *Wallace Sabine* designed the acoustics of the Boston Symphony Hall (opened 1900) to have a reverberation time of 1.9 seconds

Short reverberation times are necessary for clarity of speech, otherwise the continuing presence of reverberant sound will mask the next syllable and cause the speech to be blurred. Longer reverberation times are considered to enhance the quality of music, which will sound 'dry' or 'dead' if the reverberation time is too short. Larger rooms are judged to require longer reverberation times, as is also the case with lower frequencies of sound.

Optimum reverberation times can be calculated by formulas, such as the following *Stephens and Bate* formula.

$$t = r\left[0.0118 \sqrt[3]{V} + 0.107\right]$$

where

t = reverberation time (s)
V = volume of hall (m³)
r = 4 for speech, 5 for orchestras, 6 for choirs.

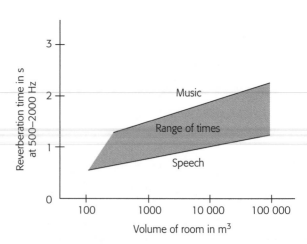

Figure 10.10 Optimum reverberation times

Ideal reverberation times can also be presented in sets of graphs, such as those shown in Figure 10.10.

Reverberation time formulas

When reverberation time cannot be directly measured, as at the planning stage for example, it can be predicted from a knowledge of the factors that affect the decay of sound. Reverberation time depends on the areas of exposed surfaces, the absorption coefficients of these surfaces, and the distances between the surfaces (or volume). If these factors are numerically related by a formula, then one of the factors can be calculated if the others are known.

Sabine formula

The Sabine formula assumes that the reverberant decay is continuous and it is found to give reasonable predictions of reverberation time for rooms without excessive absorption.

Spreadsheet programs are useful for reverberation calculations

$$t = \frac{0.16\,V}{A}$$

where

t = reverberation time (s)
V = volume of the room (m³)
A = total absorption of room surfaces (m² sabins)
  = Σ(surface area × absorption coefficient).

Calculation guide

The worked examples illustrate common types of calculation where the following guidelines are useful.

- Do not directly add or subtract reverberation times with one another. Use the Sabine formula to convert reverberation times to absorption units, make adjustments by addition or subtraction of absorption units, then convert back to reverberation time.
- Surfaces that are not 'seen' do not usually provide absorption – an area of floor covered by carpet, for example.
- Use a sketch of the enclosure with dimensions to help identify all surfaces and their areas.
- Lay the calculations out in a table, as shown in the worked examples. This method encourages clear accurate working, allows easy checking, and is also suitable for transfer to a computer spreadsheet.

Eyring formula

If the average absorption in a room is high, as in a broadcasting studio, the reverberation times predicted by the Sabine formula do not agree with actual results. A more accurate prediction is estimated by the Eyring formula given below.

$$t = \frac{0.16\,V}{-S\,\log_e(1-\bar{a})}$$

where

t = reverberation time (s)
V = volume of the room (m³)
S = total area of surfaces (m²)
$\bar{a}$ = average absorption coefficient of the surfaces.

Calculation of reverberation time

Reverberation times are calculated by finding the total absorption units in a room and then using a formula such as the Sabine formula. The absorption of materials varies with frequency, and the reverberation time predicted by the formula is only accurate for the frequency at which absorption coefficients are valid.

If the same value of reverberation time is required at different frequencies then the total absorption must be the same at each frequency. This can be achieved by choosing materials or devices that provide absorption at certain frequencies and not at others. For example, the absorption of porous materials increases with higher frequencies while the absorption of the panel decreases with higher frequencies.

Worked Example 10.1

A hall has a volume of 5000 m³ and a reverberation time of 1.6 s. Calculate the amount of extra absorption required to obtain a reverberation time of 1 s.

Know

$t_1 = 1.6$ s, $A_1 = ?$
$t_2 = 1.0$ s, $A_2 = ?$
$V = 5000$ m³

Using the Sabine formula

$$t = 0.16V / A$$

and substituting for t_1

$$1.6 = \frac{0.16 \times 5000}{A_1} \quad \text{so } A_1 = \frac{0.16 \times 5000}{1.6} = 500 \text{ sabins}$$

and substituting for t_2

$$1.0 = \frac{0.16 \times 5\,000}{A_2} \quad \text{so } A_2 = \frac{0.16 \times 5\,000}{1.0} = 800 \text{ sabins}$$

Extra absorption needed $= A_2 - A_1 = 800 - 500$
$= \textbf{300 m}^2 \textbf{ sabins}$

Worked Example 10.2

A lecture hall with a volume of 1500 m³ has the following surface finishes, areas and absorption coefficients at 500 Hz.

	Area	Abs. coeff.
Walls, plaster on brick	400 m²	0.02
Floor, plastic tiles	300 m²	0.05
Ceiling, plasterboard on battens	300 m²	0.10

Calculate the reverberation time (for a frequency of 500 Hz) of this hall when it is occupied by 100 people.

Tabulate information and calculate the absorption units using

Absorption = Area × Absorption coefficient

Surface	Area/quantity	500 Hz Absorption coefficient	Abs. unit (m² sabins)
Walls	400 m²	0.02	8
Ceiling	300 m²	0.10	30
floor	300 m²	0.05	15
Occupants	100 people	0.46 each	46
		Total A	99

Using the Sabine formula

$$t = \frac{0.16\,V}{A} = \frac{0.16 \times 1500}{99} = 2.42$$

So reverberation time $= \textbf{2.42 s at 500 Hz}$

Worked Example 10.3

The ideal reverberation time required for a hall in is 1.5 s. Calculate the area of acoustic tiling needed on the walls to achieve this reverberation time (absorption coefficient of tiles = 0.4 at 500 Hz).

The areas of tiles will change the original area of plain walls. The areas can be found by trial and error, or by algebra as shown here.

Surface	Area/ quantity	500 Hz Absorption coefficient	Abs. units (m² sabins)
Tiles	S m²	0.40	0.4S
Walls	400 − S m²	0.02	8 − 0.02S
Ceiling	300 m²	0.10	30
Floor	300 m²	0.05	15
Occupants	100 people	0.46 each	46
		Total A	99 + 0.38S

Using

$$t \quad \frac{0.16\,V}{A}$$

$$1.5 = \frac{0.16 \times 1\,500}{99 + 0.38S}$$

Rearranging the formula

$$0.38S = \frac{0.16 \times 1\,500}{1.5} - 99 = 160 - 99 = 61$$

$$S = \frac{61}{0.38} = 160.5$$

So area of tiles = **160.5 m²**

EXERCISES

1. Draw a scaled plan and a section of a hall similar to that in Figure 10.4, or use any suitable drawings of an auditorium. Choose a sound source situated on the centre of the stage and draw geometrically accurate sound path diagrams to show the reflections off the ceiling and off the walls. Comment on the distribution of sound in the hall and suggest remedies for any areas where reflections might cause acoustic defects.

2. A room of 900 m^3 volume has a reverberation time of 1.2 s. Calculate the amount of extra absorption required to reduce the reverberation time to 0.8 s.

3. Calculate the actual reverberation time for a hall with a volume of 5000 m^3, given the following data for a frequency of 500 Hz:

Surface area	Absorption coefficient
500 m^2 brickwork	0.03
600 m^2 plaster on solid	0.02
100 m^2 acoustic board	0.70
300 m^2 carpet	0.30
70 m^2 curtain	0.40
400 seats	0.30 units each

4. If the optimum reverberation time for the above hall is 1.5 s then calculate the number of extra absorption units needed.

5. A large cathedral has a volume of 120,000 m^3. When the space is empty the reverberation time is 9 s. With a certain number of people present the reverberation time is reduced to 6 s. Calculate the number of people present, if each person provides an absorption of 0.46 m^2 sabins.

6. A rectangular hall has floor dimensions of 30 m by 10 m and a height of 5 m. The total area of windows is 50 m^2. The walls are plaster on brick, the ceiling is hardboard on battens and the floor is wood blocks on concrete. There are 200 fabric seats. The reverberation time required for the hall, without audience, is 1.5 s at 500 Hz. Use Table 10.1 to help calculate the area of thick pile carpet needed to achieve the correct reverberation time.

Answers are on page 326.

Electricity Supplies

CHAPTER OUTLINE

Electricity powers many aspects of our modern lives and a supply of electricity in a building is essential for creating and controlling the environment. Systems for heating, cooling, ventilating and lighting all use electricity, because of its energy content and for the ease with which it can be controlled. The electronic equipment of modern offices and intelligent buildings requires convenient supplies of electricity. On a larger scale, electricity also provides the large amounts of energy needed for industry and transport, and for pumping public water and drainage systems.

However, the production and distribution of large-scale supplies of electricity energy also involves technical, aesthetic and ethical considerations such as the following:

▶ the siting of power stations and transmission lines
▶ the use of high-voltage cables near communities
▶ the carbon emissions of power stations
▶ the use of renewable energy sources for electricity
▶ the use of nuclear power stations.

To give you a better basis for considering these matters this chapter explains the features of electricity supplies and their alternatives. The text will enable you to:

▶ appreciate how electricity systems make use of the principle of induction
▶ understand the working of practical motors, generators and transformers

→

▶ understand the properties and terms associated with alternating current electricity

▶ appreciate the role of single and three-phase supplies in electricity systems

▶ understand the nature of various types of power station

▶ appreciate the possibilities of modern generation techniques

▶ understand the features of traditional national systems for the generation, transmission and distribution of electrical energy via a grid

▶ appreciate the benefits of small local generation and distributed networks.

The principles and options described in this chapter provide a useful basis for considering energy and environmental issues put forward in Chapter 14. The Resource 4 section also outlines the science of current electricity, magnetism and induction if you need to remind yourself of basic principles.

ELECTRICAL PRINCIPLES

In the 1830s, while working at the Royal Institution in London, Michael Faraday experimented with various electrical devices that led him to demonstrate and explain the principles and effects of electromagnetic induction. For example, if a magnet is moved near a wire or coil then an electric current appears in the wire, or is 'induced'. Faraday explained how all such effects share a common mechanism: that the magnetic field is *changing*. The change of magnetic field, and hence the generation of electric current, can be produced by moving the magnet, or moving the coil; or by switching a system on or off.

Electromagnetic induction is the principle behind many important devices used for the generation, the transmission and the application of electricity. Although Faraday explained the principles in the 1830s, it was not until the 1880s that the principles were put to use in practical electrical supply systems. Some examples of applications of induction are listed below and described in the following sections:

• generators for the production of electricity
• transformers for changing voltage
• electric motors
• ignition coils for producing spark in car engines
• microphones
• linkages in electronic circuits
• heating elements in kitchen cookers.

Generators

A generator is a device that converts mechanical energy to electrical energy by means of electromagnetic induction. Most of the electricity used in everyday life is generated in this manner. The change in magnetic field necessary for induction is produced by moving a coil through a magnetic field, or by moving a

See also:

section on *Carbon and energy management* in Chapter 14 and Resource 4

See also:

section on *Induction* in Resource 4

See also:

section on *Current electricity* in Resource 4

magnetic field past a coil. Rotational motion is usually employed. Some types of generator may be known as dynamos or alternators.

Simple AC dynamo

The simplest type of generator is the AC dynamo, shown in Figure 11.1. The generator coil is set on an axle in a magnetic field and connected to metal slip rings on the axle. The slip rings make sliding contact with the stationary brushes, which carry current away from the generator. The axle is rotated by a source of mechanical energy, such as a motor, and the coil moves through the magnetic field. The coil then experiences a changing magnetic field so that an electromotive force (emf or 'voltage') is induced in the coil, and therefore a current flows. The current is led away from the generator via the slip rings and the brushes.

emf is the 'voltage' that drives current in a circuit

Maximum current is generated twice per revolution as the coil cuts the field at right angles, which happens when the coil is horizontal. Zero current is generated when the coil is vertical. The current varies from zero to maximum depending upon the angle at which the coil cuts the magnetic field. As the coil rotates through one revolution it cuts the magnetic field in two different directions and the induction reverses direction.

The emf and the current produced by this generator is therefore continuously 'alternating' with every revolution of the coil. The output of this 'alternator' is in the shape of a sine wave, as shown in Figure 11.3.

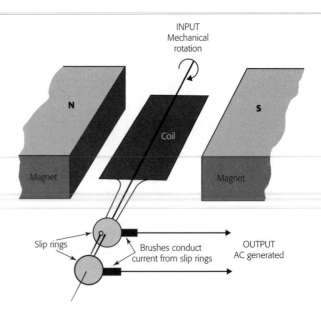

Figure 11.1 Simple AC dynamo

DC generator

The current from a battery has a steady value in one direction. To obtain this sort of 'direct current' from a generator it is necessary to replace the slip rings by a commutator which reverses the connections every half turn and gives a current flowing in one direction only. This current will still fluctuate in value, and to obtain a virtually steady flow of current the commutator is split into many sections and connected with up to 30 different coils. The construction is the same as for the DC motor, shown in Resource 4, and some machines can act as both generator and motor, at different times.

Practical AC generators

A large practical AC generator or alternator works on the same general principle as the simple generator but its construction is different, especially when it is scaled up in size for use in power stations.

Applications of dual motor-generators: Electric trains for braking Hybrid cars for charging batteries

   The *stator* is the stationary frame to which are fixed the output windings that produce the current. No moving contacts are necessary to lead the current from the generator so the construction is suitable for large supplies. The *rotor* turns inside the stator and contains the magnetic field coils. A separate self-starting excitor dynamo, run on the same axle, supplies the DC current necessary for these electromagnets.

Transformers

A transformer is a device that uses the principle of electromagnetic induction for the following purposes:

- step-up or step-down voltage
- isolate a circuit from an AC supply.

A major reason for using AC in electricity supplies is the relative simplicity and efficiency of transformers for changing voltage, as explained in the section on power transmission. A transformer does not work on DC supply. The construction of a common type of transformer is shown in Figure 11.2.

Operation of transformer

The primary coil is connected to an AC supply and sets up a magnetic field which is continuously changing with the alternating current. The secondary coil experiences this changing magnetic field and produces an induced emf which can then be connected to a load. The induced emf alternates with the same shape and frequency as the emf of the supply frequency, but the ratio of the two emfs is proportional to the ratio of the turns on the two windings.

Transformer equation

$$\frac{V_s}{V_p} = \frac{N_s}{N_p}$$

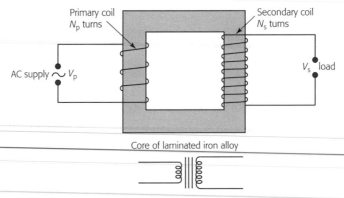

Figure 11.2 Double-wound step-up transformer

where

V_s = emf induced in the secondary coil (V)
V_p = emf applied to the primary coil (V)
N_s = number of turns on the secondary coil
N_p = number of turns on the primary coil.

A transformer with twice as many secondary turns as primary turns will, for example, 'step-up' the secondary voltage to twice the primary voltage. If the connections to such a transformer are reversed the transformer could be used as a 'step-down' transformer which then halves the applied voltage. In an 'isolating transformer' the primary and secondary voltage may be the same.

Transformer efficiency

Transformers are efficient machines and their power output is almost equal to their power input. If, for simplicity, it is assumed that there are no power losses then the following relationship is true:

- Output power = Input power.

The output power is from the secondary coils and the input power is from the primary coils. So using the power formula $P = I V$ we can write the transformer power relationship as follows:

$$P_{secondary} = P_{primary}$$

or

$$I_s V_s = I_p V_p$$

See also:

section on
Electrical power
in Resource 4, p.
379

This relationship shows that if a transformer increases the voltage then the current must decrease by the same ratio, so as to conserve energy. In practice, some energy is lost in a transformer by heating in the coils and in the core. *Eddy currents* are circulating currents which are induced in the core. These currents are minimised by constructing the core from separate laminations so that the core presents a high magnetic and electrical resistance.

For the practical case where the transformer is less than 100 per cent efficient, the following formula is used.

$$I_s V_s = I_p V_p \times \text{efficiency factor}$$

where
I = current in secondary coil (A)
V_s = emf of secondary coil (V)
I_p = current in primary coil (A)
V_p = emf of primary coil (V).

Worked Example 11.1

A transformer has 600 turns on the primary coil and 31 turns on the secondary coil. An emf of 230 V is applied to the primary coil and a current of 250 mA flows in the primary coil when the transformer is in use.

(a) Calculate the emf of the secondary coil.
(b) Calculate the current flowing in the secondary coil. Assume that the transformer is 95 per cent efficient.

(a) Know N_p = 600, V_p = 230, N_s = 31, V_s = ?

Using transformer equation

$$\frac{V_s}{V_p} = \frac{N_s}{N_p}$$

or

$$\frac{V_s}{230} = \frac{31}{600}$$

Rearranging the expression

$$V_s = \frac{31}{600} \times 230 = 11.88 \text{ V}$$

So secondary emf = **12 V**

(b) Know V_p = 230 V, I_p = 250/1 000 = 0.25 A, V_s = 12 V, I_s = ?

Using efficiency expression

$$I_s V_s = I_p V_p \times 95/100$$

$$I_s \times 12 = 0.25 \times 230 \times 95/100$$

$$I_s = \frac{0.25 \times 230}{12} \times \frac{95}{100} = 4.55 \text{ A}$$

So secondary current = **4.6 A**

Alternating current properties

The emf induced in an AC generator is constantly changing and reversing, and so the current produced by the emf also changes to give a pattern as shown in Figure 11.3. Some additional terms are needed to describe the nature of the alternating output.

Frequency

- **Frequency** (f) is the number of repetitions, or cycles, of output per second.

    Unit: hertz (Hz)

For public supplies in Britain and other countries in Europe the frequency is 50 Hz; in North America the frequency is 60 Hz.

For an AC supply at 50 Hz:
the voltage, and current, is momentarily zero 100 times per second

Peak value

The peak value is the maximum value of alternating voltage or current, measured in either direction. The peak values occur momentarily and only twice in a complete cycle, as shown in Figure 11.3.

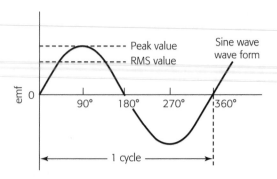

Figure 11.3 Alternating current output

RMS value

The simple mathematical average of a sine wave output is zero. But an alternating supply does produce an effective voltage or current and this is measured by RMS values of voltage or current.

- A **root mean square** (RMS) value of alternating current is that value of direct current that has the same heating effect as the alternating current.

> RMS is a form of average based on energy

A 1 kW fire, for example, produces the same heating effect using 230 V alternating current as it does using 230 V direct current. The RMS value is found mathematically by taking many instantaneous peak values, squaring them, taking the mean (average) of the squares, then taking the square root. The relationship between the values is found to be:

- RMS value = 0.707 Peak value

> RMS = 0.707 peak

The European domestic supply AC voltage of 230 V RMS has a peak value of 325 V. The values of emf and current quoted for AC supplies are assumed to be RMS values, unless it is stated otherwise.

Power factor

The voltage and the current of an AC supply are *in phase* when both have their peak values and zero values occurring at the same time. The power used by an AC circuit that is in phase is calculated as the product of the RMS values of current and voltage – similar to DC power.

Some AC circuits contain components that cause a *phase shift*, where either the voltage or the current leads or lags behind the other. In such a circuit some of the energy supplied as *apparent power* is lost in heating the circuit and does not appear as effective power or *active power*. The power factor expresses the ratio between the two forms of power.

$$\textbf{Power factor (PF)} = \frac{\text{Active power (watts)}}{\text{Apparent power (volt-amperes)}}$$

$$= \text{cosine } \theta$$

where θ is the phase angle between current and voltage.

The power factor has a maximum value of 1. Most AC equipment is rated by its apparent power together with a power factor. Supply authorities have to generate the apparent power but can only charge for the active power which appears on the meters, so they set minimum allowable values for the power factor (0.85 for example).

Devices with simple resistance, such as a heater, have a power factor of 1 and their active power is the same as their apparent power. Devices with induction

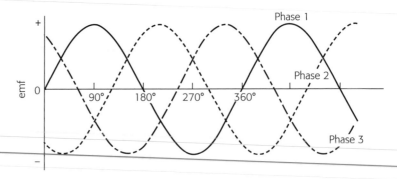

Figure 11.4 Three-phase alternating current

coils, such as motors, have power factors less than 1. Low power factors can be improved by adding a capacitance device in parallel to balance the inductance.

Three-phase supply

> For delivering energy single-phase AC can be considered 'lazy'

The output from a simple AC generator, shown in Figure 11.3, is a *single-phase* supply. Practical supplies of electricity are usually generated and distributed as a *three-phase* supply, which is, in effect, three separate single-phase supplies each equally out of step, as shown in Figure 11.4. The generator is essentially composed of three induction coils instead of one, the coils being displaced from one another by 120°. The three phases can be interconnected in different ways for different purposes, as shown in Figure 11.5.

A *delta* connection is used for larger loads, such as three-phase motors. The voltage across any two phases (the line voltage) is 1.73 (or √3) times the voltage between any one phase and earth.

A *star* connection is used for relatively small loads, such as households. It is connected to one phase and to the neutral to provide a 230 V supply. If the single-phase loads are evenly balanced then the return current in the neutral cable is zero.

Three-phase supplies are economical in their use of conductors and can supply more power than single-phase supplies. Only three cables need to be used for the

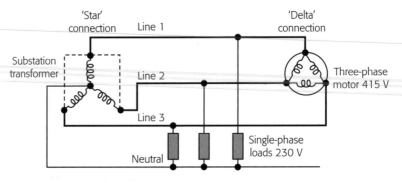

Figure 11.5 Three-phase, four-wire AC supply

three-phase supply, instead of the six cables needed for three separate single-phase supplies. A device such as a motor which is designed to make use of the three phases receives more energy per second than those devices wired to a single phase; the three-phase motor is also smoother in its operation.

POWER SUPPLIES

A public supply of electrical energy is one of the most important services in the built environment. Manufacturing, building services, transport and communications are all dependent upon electricity supplies. Large electric power systems require significant investments in construction. They also have a considerable impact upon the appearance of the landscape, and any breakdown in the system greatly disrupts everyday life. Patterns of electrical energy use in buildings will tend to change in the future and the design of electric power systems will also need to evolve.

A complete power system is a collection of equipment and cables capable of producing electrical energy and transferring it to the places where it can be used. It is made up from the following three main operations.

- **Generation**: the production of the electricity.
- **Transmission**: the transfer of electrical energy over sizeable distances.
- **Distribution**: the connection of individual consumers and the sale of electricity.

Power stations

All power stations generate electrical energy by using electromagnetic induction where an emf is produced in a coil which experiences a changing magnetic field. The mechanical energy required can be obtained using heat energy from burning fossil fuels, from nuclear reactions, from the energy of moving water or from renewable sources such as wind and waves. A *turbine* is a device that produces rotational motion from the movement of steam, running water or wind, and turns the axle of the generator.

Practical AC supply system: proposed in 1888 by Nikola Tesla

A typical generator in a steam turbine power station turns at 3000 revolutions per minute and has an excitor dynamo mounted on the same shaft. The output varies but a large 500 MW generator set commonly generates 23 000 A at 22 kV. The cooling of the generators, by liquid or gas, is an important part of their engineering.

Generator size (typical output in kW):

Nuclear	1 000 000
Steam	500 000
CCGT	250 000
CHP	50 000
Wind	1 000
Solar panel	3

Steam turbine power stations

Traditional large 'thermal' power stations use heat energy to drive the generators. The heat is obtained by burning fossil fuels such as oil, coal or gas. The components of a thermal station are outlined in Figure 11.6. The boiler burns the fuel and heats water to produce high-pressure steam at high temperature. The steam is directed

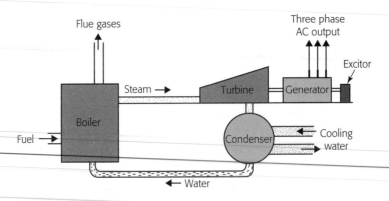

Figure 11.6 Thermal power station scheme

onto the blades of a high-speed turbine which produces mechanical energy and turns the generator. The steam is condensed and the water returned to the boiler.

The efficiency of the boiler is the main limitation of the system, but modern techniques, such as fluidised-bed combustion, offer some improvements. A thermal power station can, however, use low-quality coal and the ash recovered from the boiler can be used as a building aggregate. The condensation of the low-pressure steam from the turbines requires large quantities of cooling water from a river, or else a system of large and usually unsightly cooling towers. The process involves discarding large amounts of heat energy which would preferably be put to use, as is done in combined heat and power (CHP) units described below.

The potential advantages of thermal power stations include the fact that they can be sited relatively close to the demand for their electricity and so can save on the cost of transmission lines. It is also possible for some of the waste heat from the condensation process to be used for industrial purposes or for district heating of buildings, as described below in the section on CHP. The disadvantages of traditional thermal power stations include their use of the world's finite supplies of fossil fuels, and their emissions of carbon dioxide and air pollution components. There are however some advances in the technologies associated with thermal power stations and some of these are outlined below.

See also:

section on *Carbon and energy management* in Chapter 14

Gas turbines

Gas turbines are similar to those found in jet aircraft engines. Natural gas or gas oil is burnt in combustion chambers and the pressurised combustion gases are used to turn a series of turbine fan blades. The turning shaft of the turbine provides the mechanical energy to drive the generator and produce electrical energy. Gas turbine engines have the versatility of rapid start-up and shut-down, compared with the hours needed for a steam turbine.

Combined cycle gas turbine (CCGT)

CCGT power stations combine the use of gas and steam turbines to drive one or more generators. The gas turbine drives a generator and supplies waste heat in the form of hot exhaust gases. These gases heat a boiler to produce steam, which drives a conventional steam turbine and generator. The thermal efficiency of these stations can be as high as 60 per cent compared with a maximum of 40 per cent from conventional coal-fired steam turbine stations.

Combined heat and power (CHP)

CHP is the generation of usable heat and power in a single process. The heat is usually in the form of steam or hot water. For example, large isolated thermal power stations have to get rid of large amounts of steam energy and their tall steaming cooling towers are monuments in a way to wasted energy. CHP is a system which uses such heat from electricity generation for industrial processes or for community use such as district heating (DH).

> CHP in the UK: 7 per cent of total generation (2009)

The design of a combined heat and power plant can be optimised for total energy efficiency and achieve efficiencies of up to 90 per cent. CHP plants can use steam turbines, gas turbines or combinations of the two. Typically a CHP plant is much smaller than a large traditional power station and needs to be attached to the site which consumes the heat and a large proportion of the electrical power produced. This arrangement for the electricity avoids significant transmission and distribution losses via a grid.

> *Other terms for CHP:*
> Co-generation
> Total energy

Clean coal technologies (advanced coal)

Many countries of the world are rich in coal and wish to make use of these resources, which are likely to last longer than supplies of oil and natural gas.

Older coal-fired thermal power stations can be replaced by more advanced plants that use new techniques to be more efficient.

- **Fluidised bed technology** burns coal at lower temperatures and uses limestone to capture emissions. The technique is about 50 per cent cleaner than current coal-burning power plants.
- **Gasification** breaks coal into a mixture of carbon monoxide, hydrogen and other gases that can be used for fuel. A gasifier exposes coal to hot steam together with controlled amounts of air or oxygen under high temperatures and pressures.

These advanced techniques of burning coal can be combined with CCGT generation of electricity, as described above. Gasification also offers the possibility of *carbon capture and storage* (CCS), which involves capturing carbon after it has been used in a boiler and storing it underground in a way that prevents it entering the atmosphere.

Nuclear power stations

Nuclear power stations are thermal stations where the heat energy is released from a nuclear reaction rather than by burning a fossil fuel. Radioactive elements such as uranium and plutonium have unstable nuclei which emit neutrons. These neutrons split neighbouring atoms, thus releasing other neutrons and producing heat. This *fission* reaction is controlled and prevented from becoming a chain reaction by using moderator materials, such as carbon, which absorb neutrons without reaction.

A nuclear reactor consists of a central container holding the radioactive fuel elements, which are surrounded by control rods of the moderator material. The withdrawal of some control rods starts the reaction and the insertion of extra control rods can stop it. The heat generated by the reaction is carried by a coolant from the core to a heat exchanger, which then produces steam for the turbines. Because the core and the coolant emit dangerous radiation, they must be well protected.

> Apart from heat production, nuclear stations use the same generation technologies as other thermal stations

The *pressurised water reactor* (PWR) is a common type of reactor in practical use around the world. The coolant in the reactor is water which is kept at high pressure to prevent it boiling. The *advanced gas-cooled reactor* (AGR) is an earlier UK design which uses carbon dioxide gas as the coolant.

The advantages of nuclear power stations are that their fuel costs are low, their emission of greenhouse and pollutant gases is negligible, and that they reduce dependence upon fossil fuels. Their disadvantages include the harm that escaped radioactive fuels, coolants and waste products can cause to people and the environment. The dangers associated with the disposal and storage of radioactive waste from nuclear power stations will continue for thousands of years. The real price of nuclear-generated power therefore depends upon how the costs of radioactive waste management and disposal and the decommissioning of plant are taken into account.

The safety of nuclear reactors depends upon the reliability with which they can be built, operated and decommissioned under practical conditions, which include the possibility of operator errors and periodic natural events such as earthquakes.

Hydroelectric power stations

Hydroelectric power stations use the kinetic and potential energy of running water to drive the generators. A large quantity of water at a height is needed to provide enough continuous energy. It is worth remembering that the original source of this energy is sunshine which lifts the water molecules by means of evaporation. A dam usually provides both the head of water and a reservoir of stored water. The water flows down *penstock* pipes or tunnels, and imparts energy to the water turbines which turn the generators.

The advantages of hydroelectric power stations are that their energy source is free, they have a long operating life and they need few staff. Their disadvantages include the very large capital investment on civil engineering work, and the disruption to the environment by forming dams and lakes. Suitable sites for such stations are limited and usually need long and expensive transmission lines to transfer energy to the consumers.

Wind turbines

A wind turbine is a form of windmill that converts the kinetic energy of wind to electrical energy by rotating an electrical generator. The axis of rotation can be either vertical or horizontal. A common type of large-scale wind turbine is seen as a horizontal nacelle unit set at the top of a tall steel tower with two or three propeller-like blades attached. Wind passing over the blades exerts a turning force on the shaft, which is attached to a gearbox to increase the speed of rotation for the generator, which converts the mechanical energy to electrical energy.

The electricity is generated at a voltage of around 700 V and fed to a transformer which converts it to the higher voltages of the distribution system, such as 33 kV and 132 kV. The rated power output of a single modern wind turbine is typically 2 to 10 megawatts (MW). The turbines produce this power when the wind speed is around 15 metres per second (30 mph/50 kph).

Wind farms generate commercial quantities of electricity using collections of many wind turbines set in suitable locations such as hillsides, mountain passes and shallow offshore waters. A near constant flow of non-turbulent wind is an ideal location for a wind farm, but must be balanced with other factors such as community acceptance of wind turbines, local demand for electricity and the capacity of transmission systems.

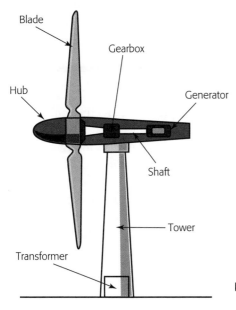

Figure 11.7 Wind turbine layout

Other sources of electricity supplies

The previous sections of this chapter describe popular methods of generating electricity on a large scale, by thermal power stations for example. There are a variety of other methods for generating electricity, and some of them are outlined below. It can take time for a new technology to develop and for the costs of using that technology to reduce to prices that make projects feasible. Some environments are particularly suited for alternative methods of generating electricity, such as by using solar energy, geothermal energy, and tidal energy.

Solar energy sources

The Earth receives energy from the Sun in the form of electromagnetic radiation. This energy is currently collected by two principal methods:

- reflecting and focusing solar radiation onto tubes of a fluid which then acts as a heat source for a steam turbine to generate electricity
- converting solar radiation directly into electricity by the use of semiconductor materials that employ the *photovoltaic (PV)* effect in a solar cell.

A *solar cell* contains thin layers of semiconductor materials, such as silicon compounds, arranged to form an electric field. When photons in the sunlight reach the semiconductor they knock electrons into higher energy states. This effect creates a voltage which drives a current when the cell connected to an electric circuit. Individual solar cells are combined together into modules or solar panels.

Solar panels can be mounted onto the roofs and walls of buildings. Collections of solar panels can be organised into solar arrays for larger-scale generation of electricity. Costs of PV devices are falling and this form of electricity generation is growing fast.

See also:

section on *Renewable sources of energy* in Chapter 14

Geothermal energy sources

The interior of Planet Earth contains thermal energy from various sources including heat left from the original formation of the planet and the continuing radioactive decay within some rocks. There is a continuous flow of this heart from the earth's core to the surface, and in some places, such as volcanic areas of the Earth, there is relatively easy access to this geothermal energy.

Geothermal electricity is generated from geothermal energy sources using steam turbines or heat engines. The thermal efficiencies of geothermal devices are lower than high-pressure steam turbines but geothermal sources are considered to be renewable and sustainable because the heat extraction is very small compared with the natural emissions of geothermal energy.

Tidal energy sources

The tides are movements of ocean waters caused by variations in the gravitational attraction exerted on the Earth by the Moon and, to a lesser extent, by the Sun. The kinetic energy of the moving seawater can be used to turn turbines and to generate electricity. Tidal energy technologies are in various stages of development, and tend to be used in places where vigorous tidal flows are easy to access. The tides are predictable so supplies of tidal power are more predictable than other renewable sources of electricity such as wind power and solar power.

Transmission systems

Electrical energy has the useful property of being easily transferred from one place to another. A transmission system, as shown in Figure 11.8, consists of conducting cables and lines, stations for changing voltages and for switching, and a method of control. The energy losses in the system must be kept to a minimum.

Alternating current is used in nearly all modern power transmission systems because it is easy to change from one voltage to another, and the generators and motors involved in AC are simpler to construct than those for DC. To obtain high

Transmission
losses
High currents =
high heat losses

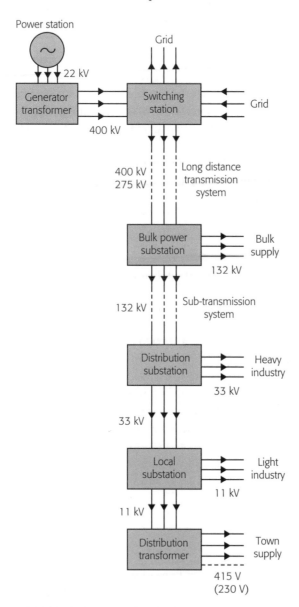

Figure 11.8 Typical power distribution

transmission efficiency the current needs to be kept low because the heating losses in a line increase with the square of the current ($P = I^2R$). Large currents also require thick conductors, which are expensive and heavy. To transmit large amounts of power at low current there must be a high voltage ($P = IV$). Transformers are used for obtaining the necessary high voltages and wide air gaps are used to supply the high insulation that is needed at high voltage.

Transmission lines

The *conductors* for overhead transmission lines are made of aluminium with a steel core added to give strength. The transmission of three-phase supply requires three conductors, or multiples of three, such as six conductors.

Transmission towers or 'power pylons' are needed to keep the lines spaced in the air. Air is a convenient cheap insulator but higher voltages require larger air gaps in order to prevent short circuits through the air. The lines are suspended from the towers by solid insulators made of porcelain or glass. Transmission lines operate at voltages of 132 kV and above. The UK supergrid is at 400 kV and some countries have 735 kV systems.

> UK grid voltage is 400 kV

In a buried cable the conductor must be insulated for its whole length and protected from mechanical damage. High voltage underground cables need special cooling in order to be efficient. It is possible to bury 400 kV transmission cables but the cost is 10 to 20 times greater than for overhead lines.

Substations and switching

Power transmission systems need provision for changing voltages, for re-routing electricity and for protecting against faults.

At *substations* the voltages are changed up or down as required, using large transformers which are immersed in oil for the purpose of cooling as well as for insulation. At a power station, for example, a 1000 MV transformer might step-up 22 kV from the generators to 400 kV for the transmission lines. At distribution substations the 400 kV is stepped down to 132 kV, 33 kV and 11 kV.

Switching substations are substations where a number of transmission lines are interconnected, enabling electrical energy to be routed to where it is required. When lines carrying large currents are disconnected, the currents form arcs of flame which melt contacts. *Circuit breakers* are used to connect or disconnect the transmission lines, and arcs are extinguished by immersion in oil or by blasts of air. Protection systems are devices which act as fuses by sensing faults on a line and then immediately isolating the line.

Transmission networks

An electrical supply system of the simplest type might consist of a hydroelectric power plant at a remote lake, transmission lines to a large city, and distribution around the buildings. If the water in the dam runs out, then there is no electricity. However the electrical supply system in developed countries usually involves many power stations and consumers, interlinked by an electric power *grid*. The output from any power station is not dedicated to just one area but can be distributed to other areas as required. Adjacent countries can also exchange electrical energy to mutual advantage when their patterns of demand differ.

Advantages of supply grids

- Larger power stations have economies of scale which can give lower generation costs.
- Sudden local demands for power can be supplied by a number of power stations.
- The effects of breakdowns in generating plant and transmission lines can be minimised.
- Periods of low demand, such as night-times, can be supplied by those plants with the lowest operating costs.

Disadvantages of grids

- Large transmission systems have financial and environmental impacts.
- Long transmission lines waste energy.
- The raw energy supply to remote power stations system may need long systems of railways or pipelines.
- It is technically challenging to keep all the stations synchronised at the same frequency and to control the grid.
- Surplus heat from large isolated thermal power stations is usually wasted.
- Initiatives for efficient generation are not possible at the local level.

Distributed energy networks

The circumstances of the twentieth century led to the establishment of large power stations and distribution grids as described above. However, the changed supply and economics of energy sources, together with the awareness of carbon emissions, have drawn attention to the benefits of generating energy locally. When power stations are close to users it is possible to use techniques such as combined heat and power (CHP) to give users the benefit of heat energy which is normally wasted at an isolated power station.

New sources of energy such as waste and biomass are likely to be available locally, and there are further energy efficiencies from reducing the amount of electrical energy lost in transmission and distribution across large distances. The idea of *distributed energy* can include a range of technologies that do not depend upon the high-voltage transmission network, or upon other energy grids such as gas pipelines. Typical features include the following:

- combined heat and power plants that supply surplus heat to the local community
- small-scale plant supplying electricity to a building, industrial site or community

- 'microgeneration' using small installations of solar panels, wind turbines, biomass and waste burners that supply one building or small community
- the ability to sell surplus electricity back into the local distribution network
- 'micro-chip' plants in homes that supply both electricity and heat.

Balancing supplies

All electricity networks need to balance the demand from their consumers with their sources of electrical power. The following factors contribute to the complexity of this balance:

- most consumers want their electricity at the same peak times during the day
- some power sources, such as thermal and nuclear, are most efficient when generating all the time, including at night
- power sources using fuels such as oil and gas are more expensive to run
- the output from wind and solar power sources depend upon weather conditions.

Operators of electricity supplies sometimes store electrical energy to help even out supply and demand. Batteries are an everyday example of electricity storage, in chemical form, but are difficult to use on a large scale. Other techniques for storing electrical energy include rotating flywheels, compressed air, hydrogen storage and pumped water.

All techniques for storing electrical energy have losses in efficiency at each conversion of energy, such as from electrical energy to chemical energy and then back from chemical to electrical. *Pumped storage* installations are specialised hydroelectric stations which use off-peak electricity to pump water up to a reservoir for later use. Other hydroelectric sources can control the flow of water from their dammed reservoirs and use this ability as a form of storage.

United Kingdom electrical supply

The national electricity supply of the United Kingdom is a useful example of a large national supply system. Most of the electricity for the United Kingdom is generated by steam and gas turbine power stations which use coal or gas as their fuel. Some nuclear energy is generated, and the UK system sometimes imports electricity from France via a cross-Channel link. The proportion of electricity generated from renewable energy sources is increasing each year as part of a national commitment to the reduction of greenhouse gases. The proportion of hydroelectric power is fixed, but wind power sources have increased and generation from other renewable sources, such as tidal and biomass, is being developed.

Large steam turbine power stations in the system have capacities in the range of 2000 to 4000 MW, and the trend towards larger units has raised the average thermal efficiency of such stations to around 35 per cent. This efficiency is measured as the ratio of the net output of electrical energy to the total input of heat energy. The total generating capacity of UK power stations is above 70 GW, although maximum loads are typically around 60 GW.

Fuels used for UK electricity:	
Gas	45%
Coal	28%
Nuclear	17%
Renewables	8%
Other	2%

(2009 data. This balance of fuels changes over the years.)

Electricity consumption is measured by the power load multiplied by total hours. The yearly totals of electrical energy used in the United Kingdom are of the order of 350,000 gigawatt hours (GWh) which can also be expressed as 350 tera-watt hours (TWh). Domestic users and industry each take about a third of this total. About one-fifth of the domestic electricity is used for space heating in buildings and about one-sixth for water heating.

The *National Grid* is the system of interconnected transmission lines which links generating stations and consumers in England and Wales. It is one of the largest power networks under unified control in the world, although there are a variety of organisations owning different parts of the generation and distribution system.

General distribution

At suitable junctions or at the ends of transmission lines, the voltage is stepped down, as indicated in Figure 11.8. In the British system the voltage is reduced from 400 kV (or 275 kV) to 132 kV. The electricity is distributed at this voltage by a sub-transmission system of overhead lines to the distribution substations. At these stations the voltage is reduced to 33 kV and 11 kV for distribution by underground cable.

European consumer voltage: 230 V RMS nominal but can vary within agreed ranges

Large industrial consumers are supplied at 33 kV while smaller industrial consumers receive 11 kV. Small transformer stations in residential and commercial areas step the voltage down to the final 415 V three-phase, 230 V single-phase supply, which was explained in an earlier section of this chapter.

The three-phase supply is distributed by three phase cables (red, yellow and blue) plus a neutral cable. Some commercial consumers are connected to all three cables of the supply. Households are connected to one of the phase cables and the neutral. Consumers are balanced between the three phases as evenly as possible, for example by connecting consecutive houses to different phases in turn.

Because perfect balance is not achieved, the neutral cable carries a small amount of return current and is earthed at the distribution transformer. To ensure true earth potential, each consumer is supplied with an extra earth cable for attaching to the metal casing of electrical appliances. If an insulation fault causes a connection to this earth then a large current flows, immediately trips the fuse, and protects the user.

Building installations

The electricity supply from the street is taken into a house via protected cables which connect to an electricity meter and then to a consumer unit. This unit contains an isolating switch and the means to supply the various sub-circuits in the building via suitable fuses or circuit breakers. The system for distributing electricity within a building needs to take account of the following factors:

Switchboard is a common name for a consumer unit

- sufficient capacity for purpose
- minimum wastage of current in the cables

Table 11.1 Typical features of domestic electrical installation

Installation feature	Purpose
Main service fuse	Emergency isolation from street supply
Meter	Measurement of energy consumption
Consumer main switch	Isolation of circuits in building
Consumer control unit, bus bar	Connection for individual circuits
Circuit breakers, circuit fuses	Emergency isolation of individual circuits
Earth connection	Prevention of shock
Ring main circuits, 30 amp	Power appliance
Radial circuits, 5 amp	Lighting
Radial circuit, 45 amp	Cooker, water heater
Outlet socket	Access to supply
Fuse in plug or appliance	Additional protection for appliance
	Prevention of shock

- protection of people from electrical shock
- prevention of fire
- means of isolation
- compliance with regulations.

A typical UK house uses a *ring main* to supply power for the appliances, such as televisions, toasters or hair driers, that we connect via sockets that are usually on the wall. The sockets can be installed at any place on the circuit, which is connected to the mains supply at each end and so forms a 'ring'. Current can be supplied to the sockets from each end of the ring main, and this arrangement allows the wiring to be of a lower current rating and smaller diameter than is required for a simple 'radial' circuit connected to the supply at one end only. To prevent a ring main being overloaded, there are limits on the number of sockets and the area of building that it may serve. It is common for a house to have a separate ring main for each floor.

Within a building there is an extra conductor in the cables which does not need to exist in the street supply. This earth conductor provides a connection between the metal casing of any appliance and the ground to help protect people from electrical shock. In the case of a fault where the live supply becomes accidentally connected to the metal casing of an appliance, the *earth connection* provides a path to ground which is more efficient than any human body also touching the appliance. Large currents will instantly flow and cause the overloaded fuses or circuit breakers to react and isolate the appliance or circuit from the supply.

A more modern method of disconnecting supplies when there is danger is by the use of *residual current* devices. These use electronic circuits to monitor the current flow to and from the appliance and quickly disconnect the circuit if an incorrect flow is detected. They can be installed at the consumer unit or used as small plug-in devices when using outdoor electrical equipment such as lawnmowers and hedge clippers.

EXERCISES

1. A transformer with 200 turns in the primary winding is to step-up voltage from 12 V to 230 V. Assume that the transformer is 100 per cent efficient.
   (a) Calculate the number of turns needed in the secondary winding.
   (b) Calculate the current flowing in the primary winding when a 100 W lamp is connected to the output.

2. The apparent power rating of an AC motor is 4000 VA and it has a power factor of 0.85.
   (a) Calculate the output power of the motor.
   (b) Calculate the current drawn from the 230 V mains.
   (c) Calculate the peak value of this current.

Answers are on page 326.

12

Water Supplies

CHAPTER OUTLINE

Living in a modern town or building, it is easy to forget that a fundamental factor when choosing a location to live is the availability of drinking water. Settlements are therefore traditionally established near a water supply, but at the same time they need to be dry and safe from the risk of flooding by water. The supply of water and the drainage of waste water is therefore a major design feature of our buildings and towns. Large financial and engineering investments are needed for the systems of collection, storage, treatment, distribution and disposal of water. The pumping of water necessary in a supply system also requires significant amounts of energy.

Humans need a small amount of essential drinking water, but much greater amounts are used for washing and waste disposal in homes, industry and commerce. According to United Nations Human Development reports, the average daily consumption of water per person ranges from around 40 litres (Nigeria) to over 500 litres (USA); the figure for the United Kingdom being about 150 litres.

The total supply of natural water in the earth is enormous and should be adequate for our needs, but local shortages do nonetheless occur, especially when droughts are combined with poor management of resources. Conservation of water used in buildings now plays an important part of strategies and codes for sustainable buildings. The water consumption in a high performance home should be less than 80 litres/person/day.

This chapter looks at the sources of water that are used for community water supplies, and describes the treatment and qualities that can be expected. By studying this chapter you will:

→

- ▶ know about natural water and the hydrological cycle
- ▶ understand the characteristics of water and possible impurities
- ▶ appreciate the links between the origins and properties of natural water supplies
- ▶ know about the hardness of water and methods of reducing hardness
- ▶ understand the methods used for the treatment of public water supplies
- ▶ know about methods for distributing water supplies to a community and within buildings.

The Resource 5 section of this book has useful background information about the concepts, terms and principles of water technology which are relevant to water supplies for towns and buildings. Topics that you can review in Resource 5 include:

- ▶ **Fluids at rest**: pressure, pressure gauges, pressure, forces caused by pressure.
- ▶ **Fluid flow**: laminar and turbulent flow, flow rate.
- ▶ **Fluid energy**: Bernoulli principle, flow measurement, venturimeter, flow in pipes, flow in open channels.

NATURAL WATERS

The world possesses a fixed amount of water, which is found in various natural forms, such as oceans, lakes, rivers, underground waters, ice caps, glaciers and rain. This water plays an important part in maintaining the balance of the world's weather, especially through the presence of water vapour in the atmosphere. Water is also essential for the growth of vegetation such as trees and food crops.

See also:

Resource 5

The hydrological cycle

A certain proportion of the world's natural water is involved in a continuous cycle of rainfall and evaporation. This hydrological water cycle, illustrated in Figure 12.1, is made up of the following stages.

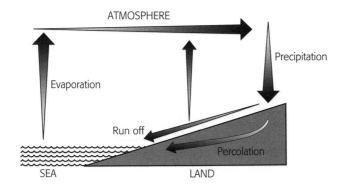

Figure 12.1 Water cycle

- **Evaporation**: when the sun shines upon a surface of water, such as the ocean, some of the water evaporates from the liquid state to water vapour. The sun also warms this water vapour which, becoming less dense than the cold air above it, rises into the atmosphere where it is circulated by weather patterns.
- **Condensation**: when the water vapour becomes cooled below its dew-point, it condenses and the liquid water appears as clouds of droplets.
- **Precipitation**: when the water droplets in a cloud are large enough, they fall as rain, snow or hail.
- **Run-off**: a proportion of the precipitated water falls on the land, where most of it flows towards the sea by two main routes: as surface water such as streams and rivers, and as groundwater which percolates through the materials below the surface of the land.

Plants participate in the hydrological cycle by absorbing moisture for the processes of their growth and by emitting water vapour in the process of transpiration. The energy for the hydrological cycle is provided by the Sun, which supplies the latent heat for evaporation and then raises huge masses of water upwards against gravity. The many components and mechanisms of the cycle are interdependent on one another and achieve a complex natural balance which could, perhaps, be upset by major changes in some parts of the cycle.

Chemical effects

Chemical terms

See also:

section on
*Chemical
processes* in
Resource 1

Some references identify chemicals by their common names, or by outdated chemical names. Table 12.1 lists equivalent names for some of the substances which are commonly referred to in descriptions of water supplies and water treatment.

Acids and alkalis

Acids and alkalis are classes of chemical compounds which are corrosive and dangerous when they are very strong, but many substances are weakly acidic or alkaline when they are dissolved in water.

- An *acid* is a substance which contains hydrogen that can be chemically replaced by other elements.
- An *alkali* or *base* is a substance which neutralises an acid by accepting hydrogen ions from it.

pH value

Acids have the ability to produce hydrogen ions, H+, when dissolved in water. The pH value is a measure of the acidity or alkalinity of a solution, rated on a scale that is related to the concentration of the hydrogen ions present.

pH 7 = neutral

- pH less than 7 indicates an acid.
- pH greater than 7 indicates an alkali.

Table 12.1 Chemical names

Chemical name	Chemical formula	Other names
Aluminium sulfate	$Al_2(SO_4)_3$	alum
Calcium carbonate	$CaCO_3$	chalk, limestone
Calcium hydroxide	$Ca(OH)_2$	lime (slaked, hydrated)
Calcium hydrogen carbonate	$Ca(HCO_3)_2$	calcium bicarbonate
Calcium sulfate	$CaSO_4$	gypsum
Magnesium hydrogen carbonate	$Mg(HCO_3)_2$	magnesium bicarbonate
Magnesium sulfate	$MgSO_4$	Epsom salt
Sodium carbonate	Na_2CO_3	soda
Sodium chloride	$NaCl$	common salt, brine
Sodium hydrogen carbonate	$NaHCO_3$	sodium bicarbonate

Note: sulfate is an agreed international chemical spelling but some English references may spell it sulphate.

The pH value of a sample can be found by chemical measurements, or by using specially calibrated electrical meters which detect the number of H^+ ions present.

Characteristics of natural water

Pure water has no colour, no taste or smell, and is neither acidic or alkaline. Water can dissolve many substances and is sometimes termed a 'universal solvent'. It is rare for natural water to be chemically pure; even rainwater dissolves carbon dioxide as it falls through the air. Water which flows over the surface of the land or through the ground comes into contact with many substances and takes some of these substances into solution or into suspension.

The impurities and qualities that may be found in natural waters can be conveniently described under the headings used below.

Inorganic matter

Dissolved inorganic chemicals, such as salts of calcium, magnesium and sodium cause the hardness in water, which is discussed in a later section of this chapter.

Suspended inorganic matter includes minute particles of sand and chalk, which do not dissolve in water. The particles are small enough to be evenly dispersed as a suspension that affects the colour and clarity of the water. A characteristic of a suspension is that it can be separated by settlement of the particles.

Organic matter

Dissolved organic materials usually have animal or vegetable origins, and the products of their decay include ammonia compounds. Suspended organic matter

Organic = based on carbon, such as animal and plant materials

can be minute particles of vegetable or animal origin such as fibres, fungi, hair and scales.

Micro-organisms

Diseases in humans are caused by small organisms such as certain bacteria, viruses and parasites. Some of these organisms can be carried by water if a supply is allowed to become contaminated. Examples of water-borne diseases include typhoid, cholera and dysentery.

Pollutants

Human activities add extra impurities to natural water, mainly in the form of waste from sewage systems and from industrial processes. Domestic sewage carries disease organisms which must not be allowed to contaminate water supplies. The detergent content of household sewage can also be high and difficult to remove.

Industrial wastes that can contaminate water supplies include toxic compounds containing cyanide, lead and mercury. Increased agricultural use of nitrogenous fertilizers, which are washed from fields or seep into the underground water sources, can lead to excessive nitrate compounds in the water supply. Certain levels of nitrates are thought to be a health hazard, especially to young children.

Acidity

Pure water is chemically pure, with a pH of 7, but natural water is invariably acidic or alkaline with a pH range of 5.5 to 8.5.

Acid   pH = 0–7
Alkali   pH = 7–14

Acidity in natural water is usually caused by dissolved carbon dioxide and dissolved organic substances such as peat. Acidic waters are corrosive and also cause *plumbo-solvency* where lead, a poison, is dissolved into the water from lead tanks or pipes. Alkalinity in natural water is more common than acidity. It is usually caused by the presence of hydrogen carbonates.

HARDNESS OF WATER

Some natural water contains substances that form a curdy precipitate or scum with soap. It is not possible to form a lather until enough soap has first been used

Table of typical values of hardness

Value in mg/litre	Hardness
0–50	soft water
100–150	slightly hard water
200 plus	hard water

in the reaction with the substances producing this 'hardness'. This property of water has conflicting effects because it causes problems with water appliances such as boilers but it is also associated with *mineral water* and good health.

Measurement of hardness

- **Hardness of water** is measured by the degree in which it is difficult to obtain a lather with soap.

> Unit: milligrammes per litre (mg/litre) of calcium carbonate ($CaCO_3$) irrespective of actual salts present.
>
> Other units: 1 part per million (ppm) is approximately 1 mg/litre;
>
> 1 'degree Clarke' is approximately 1 part per 70,000.

About 40 per cent of the public water supply in the United Kingdom is between 200 and 300 mg/litre. In general, hard water comes from underground sources or from surface water collected over ground that contains soluble salts such as carbonates and sulfates, in limestone areas for example. Soft water tends to come from surface water collected over impermeable ground, such as in granite areas.

Types of hardness

There are two main types of hardness of water, defined in the following sections. The differences between types of hardness are particularly relevant for the processes of softening water, which are described in a later section.

Temporary hardness

- **Temporary hardness** is hardness that can be removed by boiling.

Temporary hardness is usually caused by the presence in the water of the following salts.

- $CaCO_3$ calcium carbonate
- $MgCO_3$ magnesium carbonate.

The scale or 'fur' found inside kettles is the by-product of removing temporary hardness by boiling.

Permanent hardness

- **Permanent hardness** is hardness in water which cannot be removed by boiling.

Methods of removing permanent hardness are described later in the chapter. Permanent hardness is usually caused by the presence of the following salts.

- $CaSO_4$ calcium sulfate
- $MgSO_4$ magnesium sulfate

> **See also:**
>
> section on *Chemical processes* in Resource 1

- $CaCl_2$ calcium chloride
- $Mg Cl_2$ magnesium chloride

Consequences of water hardness

Hardness in water has the advantages and the disadvantages listed below. Public water supplies are *not* usually treated for hardness and the suitability of an untreated supply needs to be assessed for each application of the water. Bottles of special 'mineral water' are sold at premium prices *because* they contain the minerals that cause hard water.

Disadvantages of hard water

- Wastage of fuel occurs because of scale in boilers and pipes.
- Deterioration and damage to boilers and pipes is caused by scale.
- Wastage of soap and energy occurs before a lather forms.
- Increased wear occurs in textiles which have to be washed for longer periods.
- Industrial processes are affected by the chemicals in hard water.
- The preparation and final taste of food and drinks can be affected by hard water.

Advantages of hard water

Bottled water is usually hard water

- Less toxic lead is dissolved from pipes by hard water.
- 'Better taste' is usually a feature of hard water.
- Decreased incidence of heart disease appears to be associated with hard water.

SOURCES OF WATER

Rainfall is the original source of the water used for drinking. Part of the water evaporates from the earth soon after it falls as rain, part of it drains on the surface to join streams and rivers, and some percolates into the ground to feed underground supplies.

The balance of evaporation, surface water and underground water varies with the particular climate, the district and the time of year. A typical proportion is one-third evaporation, one-third run-off and one-third soak-in. A larger proportion of the rainwater is lost by evaporation during the summer.

Sources of water supply are usually classified by the routes that water has taken after rainfall. For supplies of drinking water the main categories are listed below and are described in the sections that follow.

- **Surface water**: examples include streams, rivers, lakes and reservoirs.
- **Underground water**: examples are springs and wells.
- **Rainwater collectors**: examples include roofs and paved surfaces.

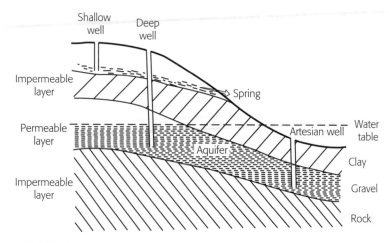

Figure 12.2 Underground water supplies

Underground water

When rain falls on soils or porous rocks such as limestone or sandstone, some of the water sinks into the ground. When this water reaches a lower layer of impervious material, such as clay or rock, it may be held in a depression or it may flow along the top of the impermeable layer. Such water-bearing layers are called *aquifers*, and a cross section of one is shown in Figure 12.2. The water table or *plane of saturation* is the natural level of the underground water. The water in some aquifers is 'confined' and held below the water table by an impermeable layer on top of the water.

Springs

A spring is a source of groundwater that occurs when geological conditions cause the water to emerge naturally, as shown in Figure 12.2. A simple land spring is fed by the surface water which has soaked into the subsoil, and its flow is likely to be intermittent. A main artesian spring taps water that has been flowing in an aquifer below the first impermeable layer.

Old farm houses on hillsides are often on the 'spring line'

The water from such a spring is usually hard, with a high standard of purity achieved because of the natural purification which occurs during the percolation through the ground. All springs need protection from contamination at their point of emergence.

Wells

Wells are a source of underground water but differ from springs as the water must be artificially tapped by boring down to the supply. Wells may be classified by the types listed below and shown in Figure 12.2.

- **Shallow wells**: shallow wells tap water near the surface.

You may not have to drill deep for a 'deep well'

- **Deep wells**: deep wells obtain water below the level of the first impermeable layer. Such a well is not always physically deep.
- **Artesian wells**: artesian wells deliver water under their own heads of pressure, because the plane of saturation is above the ground level.

The classification of wells as 'shallow' or 'deep' depends upon the sources of water and not upon the depth of their bore. Shallow wells may give good water but there is a risk of pollution from sources such as local cesspools, leaking drains and farmyards. Deep wells usually yield hard water of high purity. The construction of all wells must include measures to prevent contamination near the surface.

Surface water

Water collected in upland areas tends to be soft and of good quality, except for possible contamination by vegetation. As a stream or river flows along its course it receives drainage from farms, roads and towns and becomes progressively less pure. Many rivers receive sewage and industrial waste from towns and factories and are also required to supply fresh water to other towns. It is obviously important that the levels of pollution in rivers are controlled, and experience has shown that even rivers flowing through areas of heavy industry can be kept clean if they are managed correctly.

A flowing river tends to purify itself, especially if the flow is brisk and shallow. This self-purification is due to a combination of factors including oxidation of impurities, sedimentation of suspended material, the action of sunlight and dilution with cleaner water. Even so, water taken from a river for use in a large public supply usually needs treatment on a large scale and such water must be carefully analysed for its chemical and bacteriological content.

Rainwater harvesting

Rainwater harvesting is the collection and storage of rain from roofs of buildings or from other surface catchments. The rainwater is generally collected via a system of gutters and pipes and stored in rainwater tanks. In some countries, houses away from larger towns rely on rainwater collected from roofs as the only source of water for all household purposes. Even if there are other sources of water to a building, stored rainwater is good for flushing toilets, washing clothes and watering gardens. Modern systems of environmental assessments for sustainable buildings give credits for conservation of water by rainwater harvesting.

See also:

section on *Environmental appraisal of buildings* in Chapter 14

Rainwater collected over roofs can be contaminated with dirt and growths on the roofs, as well as by pollutants carried in the wind and the rain. The concentration of pollutants can be reduced by using mechanisms to send the first flush of rainwater to the drains. Relatively simple systems of settlement, filtration and disinfection can be used to treat stored rainwater for drinking.

WATER TREATMENT

The variety of types and qualities of natural waters described in the previous section indicates that there is a wide range of substances whose concentrations may need to be adjusted before water is used. The water for a public water supply is required to be 'wholesome', meaning that it is suitable for drinking. Good drinking water should be:

- harmless to health
- colourless
- clear
- sparkling
- odourless
- pleasant tasting.

Methods of water treatment

The principal techniques used for treatment of public water supplies are described in the following sections and can be summarised under the general headings given below.

- **Storage**: sedimentation and clarification.
- **Filtration**: slow sand filters, rapid sand filters, micro-strainers, membrane filters.
- **Disinfection**: chlorination and ozonisation.

The methods used for the treatment of a particular water depend upon whether it is in small supplies or bulk supplies, and whether it is needed for domestic or industrial use. In Britain, the entire water supply is usually made suitable for drinking, even though most of it is used for non-drinking purposes.

Many industrial processes require water with less mineral content than is acceptable for drinking water and further treatment stages, such as softening, are then necessary. The addition of chemical compounds containing metals such as copper and aluminium needs to carefully monitored and controlled.

The components of a typical water treatment works are shown in Figure 12.3.

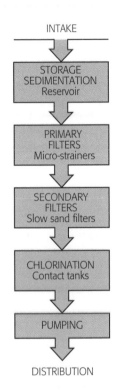

INTAKE

STORAGE
SEDIMENTATION
Reservoir

PRIMARY
FILTERS
Micro-strainers

SECONDARY
FILTERS
Slow sand filters

CHLORINATION
Contact tanks

PUMPING

DISTRIBUTION

Figure 12.3 Water treatment scheme

'potable' = drinkable

Water storage

Reservoirs are used to store reserves of water, and they are also an important preliminary stage of treatment. All contaminants in the water are diluted in their effect and different qualities of water are evened out. Pathogenic (disease-producing) bacteria tend to die when in storage because of lack of suitable food, the low temperature and the action of sunlight.

A disadvantage of prolonged storage of raw water is that it provides conditions for the growth of algae: minute plants that make subsequent water treatment difficult. Algae can be controlled by the careful addition of chemicals such as copper sulfate, and by the mixing of water in a reservoir system.

Sedimentation

Sedimentation is the gradual sinking of impurities that are suspended in the water. Simple sedimentation – the natural settling of suspended materials – takes place in reservoirs and also in specially designed settling tanks.

Clarification

Clarification is a system of chemically assisted sedimentation used for the removal of very fine suspended particles that do not settle naturally. A chemical such as aluminium sulfate (alum) produces a precipitate when it is added to the water. This precipitate coagulates with the suspended material to form a *floc*. This product of 'flocculation' then settles as a sediment, or it may be removed by mechanical collectors.

Water filtration

When water is passed through a fine material such as sand or a wire mesh, particles are removed from the water. Some filters, such as rapid sand filters, act only as a simple physical filter and the water also requires chemical treatment. Slow sand filters, however, combine a physical action with a chemical and a bacteriological action.

Slow sand filters

Slow sand filters are built in sunken rectangular basins, with 100 m × 40 m being a typical size. A cross section of a modern slow sand filter is shown in Figure 12.4. The floor of the filter bed contains a system of collector pipes and underdrains covered with a layer of graded gravel. Above the gravel is a layer of sand, about 600 mm deep, which is then covered with water to a depth of around 1 m.

Water slowly percolates downwards through the sand bed which develops a film of fine particles, micro-organisms, and microscopic plant life. It is this complex 'vital' layer (or *schmutzdecke*) which purifies the water by both physical and biological action. Because the growth of the active film reduces the rate of filtration, the head of water is gradually increased until, after a period of weeks, the bed has to be emptied for cleaning. The top 12 to 25 mm of sand is removed, washed and eventually used for replenishing the beds. A clean filter is 'charged' by slowly filling it with water from the bottom upwards and then allowing a new vital layer to form.

The slow sand filter is extremely effective and gives high-quality water, which needs little further treatment. These filters, however, occupy larger areas and work more slowly than other types of filter. The mechanical scraping and the cleaning of the sand has been made quicker and less labour-intensive by the use of machines.

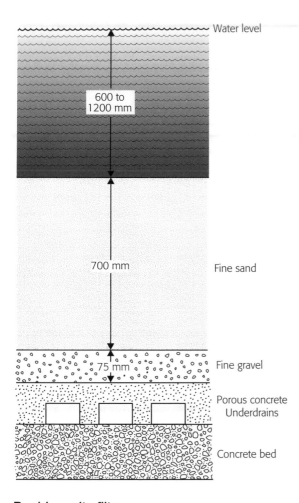

Water level

600 to
1200 mm

700 mm   Fine sand

75 mm   Fine gravel

Porous concrete
Underdrains

Concrete bed

**Figure 12.4 Slow
sand filter**

Rapid gravity filters

Rapid gravity sand filters are constructed in tanks which have dimensions up to 12 m × 3.9 m. A uniform bed of sand, about 600 mm thick, is supported on a bed of gravel of similar thickness. Beneath the gravel is a system for collecting water and also for passing air and water upwards. The filter is filled with water to a depth of 2 to 3 m.

The water passes downwards through the sand and the filtration is mainly by physical action. The filters are cleaned by draining and then blowing a strong scour of air followed by a backwash of clean water from below. Washouts are required at intervals of 12 to 48 hours but the procedure can be automated.

Rapid gravity filters work at a rate that is some 20 to 40 times faster than slow sand filters. Their construction is more compact than slow filters and they can be cheaper to install and to operate. A rapid gravity filter is much less effective than a slow sand filter and is usually used in conjunction with flocculated water. Rapid filters are also used as 'primary filters' which reduce the load on the slow sand filters which follow them.

Pressure filters

Pressure filters are contained in steel pressure vessels. The construction of the sand filter inside the vessel is similar to the open rapid gravity filters and the backwashing is carried out in a similar way. The whole of the pressure cylinder is kept filled with water so that pressure is not lost during filtration, and this type of filter can be inserted anywhere in a water main.

Micro-strainers

Micro-strainers are revolving drums with a very fine mesh of stainless steel wire or other material which is cleaned by water jets. The strainers are useful for screening stored waters which do not contain large amounts of suspended matter. By removing microscopic-sized particles of algae they can decrease the load on sand filters. Sometimes micro-strainers can produce water pure enough for sterilisation, without the need for filtration.

Membrane filters

Newer methods of micro-filtration or nano-filtration involve the use of membranes whose structure is so fine that they can physically sieve out bacteria and other organic pollutants such as herbicides. A membrane cartridge, for example, contains bundles of polymer fibres whose surfaces have tiny pores which are small enough to trap impurities greater than 0.1 micron. As the molecules of most bacteria are between 0.5 and 1.0 micron in diameter, this micro-filtration is capable of providing safe drinking water. Untreated water is pumped under pressure through collections of membrane cartridges arranged in banks. Like other filters the membranes have to be cleaned by backwashing the surfaces with water, and this process can be automated.

> 1 micron is 1 millionth of a metre

Disinfection

The disinfection of water supplies is intended to reduce harmful organisms, such as bacteria, to such very low levels that they are harmless. This safe quality needs to be maintained while the water is in the distribution system, including any reservoirs that store purified water. Disinfection can be achieved by a number of agents, but chlorine and ozone are usually employed to treat public water supplies.

Chlorination

Chlorine is a powerful oxidising agent which attacks many organic compounds and has the property of killing bacterial cells. The rate of disinfection of a particular type of water depends upon the chemical form of the chlorine and upon the length of time it is in contact with the water. The presence of ammonia in the water reduces the effectiveness of the chlorine, but the addition of ammonia after disinfection can be useful for reducing possible objectionable tastes produced by the chlorine.

The chlorine is usually kept liquefied in tanks from where it is injected into the water at a controlled rate. Because the sterilising effect is not instantaneous the water and the chlorine are held in contact tanks for a period of time, such as 30 minutes.

Ozone treatment

Ozone, O_3, is a form of oxygen with molecules that contain an extra atom. This form of oxygen is unstable and reactive, so ozone is a powerful oxidising agent which rapidly sterilises water. Ozone treatment is more expensive than chlorination as the ozone must be made at the treatment station. Air is dried and subjected to a high voltage electric discharge, and the ozone produced is then injected into the water.

Other water treatment techniques

Other techniques available for water treatment, usually on a small scale, include the ones outlined below. Some communities make use of distillation (desalination) or reverse osmosis to treat seawater.

- **Carbon filtering** uses the charcoal form of carbon, which has the property of 'adsorbing' many chemical compounds, including some toxins. Activated charcoal is commonly used in household water filters.
- **Boiling water** when at or near sea level eliminates or inactivates most micro-organisms that exist at room temperature. However, boiled water without disinfection agents added can acquire new micro-organisms during storage.
- **Distillation** involves boiling the water to produce water vapour which then condenses as a liquid when it contacts a cool surface. The impurities dissolved in the water are not normally vaporised and remain behind.
- **Reverse osmosis** applies mechanical pressure to untreated water and forces pure water through a semi-permeable membrane material that discriminates about which molecules and ions are passed.

Softening of water

Hard water is satisfactory, and even desirable, for drinking, but it also has disadvantages as described previously. It is *not* usual practice to soften public supplies of water but softening may be necessary for industrial supplies. There are several different principles used for changing hard water into soft water.

- **Precipitation**: precipitation methods act by completely removing most of the hardness compounds that are present in the water. Chemicals are added to the hard water to form insoluble precipitates which can then be removed by sedimentation and filtration.
- **Base exchange**: base exchange methods act by changing hardness compounds into other compounds which do not cause hardness.
- **Demineralisation**: demineralisation is the complete removal of all chemicals

dissolved in the water. This can be achieved by an ion-exchange, which is a more complete process than the base-exchange.

Lime–soda treatment

'Lime' and 'soda' processes are precipitation methods of water softening. They depend upon chemical reactions to make the calcium and magnesium content of the hard water become insoluble precipitates which can then be removed. The following softening hardness is treated by *lime*

See also:

section on *Chemical processes* in Resource 1

$$Ca(HCO_3)_2 + \underset{\text{slaked lime}}{Ca(OH)_2} \longrightarrow \underset{\text{precipitate}}{2CaCO_3} + 2H_2O$$

Calcium permanent hardness is treated by *soda*

$$CaSO_4 + Na_2CO_3 \longrightarrow CACO_3 + Na_2SO_4$$

or

$$CaCl_2 + \underset{\text{soda ash}}{Na_2CO_3} \longrightarrow \underset{\text{precipitate}}{CACO_3} + 2NaCl$$

Magnesium temporary hardness is treated by *lime*

$$Mg(HCO_3)_2 + \underset{\text{slaked lime}}{2Ca(OH)_2} \longrightarrow \underset{\text{precipitate}}{2CaCO_3} + Mg(OH)_2 + 2H_2O$$

Magnesium permanent hardness is treated with *lime and soda*

$$MgSO_4 + Ca(OH)_2 \longrightarrow Mg(OH)_2 + CaSO_4$$

or

$$MgCl_2 + \underset{\text{slaked lime}}{Ca(OH)_2} \longrightarrow \underset{\text{treat as for calcium}}{Mg(OH)_2} + CaCl_2$$

In a lime–soda softening plant the chemicals are measured and added to the water, either dry or in a slurry. The solids formed in the precipitation are removed by settlement in tanks, or by flocculation, and the water is then filtered. Lime–soda water softening processes have relatively low running costs but produce large quantities of sludge which require disposal.

Base-exchange method

In the base-exchange method of water softening, the hardness-forming calcium and magnesium salts are converted to sodium salts which do not cause hardness. *Zeolites* are the special materials that act as a medium for the exchange of ions.

calcium sulfate + sodium zeolite $\longrightarrow$ calcium zeolite + sodium sulfate
(hard)   (soft)

The exhausted zeolite is 'regenerated' with sodium from a salt solution.

calcium zeolite + sodium chloride $\longrightarrow$ sodium zeolite + calcium chloride

Natural zeolites are obtained from processed sands and synthetic zeolites are made from organic resins such as those of polystyrene. The water softener is usually a metal cylinder, constructed like a pressure filter, in which the water passes downwards through a bed of zeolite. When regeneration is necessary dilute brine is passed through the bed, followed by a freshwater wash.

The base-exchange process can be operated simply and automatically but its costs depend upon the availability of salt. The process is used for commercial supplies such as for boilers and laundries, and for household water softeners.

Scale prevention

The build-up of scale or 'fur' within boilers and pipes can be minimised by techniques which do not actually soften the hard water. Instead, the crystalline growth of the scale is inhibited or interrupted so that it does not form on surfaces.

These techniques include the use of chemicals, usually phosphates, such as those added to the closed circuit of a hot-water heating system. Non-chemical methods include the use of mechanical vibrations, electrical and magnetic effects.

Magnetic water treatment

In magnetic water treatment the hard water flows through a 'water conditioner' which applies a strong magnetic field to the water. When this water is heated the hard water products remain as microscopic particles suspended in the water instead of forming scale. The products are then carried by the movement of the water or else form a movable sediment.

WATER INSTALLATIONS

Distribution to buildings

After water has been treated it is pumped to the start of a local distribution system, which is often a high-level storage reservoir or a water tower. Water can then be supplied by gravity through iron pipes or polymer pipes (coloured blue) beneath the streets. These pipes flow full of water under pressure, so secure joints are important in order to prevent leakage. Damage to pipes caused by age, ground movement or heavy traffic can cause a serious wastage of water. In general, modern systems use pressure energy to bring water supplies into buildings using pipes under pressure, and use gravity to take waste water away in drain pipes which are not under pressure.

Domestic water installations

The system for distributing water within a building needs to take account of the following factors:

- sufficient capacity for purpose
- leakproof pipework
- means of isolating pipework appliances
- means of draining pipework and appliances
- arrangements for overflows
- prevention of back pollution to the public supply
- compliance with regulations.

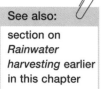

See also:

section on *Rainwater harvesting* earlier in this chapter

Modern regulations and guides, such as the *Code for Sustainable Homes*, also aim to reduce the consumption of potable water in buildings through the use of efficient water fittings and appliances and by the use of water recycling systems. Techniques include low-flush WCs, flow restrictors in pipes or taps, rainwater

Table 12.2 Features of water installation

Installation feature	Purpose
Street valve	Emergency isolation from street supply
Meter	Measurement of water consumption
Storage cistern	Constant pressure
	Continuity of supply
	Isolation from public supply
Drain-off valve	Means of draining system
Overflow pipe	To divert overflow water outside
22 mm and 15 mm diameter service pipes	To provide appropriate capacity
Isolating valves	To service appliances
Indirect cold water supply	*Direct cold water supply*
water stored in cistern at higher level	no storage of water involved
mains is fed to storage cistern	mains supply is fed directly to all outlets
cold taps not for drinking	all cold taps suitable for drinking
demand from mains is smoothed	higher peak demands on mains
building is protected from mains failure	risk of back-syphonage to mains
more plumbing installation and higher costs	less plumbing installation and lower costs

recycling, and *grey-water recycling* which reuses bath and shower water for toilet flushing.

Once inside the building the cold water can be distributed to various points by two main methods: *direct* systems, where water is taken straight 'up' from the mains, and *indirect* systems where it is taken 'down' from storage tanks, often in the roof space. The features of the different methods of water supply are listed in Table 12.2.

13

Waste Water

CHAPTER OUTLINE

Waste water or sewage is the material that we drain into the sewers and take away for treatment. In addition to the material flushed from toilets, waste water includes the material from other drains in houses, businesses and industries. Although sewage is usually over 99 per cent water, it also contains a variety of natural and synthetic materials that are damaging to health and the environment unless they are treated. The aim of waste water systems and sewage treatment is therefore to take the waste safely away from buildings, to convert it to materials that are not harmful, and to dispose of those materials safely.

This chapter describes the nature of waste water and the systems currently used to carry it and to treat it. By studying this chapter you will:

▶ know the terminology and technology options for drainage systems
▶ understand the significance of drainage design options
▶ appreciate the purpose and principles of sustainable drainage systems
▶ know the chemical and biological components of waste water
▶ be familiar with the chemical and biological processes relating to waste water
▶ understand the aims of waste water treatment and technology options available
▶ be familiar with the operating principles of large water-treatment plants
▶ understand the options for smaller-scale and natural approaches to waste water treatment.

Traditional systems for waste water treatment are large and expensive, and create materials that require special disposal. It should be noted that some communities question the need for large-scale treatment systems and are seeking and implementing sustainable alternatives. The final sections outline some of the alternative options for the treatment of waste water.

The Resource 5 section of this book has useful background information about the concepts, terms and principles of water technology which are relevant to water supplies for towns and buildings. Topics that you can review in Resource 5 include:

▶ **Fluids at rest**: pressure, pressure gauges, pressure, forces caused by pressure.
▶ **Fluid flow**: laminar and turbulent flow, flow rate,
▶ **Fluid energy:** Bernoulli principle, flow measurement, venturimeter, flow in pipes, flow in open channels.

DRAINAGE SYSTEMS

Fresh water is brought into modern buildings by a system of pipes which are full of water under pressure; the pressure is supplied by central pumping stations which use large amounts of energy. To take waste water away from buildings we normally use drains where the water flows by gravity and carries along small amounts of solids. There is usually no pumping of this waste water except at some specialised places in the sewerage system.

When liquid flows in drains, the pipe or channel is only partly full and not subject to the high pressures of supply pipes. Therefore joints between parts of drains are easier to make, but the gravity flow requires a constant sloping gradient which needs careful design and installation. In addition to correctly moving the waste water, a drainage system also needs to satisfy the following technical specifications:

- protection from health risk
- protection from foul air
- sufficient capacity for maximum design flow
- restriction of pressure effects such as suction and compression
- encouragement of smooth non-turbulent flow
- self-cleansing of system by normal flows
- minimal chance of blockage
- easy access to all parts
- long-term resistance to effects of waste water
- long-term resistance to leakage
- protection from effects of ground pressures and movement
- protection from extremes of weather.

Drainage systems also need to manage water draining from the ground surfaces around buildings, which in developed areas are generally hard surfaces that do not absorb water. During high rainfall events these impervious surfaces can pass surface water into the drains so rapidly that the drainage capacity of an area is exceeded, and serious flooding of property occurs. The section about sustainable

See also:
Resource 5

See also:
section on *Water treatment* in Chapter 12 and on *Fluid flow* in Resource 5

COMBINED DRAINAGE SYSTEM

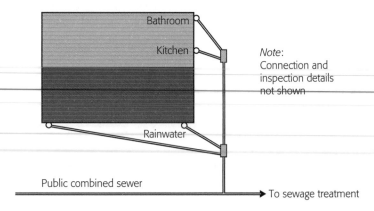

Bathroom

Kitchen

Note:
Connection and
inspection details
not shown

Rainwater

Public combined sewer

To sewage treatment

SEPARATE DRAINAGE SYSTEM

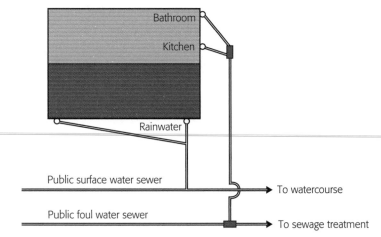

Bathroom

Kitchen

Rainwater

Public surface water sewer

To watercourse

Public foul water sewer

To sewage treatment

Figure 13.1 Drainage systems

Drainage terms
Drain:
a below-ground
pipe or closed
channel where
water flows by
gravity
Soil water:
waste water from
WCs
Foul water:
mixture of soil
water and other
waste water
*Surface or storm
water*:
clean water
collected from
roofs and paved
surfaces

drainage systems later in this chapter outlines how this surface water can be better controlled.

Types of drainage system

The drainage system in and around buildings in built-up areas usually connects to a community drainage system. In country areas which do not have any system of public sewerage, the rainwater is usually directed straight to *soakaways* in the ground, while the foul water is directed to a *septic tank*, as described later in this chapter.

Traditionally, community drainage has been of two main types of system depending on whether rainwater and surface water is combined with foul water or kept separate from it. The features of these two systems are described below and illustrated in Figure 13.1. Some communities may have a foul water sewer and leave buildings to deal locally with rainwater and surface water. When systems are correctly designed, this is a more sustainable approach.

Separate drainage system

The features of a separate drainage system are:

- Surface water is kept in separate drains and sewers from foul water.
- Roadways contain two sewers: foul water and surface water.
- Surface water remains clean and needs no treatment.
- Surface water can be managed separately in a sustainable fashion.
- The cost of building a double system of drainage is higher.
- There are lower volumes and costs at sewage treatment works.

Combined drainage systems

The features of a combined drainage system are:

- A historical system developed for crowded cities.
- Surface water shares the same drains and sewers as foul water.
- Installation at the building is cheaper and easier.
- Roadways contain one combined sewer.
- High rainfall events cause overflows that contaminate streets and waterways with foul water.
- High rainfall events overburden sewage treatment works.
- There are higher volumes and costs at sewage treatment works.

Sustainable drainage systems

Sustainable drainage systems (SuDS) are an approach to managing surface water by management practices and installations which reduce the volume and the runoff rate of surface water. Such systems help minimise the impact of the built environment on the natural environment and also play a major role in flood control in the wider area. Systems for assessing *sustainable buildings* have criteria for surface water management. Some principles and features of sustainable drainage systems are given below.

- **Local management**: the surface water is disposed of near the building. Techniques include rainwater harvesting, soakaways, garden use, porous/pervious paving, green roofs.
- **Detention:** the surface water is held locally until the high rainfall event has

> *Drainage terms*
> *Sewage*:
> mixtures of foul and surface waters which need to be drained from a building
> *Sewer*:
> form of drain which collects foul water from a number of drains
> *Sewerage*:
> a network of sewers that disposes of sewage from a community

> The construction of *combined systems* is now uncommon, or not allowed

Drainage *swales* are depressions in the ground designed to hold and absorb surface water.

passed. Techniques include attenuation tanks at the building, ponds, swales (depressions), detention basins and wetland areas.

NATURE OF WASTE WATER

Waste water or sewage is the liquid material which we take away from buildings and their surroundings. The typical sources of waste water from a building are:

- bathroom waste
- kitchen waste
- industrial waste.
- runoff from yards and roads
- rain from roofs and yards.

Sewage is therefore mainly composed of water but it also contains a variety of natural and synthetic materials that are damaging to health and the environment unless they are treated. The solid materials in raw sewage are usually a mixture of faeces, food, grease, oils, soap, detergents, metals, plastic and grit. A typical percentage composition is shown in Table 13.1 where the figures confirm the fact that sewage is mainly water.

The 'strength' of sewage depends upon local conditions and habits, and it changes during the day and the year. The liquid is normally grey to brown in colour with small but visible particles of organic material. Darker sewage usually indicates that anaerobic processes have started and these processes give pungent odours.

Table 13.1 Composition of typical raw sewage

Water 99.9%	Water proportion varies with time and local conditions, but is always high	
Solids 0.01%	Organic 70%	Proteins
		Carbohydrates
		Fats
	Inorganic 30%	Salts
		Metals
		Grit

Biological factors

See also:

Chemical processes in Resource 1

Small organisms or microbes are present in most natural water systems such as streams and ponds. They are present in greater amounts in waste water and are put to good use during biological stages in the treatment of sewage. However, some types of micro-organisms cause disease and must not be allowed to contaminate living areas or clean water supplies.

Micro-organisms such as bacteria react readily with the organic matter in water and therefore break down the remains of dead plants and animals. When organic pollution increases, the micro-organisms also increase their growth and metabolism. This biological reaction allows natural waters to purify themselves and is also used in sewage treatment, where conditions are provided to encourage maximum reactions. Microbes are also capable of breaking down toxic waste, such as an oil spill, if it has an organic origin.

The metabolism of a micro-organism requires sources of energy and carbon in order to make new cells, and the organic pollutants in water can supply these nutrients for the micro-organism, as shown in the equation below. Some micro-organisms take their nutrients by reacting with inorganic matter and using carbon dioxide as a source of carbon.

Pathogen:
organism that
causes a disease

$$\text{organic matter} + \text{oxygen} \xrightarrow{\text{(Micro-organisms)}} \begin{array}{l} \text{aerobic micro-organisms} \\ + \text{ carbon dioxide} \\ + \text{ water} + \text{ammonia} \end{array}$$

Organic: based on
carbon
Inorganic:
non-carbon
based

Oxidation processes

The process of *oxidation* is common to the various biochemical reactions involved in waste water treatment. This process may use free oxygen, such as the oxygen dissolved in the water, or use oxygen chemically present in other compounds such as sulfates (sulphates) (SO_4) or nitrates (NO_3).

- **Aerobic processes** take place in the presence of free oxygen.
- **Anaerobic processes** take place in the absence of free oxygen.

Water pollution

Various measures are available to assess water quality and to monitor pollution levels. Biological indices use observations of the presence or absence of certain biological species and the diversity of certain animal and plant communities. Physical and chemical qualities of water can be measured by many indices and oxygen content, described below, is an important chemical indicator.

Oxygen content

The quality of water in a river or other body of water depends greatly upon the oxygen content of the water. If organic pollutants enter the water they use dissolved oxygen from the water and create an oxygen deficit. Natural waters will recover oxygen levels if given time, but not if they are consistently overloaded with pollutants. As pollution increases, the levels of dissolved oxygen decrease and eventually fish, which breathe the dissolved oxygen, cannot live in that water. When fish return to city rivers, like the Thames in London, it is a sign of improved water quality and recovery from pollution.

The oxygen content of a water sample can be assessed by measurements that use the reactions of micro-organisms with oxygen. The following indices are common, and typical levels are given in Table 13.2.

Table 13.2 Typical BOD levels

Surface	BOD mg/l
Farm slurry	50 000
Raw domestic sewage	300
Treated domestic sewage	30
Polluted river	20
Large lowland river	5
Upland stream	1

Note: these figures vary with sources and dilution.

Higher BOD readings mean higher pollution

- A **biochemical oxygen demand** (BOD) test is a laboratory measurement that indicates the amount of dissolved oxygen used by the sample under standard laboratory conditions.
- A **chemical oxygen demand** (COD) test is a laboratory measurement that includes non-biodegradable compounds present in a sample, and can be used to supplement the BOD test.

WASTE WATER TREATMENT

Engineer *Joseph Bazalgette* designed the London sewerage system following 'The Great Stink' of 1858

The system of sewage collection and treatment for a town or district usually involves careful consultation, appropriate design and a major community investment. The purpose of sewage treatment can be summarised by the aims given below:

- to protect public health
- to protect the environment
- to convert waste water into stable end-products
- to dispose of the end-products in a safe manner
- to recover and recycle materials if possible
- to provide a service which is reliable and regular
- to operate without nuisance or offence
- to provide an economic system
- to comply with appropriate standards and legislation.

Some of the technical and economic factors may need to be balanced against each other while maintaining absolute safety about health. Various sets of national and international standards and legislation reinforce the health and environmental aspects of sewage treatment.

Sewage treatment processes

The techniques and processes available for sewage treatment can be grouped in terms of their operating principles under the following types.

- **Physical processes**. Examples: screening, sedimentation, flotations, filtration, centrifugation, reverse osmosis, micro-filtration.
- **Chemical processes**. Examples: neutralisation, precipitation, oxidation-reduction, ion-exchange.
- **Biological processes**. Examples: biological filtration, activated sludge, stabilisation ponds, anaerobic digestion.

Sewage treatment plants

The practical arrangements involved in large-scale sewage treatment are described in the following sections, and a typical scheme is shown in Figure 13.2. A sewage treatment plant is a combination of units that are installed to carry out particular treatment processes. The exact combination of treatment plant units depends upon the nature of the input and the objectives for quality of the water system that is receiving the effluent from the treatment plant.

> *Sewage*: the material
> *Sewer*: the drain
> *Sewerage*: the network

Preliminary treatment

Preliminary treatment is needed to remove large solids and grit which may cause blockages and damage to other parts of the treatment plant. Screens made of steel bars about 25 mm apart are commonly used to remove solids such as rags, paper, wood and plastic. Finer screens, when needed, may be made of stainless steel wire mesh.

Grit, such as sand, gravel and small fragments of glass and metal, is usually removed by a process of rapid sedimentation which allows the smaller solids to remain suspended and pass to the next stage. The process is carried out by equipment such as *vortex separators* which use centrifugal force to throw the grit to the sides of a circular chamber.

Other forms of preliminary treatment may remove oil and grease if they are present in large quantities. Outputs from some industries require pre-treatment before they enter a treatment system.

Primary treatment

Primary treatment uses sedimentation tanks in which the waste water is moved at a velocity which allows the fine solids to fall out of suspension by gravity. These solids form a removable sludge at the base of the tank. Sedimentation tanks also allow floatable materials, such as oil and grease, to form a scum on the surface. Scrapper blades or skimming devices move across the surface of the tank and separate the scum. Figure 13.3 is a simplified view of an arrangement for a sedimentation tank.

> *Primary treatment* mainly uses physical processes

Primary treatment can reduce the BOD measurement by up to 40 per cent and reduce the suspended solids by up to 70 per cent. The output from the primary treatment tank contains only dissolved material and fine suspended (colloidal) material.

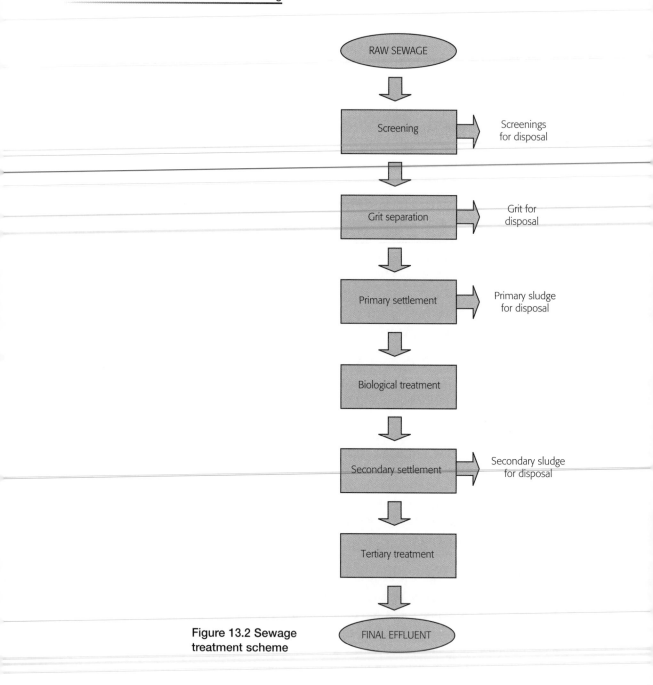

Figure 13.2 Sewage
treatment scheme

Secondary treatment

*Secondary
treatment* mainly
uses biological
processes

The aim of secondary treatment is to expose the settled sewage waste to biochemical reactions under aerobic conditions. The purified waste water is then separated by secondary settlement tanks. A variety of plants can be used for the process and two common choices are described below.

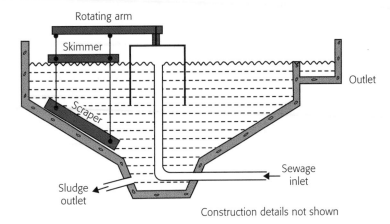

Figure 13.3 Primary sedimentation tank

- **Biological filtration**: the waste water is trickled through a bed of inert material on which a biomass of micro-organisms develops. Aeration occurs naturally.
- **Activated sludge**: the sewage is aerated in tanks by mechanical agitators which keep oxygen levels high. New biomass or sludge develops and some of it is returned to the aeration tank to seed or 'activate' the raw sewage.

In both cases the aerobic bacteria are able to grow and to purify the sewage. Secondary sedimentation tanks separate the biomass or 'humus' from the effluent, in which the BOD reading is usually less than 20 mg/l.

Tertiary treatment

A third stage of treatment is used if the quality of the effluent is not good enough to meet requirements or standards. For example, the level of suspended solids in the effluent may be too high. Methods of tertiary treatment include:

- settlement in lagoons
- irrigation onto land
- filtration through beds of sand or gravel, fine metal filters.

Effluents may also be disinfected if there is a danger of pathogens affecting nearby activities, such as bathing. Techniques used are the same as those available for the treatment of water supplies. Chlorination is a simple method of disinfection, while the use of ozone (O_3) or ultraviolet (UV) light are more complex treatments that cause less environmental damage than chlorination.

Sludge treatment

The treatment of sewage produces two major components: liquid effluent and sludge solids. After treatment to suitable standards the effluent can be safely returned to surface water systems, such as rivers, but the sludge has to have further treatment and then be transported for disposal. The costs of the sludge treatment can represent over 50 per cent of the capital and operating costs for sewage treatment.

The sludge from primary and secondary treatment processes typically has a water content greater than 90 per cent and needs treatment to reduce this water content. The exact methods of treatment depend upon the nature of the sludge and the proposed disposal method. The following processes are commonly used.

- **Thickening**: a process of reducing the volume of sludge, often by gravity settlement. As the sludge concentration increases, the volume typically decreases by three to five times.
- **Stabilisation**: a process which prevents anaerobic breakdown of the sludge and subsequent offensive odours. Anaerobic digestion is a common method which utilises the organic matter in the sludge. Chemical stabilisation is an alternative method involving the addition of lime (calcium hydroxide) to the sludge, thereby creating an alkaline environment unsuitable for micro-organisms.
- **Dewatering**: a process of reducing the water content of sludge by physical methods including drying, filtration, squeezing, centrifugal action and natural compaction.

Sludge disposal

The amount of sewage sludge in European Union countries is increasing in volume as regulations require all towns to treat their sewage. The disposal of sewage sludge remains a technical and environmental challenge that is currently met by the options described below.

Landfill

The disposal of sewage sludge directly to landfill sites has been a common practice but has the following undesirable characteristics:

- contamination of waters by leaching from the landfill
- risk to public health by pathogen transfer
- increased methane production during decomposition
- environmental pollution by odour, flies and transport.

Land disposal

Sewage sludge can be applied to farmland as a soil conditioner and fertilizer. The value of sludge depends on the treatment it receives, but it is rich in organic matter and nutrients such as nitrogen (N) and phosphorus (P). The sludge may need to be balanced with other nutrients and cannot usually be spread onto farmland all year long.

Sea disposal

In the past it was convenient for some communities near the sea to dispose of sewage sludge by loading it onto special boats and discharging it out at sea. The European Community no longer permits this method of sludge disposal. Some towns near the coast used to discharge raw sewage into the sea but they must now carry out full primary and secondary treatment of sewage, and these changes also increase the total quantities of sludge which need disposal.

Incineration

Sludge contains considerable quantities of combustible materials and, once dried, it can be burnt. During incineration of sludge the organic and volatile components, including toxic compounds, are destroyed. The inert ash that remains is a hazardous waste that still needs to be disposed of via a controlled waste management site. Incinerators have a high capital cost but are increasingly being used by large cities. There are opportunities to mix sludge with refuse material and to use energy from the burning for heat and electrical power.

OPTIONS FOR SEWAGE TREATMENT

This section describes methods of treating that can be alternatives to the large-scale treatment plants described in the previous section. These methods share the physical, chemical and biological processes used in large sewage treatment plants and may be more appropriate for the following situations:

- single buildings
- isolated groups of buildings
- buildings and communities which lack the resources for building or maintaining treatment plants
- communities who wish to use more natural methods of sewage treatment
- communities who have enough unused land to put aside for alternative treatments
- initial treatments of strong industrial waste waters
- tertiary treatment of effluent from some treatment plants.

Cesspool

A cesspool is a watertight underground container used for the storage of household sewage. No treatment of the sewage occurs and the tank must be periodically emptied and the sewage taken away for treatment. Cesspools are constructed in a variety of ways including *in situ* concrete, and prefabricated plastic and fibreglass which are usually set into concrete. They are usually used for single dwellings or small groups of houses and require no resources or monitoring apart from emptying. However, the running cost of emptying the tank is high and large underground tanks can be expensive to construct or install.

> A good sustainable principle is to treat all domestic sewage at source

Septic tank

A septic tank is a small-scale sewage treatment plant in the form of an underground tank in which treatment takes place and from which there is a continuous discharge of liquid. The tank contains separate zones which allow solids to settle and form a sludge layer where anaerobic processes decompose the sewage. Lighter materials may form a scum layer which helps prevent oxygen transfer and provides useful thermal insulation. A septic tank acts as a primary treatment and the effluent from the tank needs further treatment such as a secondary filter or percolation into the soil via a system of percolated drainage pipes.

Septic tanks are usually rectangular in shape if constructed on site, as shown in Figure 13.4, or they may be supplied as rounded prefabricated tanks in plastic or glass-reinforced plastic. Septic tanks usually require no particular maintenance apart from regular removal of sludge. After desludging, a small percentage of sludge is kept in the tank to 'seed' the new sludge and the tank is filled with water. Septic tanks are commonly used for isolated single dwellings but they can also be used for small rural communities such as a village. A reed bed might be used following a bank of septic tanks arranged so they can be maintained without interrupting operation.

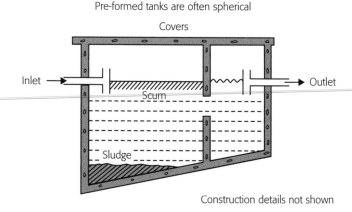

Figure 13.4 Septic tank

Stabilisation ponds

A stabilisation pond or lagoon is any enclosed body of water where organic waste is allowed to oxidise by natural activity. Ponds can be simply contained by earth embankments and used in combination if necessary. They are easiest to operate where there is plenty of sunshine and land available; they are simple to maintain and very effective at removing dangerous pathogens.

Stabilisation ponds have been used since early times, are the most common form of treatment in developing countries and are of increased interest to all

Aerobic =
presence of free oxygen
Anaerobic =
absence of free oxygen

communities. The operation of ponds can be loosely grouped as follows, according to their method of operation.

- **Anaerobic pond**s: the formation of sludge and a top crust creates good anaerobic conditions. These are particularly suitable for stronger and thicker wastes.
- **Oxidation ponds**: these are aerobic systems where the oxygen is taken from the atmosphere and also supplied by the activity of algae in the ponds. The ponds are kept shallow (1 to 2 m) to allow maximum penetration of sunlight.
- **Aeration ponds**: in these ponds the oxygen is supplied by aerators and not by the action of algae.

Wetlands and reed beds

The natural processes associated with the growth of aquatic plants can be used in the treatment of waste water. These plants are generally submerged algae, floating plants such as hyacinth and emergent plants such as reeds. The development of micro-organisms around the plants and their root systems allows both aerobic and anaerobic bacteria to develop and to 'digest' the sewage products. This vegetation can encouraged to form in artificial *wetlands* or *reed beds,* and these can then be used for the secondary or tertiary treatment of sewage, the treatment of storm water and the stabilisation of sludge.

The plants are grown in beds of soil or gravel which are retained by an impervious lining. The by-products of the biological treatment processes are taken up by the plants and removed when the plants are harvested; metals, if present, get removed into the mud at the bottom of the bed. The beds do not need daily attention but the water levels, plant growth and harvesting do need to be managed. Reed beds are being used to treat industrial waste and are also well-suited for the treatment of domestic sewage from small isolated developments.

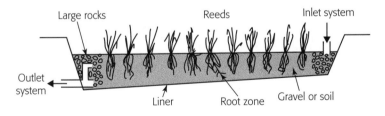

Figure 13.5 Reed bed

14

Green Buildings

CHAPTER OUTLINE

Environmental studies cover a wide range of topics and issues that involve worldwide research, government policies, academic courses and many related publications, This single chapter has to be modest and focus on linking the subject matter of this particular book about built environment topics to some of the wider topics of the total environment. The creation and improvement of buildings, townships and civil engineering projects such as transport systems do have long-lasting effects on the wider environment, in addition to their potential to improve our lives.

This chapter suggests how the correct design and use of buildings can help them exist in harmony with our wider environment. The contents of the chapter enable you to:

▶ understand the effects on buildings of macro- and microclimates
▶ make use of the terms and units for rain, wind and solar climate conditions
▶ appreciate the local and global resource and waste implications of construction activities
▶ understand the principles and techniques of conservation of energy associated with buildings
▶ appreciate the importance of the management of carbon emissions associated with buildings
▶ appreciate the various types of pollution and how construction activities can minimise polluting effects
▶ understand the causes of 'sick' buildings and cures associated with them

▶ understand the importance of sustainability and techniques for achieving sustainable construction

▶ appreciate the systems for evaluating and labelling sustainable buildings

▶ understand the concepts of zero energy and zero carbon buildings

▶ understand principles and techniques for designing and constructing satisfactory buildings for the future.

This chapter also complements the introduction to the environment given in the first chapter of the book about the built environment. The Resource sections contain supporting information that can be used for revision of principles and for extended investigation of topics.

CLIMATE AROUND BUILDINGS

A fundamental reason for the existence of a building is to provide shelter from the effects of climate, such as the cold and the heat, the wind and the rain. The climate for a building is the set of environmental conditions which surround it and link to the inside of the building by means of heat transfer.

Climate has significant effects on the energy performance of buildings in both winter and summer, and on the durability of the building fabric. Climates that are favourable to energy use and durability also make the external environment of a building attractive and useful for recreation. Although the overall features of the climate are beyond our control, the design of a building can have a significant influence on the climatic behaviour of the building. The following measures can be used to enhance the interaction between buildings and climate:

See also:

section on *The Built Environment* in Chapter 1, and Part II: Resources sections

- selection of site to avoid heights and hollows
- orientation of buildings to maximise or minimise solar gains
- spacing of buildings to avoid unwanted wind and shade effects
- design of windows to allow maximum daylight in buildings
- design of shade and windows to prevent solar overheating
- selection of trees and wall surfaces to shelter buildings from driving rain and snow
- selection of ground surfaces for dryness.

See also:

section on *Climate* in Chapter 1

Climate types

The large-scale climate of the Earth consists of interlinked physical systems powered by the energy of the Sun, and major climate types are explained in Chapter 1. The built environment generally involves the study of smaller systems for which the following terms are used.

- **Macroclimate**: the climate of a larger area, such as a region.
- **Microclimate**: the climate around a building and upon its surfaces.

A building site may have natural microclimates caused by the presence of hills, valleys, slopes, streams and other features. Buildings themselves create further microclimates by shading or drying the ground, and by disrupting the flow of wind. Further microclimates occur in different parts of the same building, such as parapets and corners, which receive unequal exposures to the sun, wind and rain.

Effects of microclimate

An improved microclimate around a building brings the following types of benefits:

- lower heating costs in winter
- reduction of overheating in summer
- longer life for building materials
- pleasant outdoor recreation areas
- better growth for plants and trees
- increased user satisfaction and value.

Climatic data

In order to design a building which is appropriate for its site, the climate of that site needs to be studied and predicted. The following climatic factors should be considered:

- temperatures
- wind speed and direction
- humidity
- sunshine hours and solar radiation
- precipitation of rain and snow
- atmospheric pollution.

These factors can vary by the hour, by the day and by the season. Some of the variations will cycle in a predictable manner like the Sun, but others such as wind and cloud cover will be less predictable in the short term. Information about aspects of climatic factors is collected over time and made available in a variety of data forms including the following information:

'Diurnal' = daily

- maximum or minimum values
- average values
- probabilities or frequencies.

The type of climatic data that is chosen depends upon design requirements. Peak values of maximum or minimum are needed for some purposes, such as sizing heating plant or designing wind loads. Longer-term averages, such as seasonal information, are needed for prediction of energy consumption. Some measurements in common use are described in the following sections.

Degree-days, accumulated temperature difference

The method of degree-days or accumulated temperature difference (ATD) is based on the fact that the indoor temperature of an unheated building is, on average, higher than the outdoor. For a traditional British construction the difference is taken to be 3°C.

In order to maintain an internal design temperature of 18.5°C, for example, the building only needs heating when the outdoor temperature falls below 15.5°C (18.5 – 3). This *base temperature* is used as a reference for counting the degrees of outside temperature drop and the number of days for which such a drop occurs.

- One day at 1°C below base temperature gives 1 degree-day.
- Two days at 1°C below base temperature gives 2 degree-days.
- One day at 2°C below base temperature also gives 2 degree-days.

The accumulated temperature difference total (degree-days) for a locality is a measure of climatic severity during a particular season, and some typical values are given in Table 14.1. These data, averaged over the years, can be used in the calculation of heat loss and energy consumption. ATD totals do not take account of extra heat losses caused by exposure to wind, or of heat gains from solar radiation.

Table 14.1 Climatic severity

Area		Degree-days
England	South-west	1800–2000
	South-east	2000–2100
	Midlands	2200–2400
	North	2300–2500
Wales		2000–2200
Scotland		2400–2600

Notes:
Accumulated temperature differences using base temperature of 15.5°C, September to May.
Increase degree-day values by 200 for each 100 m above sea level.

Driving-rain index, DRI

Rain does not always fall vertically upon a building, and when it drives sideways it can penetrate walls.

- The **driving-rain index** (DRI) is a combined annual measure of rainfall and wind speed.

Wind-driven rain (WDR) is an alternative term

The DRI is also associated with the moisture content of exposed masonry walls whose thermal properties, such as insulation, vary with moisture content. Damp walls have poorer insulation than dry walls. Driving rain is usually caused by storms but intense driving rain can also occur in heavy showers which last for minutes rather than hours. These conditions are more likely in exposed areas, such as coasts, where high rainfall is accompanied by high winds. Table 14.2 gives typical values of driving-rain index for different types of area.

Table 14.2 Driving-rain indices for British Isles

Exposure grading	Driving-rain index	Example
Sheltered	3 or less	Within towns
Moderate	3 to 7	Countryside
Severe	7 or more	West coastal areas

Note: High buildings, or buildings of any height on a hill, usually have an exposure one degree more than indicated.

Data are published, in the form of maps for example, which enable the driving-rain index to be predicted for different parts of Britain. Variations in microclimates and in types of building make an exact correlation difficult but, in general, it has been predicted that in a 'sheltered' region a one-brick-thick solid wall would not suffer from rain penetration. In addition to damp walls, rain penetration through poorly sealed windows often causes problems in areas of severe exposure.

Wind data

The main effects of wind on a building are those of force, heat loss and rain penetration. These factors need to be considered in the structural design and in the choice of building materials. The *wind chill* factor relates wind to the rate of heat loss from the human body rather than the loss from buildings. The unfavourable working conditions caused by wind chill have particular relevance to operations on exposed construction sites and tall buildings.

Wind speed

The force of a wind increases with the square of the velocity, so that a relatively small increase in wind speed produces a larger than expected force on a surface such as a building. The cooling effect of wind, measured by wind chill, also increases greatly with the speed of the wind. Typical wind speeds range between 0 m/s and 25 m/s, as described below.

- 5 m/s wind disturbs hair and clothing.
- 10 m/s wind, force felt on body.
- 15 m/s wind causes difficulty walking.
- 20 m/s wind blows people over.

See also:

section on *The Bernoulli principle* in Resource 5

The airflow around some parts of a building, especially over a pitched roof, may be high enough to provide an aerodynamic lifting force by using the Bernoulli principle described in Resource 5. This force can be strong enough to lift roofs and also to pull out windows on the downwind side of buildings.

Wind direction

The direction of the wind on a building affects both the structural design and the thermal design of the building. The directional data of wind for a location can be

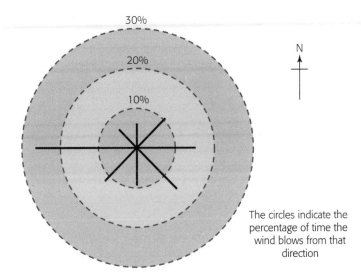

Figure 14.1 Simple directional wind rose showing dominant directions

shown by a wind 'rose' of arms around a point, with the direction north being at the top. The arms indicate the frequency or percentage of time that the wind blows from each direction.

The length of each arm of the rose can be proportional to the number of days that the wind blows from the direction shown by that particular arm. Directions with longer arms indicate prevailing winds that may affect energy consumption if, for example, they are generally colder winds. Other systems of wind roses can indicate wind speeds the thickness of the arms.

Wind effects around buildings

The existence of buildings can produce unpleasantly high winds at ground level. It is possible to estimate the ratio of these artificial wind speeds to the wind speed that would exist if the building were not present. A typical value of *wind speed ratio* around low buildings is 0.5, while around tall buildings the ratio might be as high as 2. A wind speed ratio of 2 will double normal wind speed. A maximum wind speed of 5 m/s is a suitable design figure for wind around buildings at pedestrian level.

> Wind speed
> *Typical design figure*: 5m/s
> maximum

Some general design rules for the reduction of wind effects are listed below.

• Reduce the dimensions, especially the height and the dimensions facing the prevailing wind.
• Avoid large cube-shapes.
• Use pitched roofs rather than flat roofs; use hips rather than gable ends.
• Avoid parallel rows of buildings.
• Avoid funnel-like gaps between buildings.
• Use trees, mounds and other landscape features to provide shelter.

Solar data

Some of the effects of sun on buildings are considered under the topics of *Climate* in Chapter 1, *Heat gains* in Chapter 3 and *Natural lighting* in Chapter 7. In order to predict solar effects on a building we need to know the following information about the Sun and the building:

- Sun position in the sky and the angle made with building surfaces
- quantity of radiant energy received upon the ground or other surface
- obstructions and reflections caused by clouds, landscape features and buildings.

Sun position

See also:

section on *Climate* in Chapter 1

The path that the Sun makes across the sky changes each day, but repeats in a predictable manner which has been recorded for centuries. For any position of the Sun, the angle that the solar radiation makes with the wall or roof of a building can be predicted by geometry. This angle of incidence has a large effect as the energy received as solar radiation obeys the *Cosine law of illumination*, described in Chapter 5, so that intensity is at a maximum when the radiation strikes a surface at right angles.

Figure 14.2 is a simple form of sunpath diagram showing the position of the Sun in the sky at different times of the year. Other graphical forms of sunpath diagram and 'sky maps' allow the dimensions and orientation of buildings and landscape to be plotted on the same diagram in order to predict the radiation on the building surfaces. Another design approach is to use a movable light source, known as a *heliodon*, which can imitate the movements of the Sun around a model of the proposed building.

Traditional methods of solar prediction are often replaced by *computer simulation* methods for assessing the directions and quantities of solar radiation falling

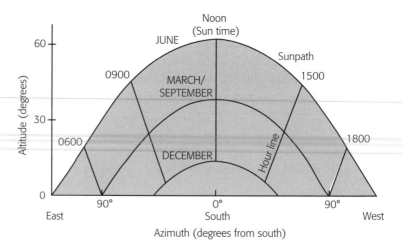

Figure 14.2 Sunpaths for southern England

upon a building. In addition to numerical results, these programs produce visualisations of the proposed building and its environment so that the effects of sunshine and shade can be seen for any time of the day or year.

Solar radiation

The intensity of solar radiation falling on a surface, such as the ground, can be measured in watts per square metre (W/m^2) of that surface. The watt is defined as 1 joule per second so this is an instantaneous measurement of the energy received per second on each square metre. When the solar energy is measured over a period of time, such as a day or year, the units will be joules or mega-joules per square metre (MJ/m^2). A typical figure for the intensity of solar radiation available on many parts of the earth's surface is 1000 W/m^2. This is a valuable potential source of energy, although only a fraction of it can usually be converted for use.

The solar radiation data for Southern England shows a June peak value of 850 W/m^2 received on a horizontal surface. When clouds obscure direct sunshine, significant amounts of diffuse radiation are still available. Annual totals of received solar radiation only vary by a factor of about 1.0 and 1.3 between the south and the north of the British Isles. Assessment procedures for energy use in buildings use solar flux values in a range between 30 and 70 W/m^2 for vertical surfaces averaged over a winter heating season.

RESOURCES FOR BUILDINGS

The activities and the materials associated with constructing the built environ-ment use the largest single portion of national resources. In established countries like the United Kingdom, for example, 5 to 8 tonnes of building materials per person per year are used in construction, and in some developing countries the figures are much higher. Most of this material is in the form of minerals from the ground, such as aggregate, and over 50 per cent of it is used for the repair and maintenance of existing buildings and infrastructure.

Resources:
land
energy
transport
minerals
timber

Mineral resources are abundant in the earth and timber is renewable so, if correctly managed, there should be raw materials available for construction in the future. However, there are bad practices in our current use of resources for construction. The manufacture and transport of materials can involve large amounts of energy, construction processes contribute to various forms of pollu-tion, and these processes also create a high percentage of the waste materials sent to landfill sites.

Although the raw materials for some building purposes are cheap and plenti-ful, they may have significant amounts of energy and pollution associated with their production. The process of making cement powder, for example, involves a high-temperature kiln, then processes for grinding and packaging the powder, followed by delivery to a building site. All these operations require an input of energy and also generate the carbon emissions associated with that energy. The

Relative energy content of construction materials:
plastics – very high
cement – high
brick and block – high
iron and steel – high
wood – medium
aggregate – low

traditional study of materials and buildings therefore needs to be extended to include all properties of construction, including the *embodied* energy of building materials.

Life cycle assessment (LCA)

Life cycle assessment (LCA), sometimes also termed life cycle analysis, is a technique for assessing the total environmental impact of construction materials by analysing their entire life cycle, from 'cradle to grave'. This type of assessment can be applied to many products, processes and services, and an International Standard (ISO 14040) defines how the life cycle assessment is made.

In the case of construction materials, a typical life cycle involves the following phases.

The BRE *Environmental Profile* of building materials and elements provides a 'cradle-to-grave' profile per tonne

- **Raw materials**: extraction and transport.
- **Production**: manufacture into building products.
- **Transport**: from factory to construction site.
- **Installation**: construction processes.
- **Service life**: impact and maintenance over lifetime of building.
- **End of life**: demolition, recycling and disposal.

At each phase of a life cycle the impacts on the environment are identified, analysed and given numerical units. Typical factors that are analysed for environmental impact are listed below.

- global warming potential
- use of metal ore, minerals and aggregates
- use of land resources
- use of water resources
- use of non-renewal energy
- use of renewable energy
- pollutants harmful to humans
- pollutants harmful to ecosystems
- treatment and disposal of waste products
- acid rain potential.

These environmental impact factors are described in various sections throughout this book.

Construction waste management

In addition to being major users of national resources, the processes of construction and demolition are major sources of waste of resources. Up to one third of the landfill waste in the United Kingdom has been generated by the construction industry. Fortunately there are also tremendous opportunities to reduce this waste, and the UK Strategy for Sustainable Construction in 2008 was able to set and achieve the ambitious target of a 50 per cent reduction in construction, demolition and excavation (CD&E) waste to landfill between 2008 and 2012.

Reductions in construction waste are achieved by the processes of reclaiming and recycling construction materials as described below.

- **Reclaimed materials**: used materials that are used again as construction materials without reprocessing. They may be cleaned or changed in size but are essentially the same material.
- **Recycled materials**: used materials that have been reprocessed to form all or part of a new product.

Table 14.3 Examples of reclaimed and recycled construction materials

	Reclaimed	Recycled
Bricks:	Bricks cleaned and used again in new construction	Bricks crushed and used for hardcore
Timber:	Timber sections or floorboards used again in new construction	Timber is chipped and used as part of a wooden panel product
Steel:	Steel cleaned and used again in new construction	Steel is used as a proportion of new steel

Site waste management plan

A site waste management plan (SWMP) is a plan that sets out the management of resources and waste control for all stages of a construction project. A SWMP includes the following information:

- Identity of those responsible for resource and waste management.
- Types of waste that will be generated.
- Identity of the contractors used remove the waste.
- Identity of the site to where the waste is taken.

For some projects in the United Kingdom it is compulsory to have a site waste management plan. However, an SWMP procedure will bring economic and environmental benefits to any project and is promoted as best practice.

CARBON AND ENERGY MANAGEMENT

Carbon (symbol C) is one of the most common elements in the universe and can be found on Earth occurring in natural states such as coal, soot, graphite and diamond. Although physically different, these forms of carbon are all chemically the same, and when carbon joins chemically with other elements it shows great chemical versatility. Carbon atoms can join together, with themselves and with the atoms of other elements, in long chains that give rise to a huge number of different chemical compounds. We call these compounds 'organic', and they are the chemical basis of all known forms of 'life' on Earth, including humans.

With carbon forming much of our bodies, we humans obviously need carbon and should perhaps be fond of carbon. Our process of breathing continually produces carbon dioxide (CO_2), which is also an ingredient in the vital photosynthesis process by which plants grow and develop. Our reliance on carbon should be kept in mind in when we discuss how best to 'manage' excessive carbon and carbon compounds like carbon dioxide.

See also:

Elements and compounds in Resource 1

Carbon management

The total amount of carbon on Earth is constant, some of it existing as pure forms of the element such as coal, but most of it existing in compounds with other elements, such as oxygen. The main 'reservoirs' for carbon in the environment are the atmosphere, the oceans, the ground and the biosphere of plants and animals, which includes humans.

See also:

Target CO₂ emission rate in Chapter 3

Carbon in the environment moves between these reservoirs by natural processes of the carbon cycle such as plant growth and decay, ocean warming and cooling. Most of this exchange is via the atmosphere where the carbon exists mainly as CO_2 gas, which is in a very small proportion compared with the main gases of the atmosphere. But this small proportion of CO_2 has a controlling effect on the greenhouse action which is warming the Earth's climate.

Our activities of burning fuels to supply energy for transport, buildings and manufacturing emit quantities of CO_2 that far exceed those generated by natural fires or volcanic activity. Controlling the quantities of carbon fuels we use, together with the methods used to burn the fuels, will also control the amount of CO_2 emitted into the atmosphere. This 'carbon management' is therefore interdependent with energy management.

Energy supplies

At present most of the energy used to heat buildings, including electrical energy, comes from fossil carbon fuels such as oil and coal. This energy originally came from the sun and was used for the growth of plants such as trees. Then, because of changes in the Earth's geology, those ancient forests eventually became a coal seam, an oil field or a natural-gas field. The existing stocks of fossil fuels on Earth cannot be replaced and, unless conserved, they will eventually run out.

Reduced fuel use = lower CO_2 emissions

In the United Kingdom, some 40 to 50 per cent of the national consumption of primary energy is used for building services such as heating, lighting and electricity. Over half of this energy is consumed in the domestic sector, mostly for space heating. Reducing the use of energy in buildings will therefore be significant for conserving energy resources, reducing carbon dioxide emissions, and hopefully reducing running costs for the occupants of buildings.

Methods of conserving energy in buildings are influenced by the costs involved and, in turn, these costs vary with the types of building and the current economic conditions. Some of the important options for energy conservation in buildings are outlined in the following sections and can be used as starting points for staying informed about the balance and interactions of these matters.

Alternative energy for buildings

It is difficult to carry on modern life without oil and coal for certain purposes, such as conversion to chemicals and making polymer products like plastics and paints. But the stocks of fossil fuels such as oil are not renewable, except by waiting for long geological epochs to produce more. It is not essential, however, to use fossil fuels for low-grade tasks such as the heating of buildings. There are also alternative sources of energy available which are 'renewable'.

Large amounts of energy are contained in the heat, wind and waves of the Earth's weather system, which is derived from the energy of the sun. Also available is solar heat stored in the oceans and heat that comes from the Earth's interior, where it is generated by radioactivity in rocks. Other sources of renewable energy are in the use of growing plants, such as wood, and in the use of other *biofuels*. This type of energy is widely available at no cost except for the installation and running of conversion equipment.

> See also:
>
> section on *Using energy* in Chapter 3

Renewable sources of energy

The following are sources of energy that do not use fossil fuels and can be considered renewable.

- **Solar heating**: energy direct from the sun captured by special panels.
- **Wind**: energy captured using windmills.
- **Waves**: energy captured using devices which rise and fall.
- **Hydro**: energy captured from falling water using turbines.
- **Geothermal**: energy from the heat of the earth, captured as hot water or steam.
- **Biofuels**: including landfill gas, sewage gas, wood, straw, refuse.

> See also:
>
> section on *Power supplies* in Chapter 11

Solar energy

All buildings gain some casual heat from the Sun during winter, but more use can be made of solar energy by the design of the building and its services. Despite the high latitude and variable weather of countries in north-western Europe, like the United Kingdom, there is considerable scope for using solar energy to reduce the energy demands of buildings.

The utilisation of solar energy need not depend upon the use of special 'active' equipment such as heat pumps. *Passive solar design* is a general technique which makes use of the conventional elements of a building to perform the collection, storage and distribution of solar energy. For example, the afternoon heat in a glass conservatory attached to a house can be stored by the thermal capacity of concrete or brick walls and floors. When this heat is given off in

> *Passive* features already exist
> *Active* features need adding

the cool of the evening it can be circulated into the house by natural convection of the air.

Energy efficiency in buildings

The total energy of the Universe always remains constant, but when we convert energy from one form to another some of the energy is effectively lost to use by the conversion process. For example, when a boiler converts the chemical energy stored in a fuel into heat energy, hot gases must be allowed to escape up the chimney flue to maintain combustion. Around 90 per cent of the electrical energy used by a traditional light bulb is wasted as heat energy rather than light energy.

Efficient equipment

See also:

sections on *Lamps and Luminaires* in Chapter 6

New techniques are being used to improve the conversion efficiency of devices used for services within buildings. Condensing boilers, for example, recover much of the latent heat from flue gases before they are released. More efficient forms of electric lamp have been described in Chapter 6 about *Artificial lighting*. Heat pumps can make use of low-temperature heat sources, such as waste air, which have been ignored in the past.

Electricity use

See also:

section on *Power stations* in Chapter 11

Although electrical appliances have a high energy efficiency at the point of use, the overall efficiency of the electrical system is greatly reduced by the energy inefficiency of large power stations built at remote locations. It is possible to greatly improve the efficiency of electricity generation by combining it with the generation of heat for industrial and domestic use. The various techniques of CHP (combined heat and power) can raise the energy efficiency of electricity generation processes from around 35 per cent to over 80 per cent. CHP techniques can also be applied on a small scale to meet the energy needs of just one building or a series of buildings.

In the absence of CHP, savings in national energy resources are made if the 'grade' of energy for a task is considered. High-grade energy such as electricity is needed to run devices such as lights, motors and electronics, but it need not be used for a low-grade requirement such as space heating. Fossil fuels, such as gas or oil, are best converted into heat at the place where heating is required so that there is an opportunity to extract maximum amounts of heat from the fuel and the exhaust gases. It is wasteful to convert fossil fuels into electricity and then convert that electricity into heat.

Thermal insulation

External walls, windows, roof and floors are the largest areas of heat loss from a building and are discussed under *Thermal insulation* in Chapter 2. Current standards of thermal insulation in the United Kingdom, as defined by Building Regulations, have scope for continued improvement.

The upgrading of insulation in existing buildings can be achieved by techniques of roof insulation, cavity fill, double glazing, internal wall-lining and exterior wall-cladding. Increasing the insulation without disrupting the look of a building is a particular challenge for countries with stocks of traditional buildings.

See also: section on *Thermal insulation* in Chapter 3

Ventilation

The warm air released from a building contains valuable heat energy, even if the air is considered 'stale' for ventilation purposes. Minimising the heat lost during the opening of doors or windows becomes a significant factor in energy conservation, especially when the cladding of buildings is insulated to high standards.

These ventilation losses are reduced by better seals in the construction of the buildings, by air-sealed door lobbies, and the use of controlled ventilation. Some of the heat contained in exhausted air can be recovered by heat-exchange techniques such as heat pumps.

POLLUTION

Environmental pollution includes the various ways that human activity affects and harms the natural environment. Everyone would like to see less pollution but unfortunately many of the activities and products associated with modern life are also associated with pollution.

The construction of buildings and other structures contributes to pollution in various ways, and the pollution we generate within our buildings is also significant, especially as we spend up to 90 per cent of our time indoors, most of it at home.

Pollution can be broadly classified as air pollution, water pollution and soil pollution, although all types are linked together by the ecosystems of the Earth. The following sections amplify some of the causes and issues associated with pollution. The term pollution can also be used to describe excess of noise (noise pollution) and excess of light in the sky from inefficient street lighting of towns (light pollution). Some of these topics are described in more detail in earlier sections of the book.

Causes of pollution: buildings transport manufacturing food production

Ecosytem: the relationships between all living and non-living things in an environment

Air pollution

The lower atmosphere, where we breathe, has a natural content of nitrogen, oxygen, carbon dioxide and small amounts of other gases and particulates. Plants and animals exchange carbon dioxide and oxygen. Natural fires and volcanoes push particulates into the atmosphere where wind and rain scatter them. However, the natural balance of gases in the air is upset by the millions of tonnes of extra gases and particulates that human activities exhaust into the

Particulates: very small particles of solids and liquids

atmosphere. These pollutants are summarised below and some of them are described further in other sections.

- **Carbon dioxide** (CO_2): produced by the burning of fossil fuels and forests and by all organic decay. Chimneys, motor-vehicle exhausts and forest fires are major sources.
- **Smog**: a brown fog of various pollutant gases and particulates. In sunlight a photochemical reaction produces droplets of organic compounds which irritate eyes, throat and lungs.
- **Ozone** (O_3): a powerful oxidising agent and a dangerous irritant to eyes, throat and lungs. It is produced as part of *smog* by photochemical reactions. In the upper atmosphere it forms a protective layer which we wish to keep.
- **Nitrogen oxides** (NO_x): various oxides of nitrogen which are mainly produced by motor-vehicle emissions.
- **Sulfur oxides** (SO_x): various oxides of sulfur which are produced by the combustion of fuels such as oil and coal.
- **Methane** (CH_4): the main component of natural gas supplies. Methane is produced by decay of organic matter and also by the digestive processes of sheep and cattle.
- **Chlorofluorocarbons** (CFCs): families of chemical compounds that have been used in refrigerators, spray cans and in insulation. Although inert at point of use, when CFCs escape to the upper atmosphere they react and deplete the protective ozone layer.
- **Heavy metals**: typically lead and mercury which are contained in the gases given off by motor vehicles and by burning waste in incinerators.

Air pollution can cause immediate problems of poor visibility, discomfort and danger to people's health. Some polluting gases are also grouped as greenhouse gases which are linked to the problems of global warming, as described in Chapter 1.

Water pollution and runoff

Water can be contaminated by pollution from homes, offices, factories and farms. A variety of substances cause this pollution, including sewage, industrial chemicals and agricultural chemicals. When the contamination contains micro-organisms such as certain bacteria, it can cause dangerous illnesses.

Another form of water pollution is caused by excessive nutrients from sewage and fertilizer. This condition of *eutrophication* causes rapid growths of algae which deprive the water of oxygen and leave streams, rivers and lakes deprived of other life.

In their natural state, the ground and vegetation in most localities deal with rainwater by letting it be absorbed by plants and by the earth, where it is filtered and slowly dispersed. However the roofs, roads and other impermeable areas of the built environment allow rain to run off the surfaces as torrents which need

See also:

section on *Refrigerants* in Chapter 4

See also:

section on *Climate change* in Chapter 1

See also:

sections on *Water supplies* in Chapter 12 and on *Waste water* in Chapter 13

to be collected and channelled away safely. This rapid runoff also collects pollutants that are quickly swept into waterways such as streams and lakes. These
effects can be reduced by using semi-permeable surfaces and by building ponds
and absorbent drainage ditches (*swales*) to hold storm water.

Soil pollution

Industrial processes, mining operations and motor vehicles can leave heavy
metals, such as mercury and lead, in the atmosphere and soil.

> *Toxic metals*:
> mercury
> lead
> cadmium
> zinc
> chromium

Heavy metals are toxic, and because they accumulate in animal organs they
can enter the food chain of animals and then humans. It is good environmental
policy to recycle land within cities and to use *brownfield* sites. However, if this land
has been used for heavy industry in the past then the soil must be carefully
analysed and, if necessary, cleaned.

Waste material

Waste material is that which is unwanted for various reasons and has been sent
for disposal. One aspect of waste management is ensuring that waste materials
are truly waste and cannot be recycled. *Solid waste* is the material we discard from
homes, offices and factories. It includes paper, plastic, glass, waste food and
garden trimmings. Some waste is in bulk form, such as the rubble in demolition
waste or spoil left over from mining and agricultural processes.

Hazardous waste is waste is made up of a substance that can threaten health
and the environment. Waste is hazardous if it is poisonous, corrosive, explosive,
catches fire easily or reacts strongly with materials. Some hazardous waste can
cause harm to humans, animals and plants.

All waste is difficult to dispose of safely. If it is buried in *landfills*, precautions
must be taken to prevent liquids seeping into the water system and avoid the
build-up of flammable gases. Burning waste in incinerators can be a useful
source of energy but the combustion can release toxic chemicals, ash and metals
into the air. *Radioactive waste*, produced by nuclear power stations and by nuclear
weapons factories, will remain hazardous for thousands of years and its
treatment and storage will always be difficult and expensive.

Radon

Radon is a colourless, odourless gas given off naturally through the radioactive
decay of uranium rocks within the earth. The natural levels of radon vary in
different parts of the country and different buildings, but rocks such as granite
are associated with higher levels. If the radioactive products of radon decay are
breathed into the lungs, this can cause or contribute to lung cancer.

Natural radon is quickly diluted by the atmosphere and becomes harmless, so
the aim is to prevent it entering a building and increasing in level. Measures to
prevent radon entering buildings include the following:

- sealing floors and walls
- providing ventilation beneath the floor and in the building
- installing a sump where radon can accumulate and be extracted
- using a ventilation system with positive pressure.

Indoor pollution

Indoor pollution sources:
tobacco smoke
particles and
   gases from gas
   burners
cooking fumes
paint fumes
carpets
glue
household
   cleaners
radon

Measurements show that some household kitchens have a more dangerous level of air pollutants than a busy street corner! These pollutants are generated by normal processes of cooking, especially cooking with gas, and harmful levels can be avoided by good ventilation. Control of ventilation is also a factor in saving energy, so the ventilation system should have recovery systems to extract and recycle the energy from the warm but polluted exhaust air. Indoor pollution is also a factor of sick building syndrome discussed in the next section.

SICK AND HEALTHY BUILDINGS

There are some buildings, especially offices, where the occupants appear to suffer ill health more often than might reasonably be expected. These 'sick building' illnesses have no readily identifiable cause. Some illnesses, such as legionnaire's disease, which can be traced to a particular cause, are *not* strictly sick building illnesses.

Alternative terms:
Sick building
   syndrome
Building-related
   illness
Tight building
   syndrome
Office eye
   syndrome

Sick building syndrome (SBS) is the most commonly used term for this phenomenon and is recognised by the World Health Authority. It has been suggested that up to 30 per cent of new and refurbished buildings have given rise to complaints of sick building illness. Although this consideration of sick building syndrome is centred around office environments, it is well to consider that the home environment can be troubled by many of the same physical causes, if not the mental ones.

Sick building effects

The illnesses related to sick building syndrome generate the following types of symptoms:

- eye, nose and throat irritations
- dryness of throat, nose and skin
- breathing difficulties and chest tightness
- headaches, nausea, dizziness
- mental fatigue
- skin rashes
- aching muscles and 'flu-like symptoms.

This wide range of symptoms includes illnesses which most people occasionally suffer while at home or work, so investigations are difficult. However, certain patterns have been found.

Sick building syndrome is linked to the size and structure of the organisation using the office. Most complaints occur in offices which contain many staff, and symptoms are more frequent in the afternoon than the morning. The people with most symptoms are those who see themselves having least control over their environment, and clerical staff are more likely than managerial staff to suffer from SBS.

Whatever the causes, a sick building results in absenteeism among staff and lower productivity while they are at work. Sick building syndrome also costs money by loss of profit, by bad publicity and, in the extreme, by closure of a building.

Buildings at risk

Despite the difficulties of investigating sick building syndrome, the following features have been identified as common to many sick buildings:

- forced ventilation, including mechanical ventilation and air conditioning
- windows and other openings sealed for energy efficiency
- lightweight construction
- carpets and other textiles used on indoor surfaces
- warm and uniform environments.

Sick building causes

The following list of factors which contribute to sick building syndrome has been arranged under three general headings of physical, chemical and microbial factors.

Physical comfort conditions

- uncomfortable temperatures
- low humidity
- low air movement and 'stuffiness'
- low ventilation rates
- insufficient negative air ions
- unsuitable lighting and decoration
- low daylight levels
- uncomfortable seating
- excessive noise levels
- electromagnetic radiation from electrical services and appliances
- low morale and general dissatisfaction.

Chemical pollutants

- cigarette smoke
- formaldehyde vapours from furniture, particle boards
- vapours (VOC) from adhesives, paints and cleaners

VOC:
volatile organic
compounds

- radon decay products from granite stone and aggregates
- ozone gas from photocopiers, laser printers and high-voltage sources.

Microbial

- airborne micro-organisms from bacteria and fungi in air-conditioning systems
- micro-organisms in drinking water and vending machines
- micro-organisms in carpets, fabrics and pot plants.

Healthy buildings

The previous lists of possible causes of sick building illnesses highlight the fact that is not possible to identify single factors and, therefore, there is no single cure. Of the physical causes, poor air quality and dirty machinery are common. The whole problem of sick buildings centres around human beings, and the study of the people concerned is as important as the surroundings.

Antidotes for sick buildings:
good design
correct installation
constant
maintenance

To help to eliminate the various causes which give rise to poor environments within buildings, attention needs to be paid to good design, correct installation and constant maintenance. These processes must be directed towards creating a healthy and pleasant working environment for the occupants of a building, and these occupants need to feel involved in the creation and control of their environment.

SUSTAINABLE BUILDINGS

Chapter 1 described a *green or sustainable building* as a building that is deliberately designed to minimise impact on the environment and to maximise efficiency when using resources such as materials, water and energy; and to maintain this efficiency over the lifecycle of the building.

When we assess the environmental impact of our buildings it should lead to questions about how long we can continue or should continue to use existing methods of constructing and using buildings. For example, we need energy to live and work, yet the use of energy at current rates will create environmental problems for the generations that come after us; unless we consider the future of the resources that we are using.

- **Sustainable development** aims to meet the needs of the present without compromising the needs of the future.

We can summarise the broad aims of sustainable construction as:

- to improve the quality of our lives
- to be acceptable to other people and future generations
- to cause minimum damage to the wider environment and its resources.

The key messages of sustainability actually link well with what are generally considered to be good construction business practices, such as producing efficient buildings, minimising waste and maximising resources. When identifying and assessing a building for sustainability the following broad areas are generally considered.

- land use
- materials use
- energy and carbon use
- water management
- ecology control
- waste management
- health and well-being
- pollution control

Environmental appraisal of buildings

The issues of the environment generate a lot of words, so it is helpful to also have some numbers or classifications based on numerical data. Various systems have evolved to assess and certify buildings for their environmental performance. The criteria used for appraisal systems include the areas associated with sustainability listed in the previous section and also discussed in various chapters of this book.

For example, the BRE organisation in the United Kingdom supervises the BREEAM system which assesses the environmental performance of both new and existing buildings. The BREEAM appraisal system is also used outside the United Kingdom. The system appraises the following building performance in the categories and properties described below.

- **Management**: includes procurement procedures, construction practices, service life planning and costing, stakeholder participation.
- **Health and well-being**: includes visual comfort, indoor air quality, thermal comfort, water quality, acoustic performance, safety and security.
- **Energy**: includes reduction of energy use and carbon dioxide emissions, energy monitoring, use of low carbon technologies, refrigeration systems, transportation systems such as lifts and escalators, energy efficient equipment.
- **Transport**: includes public transport accessibility.
- **Water**: includes consumption, monitoring, leak detection and prevention.
- **Materials**: includes life cycle impacts, responsible sourcing, insulation type, robustness of design.
- **Waste**: construction waste management, recycled aggregates, operation waste
- **Land Use and Ecology**: includes site selection, re-use of land, ecology.
- **Pollution**: includes use of refrigerants, emission of nitrogen oxides, surface water run-off, light pollution of clear skies, environmental noise attenuation.
- **Innovation**: these properties will be new, by definition!

The BREEAM appraisal system awards credits in categories according to performance appropriate for the type of building, adjusts the credits by a set of environmental weightings, and combines the credits to produce a single overall

Greenfield = new building land
Brownfield = recycled building land

score. The building is then rated on a scale of Pass, Good, Very Good or Excellent; or for some purposes a star system is used.

The Code for Sustainable Homes (the *Code*) uses practices contained in the BREEAM system and applies them to the sustainable design and construction of new homes in England, Wales and Northern Ireland. The *Code* uses a one to six star rating system to assess the overall sustainability performance of a new home against the categories of: Energy/CO_2, Water, Materials, Surface Water Runoff (flooding and flood prevention), Waste, Pollution, Health and Well-being, Management and Ecology.

PassivHaus is a design method and a standard for low-energy homes, and it can be extended to other building types. To achieve the PassivHaus standard the total energy required for the space heating or cooling of a building must not exceed 15 kWh/m² per year. This performance is achievable by using very high standards of thermal insulation, triple glazing, and air tightness. A PassivHaus building may include some *passive design* principles, such as solar gain from orientation, but it also uses the *active* technique of a mechanical ventilation and heat exchanger system that recovers maximum heat energy from outgoing air.

Comparable systems of green building appraisal established in other parts of the world include LEED (USA and Canada) and Green Star (Australia, New Zealand and South Africa). The detail and weightings of areas considered in any environmental assessment of a building will continue to change as our knowledge evolves about the risks associated with construction materials and techniques.

> *Check list for Green Building Practice:*
> fitness for purpose
> integrated design and construction
> resource conservation
> low-impact design
> sustainable construction
> health and safety
> energy conservation features
> low carbon emission
> low-pollution features
> intelligent features
> whole-life costing
> low resource maintenance

FUTURE BUILDINGS

The activities of constructing and running buildings have major effects upon the environment. Local consequences, such as the visual impact of buildings, have always been recognised but now it is also acknowledged that the built environment is a major contributor to world-wide effects such as global warming and wastage of resources. Fortunately the built environment sector has high potential for achieving large reductions in carbon emissions relatively rapidly. These reductions also bring national economic benefits, improve social well-being and improve individual quality of life.

The previous sections and chapters of this book provide details of technology that can improve the built environment. This section describes some aspects of good modern practice for designing, constructing and operating buildings that will produce buildings fit for a future that needs to be more environmentally responsible than the past.

> *Buildings are globally responsible for:*
> 30% of resource use
> 40% of energy use
> 30% of greenhouse gas emissions
> (World Green Building Council)

Intelligent buildings

A modern car has many sophisticated systems and controls that go well beyond its primary role of travelling along the road. Similarly, modern buildings are more than inert structures intended just to provide shelter from the weather; at the very least it will have controls for heating and lighting. Modern building management systems (BMS) include the control and optimisation of services such as heating ventilation and air conditioning (HVAC), lighting, closed-circuit television (CCTV), security systems, and access systems that control and report the movement of people in the building.

When a building is described as 'intelligent', it is considered to have a high proportion of the following features:

- automated building services such as those for energy management, security and fire precautions,
- information management such as telecommunication systems and computer systems for IT (information technology),
- connectivity determined by internal cabling, wireless networks, and access to external services,
- control of the environment achieved by the monitoring and control of building services and safety such as in a *building automation system* (BAS),
- premises management achieved by the controlled monitoring and scheduling of maintenance and other building functions.

Whole-life costing and performance

The cost of constructing a building is actually a small percentage of the total cost of operations in that building during its lifetime. The Royal Academy of Engineering estimates that for an office building the costs over 30 years are in the ratio of 1 for construction to 5 for maintenance to 200 for operations including staff costs.

The whole-life costing of a building assesses the performance and total costs of that building over a defined service life, such as 60 years. This concept is comparable to the life cycle assessment (LCA) of construction materials described in an earlier section. Knowing the costs of a building over its lifetime is obviously useful to the owners and reflects the performance of the building. Information about the costs of running a building later in its life can also help designers and their clients make optimum decisions about the features of the building. For example, initial design decisions about the orientation of a building towards the sun and the nature of windows will affect the energy use and carbon emissions for that building over its complete lifetime.

Whole-life costing of a building takes into account the following factors:

- capital costs of design and construction
- maintenance, replacement and repair costs
- demolition and disposal costs at the end of the building's life.

Building for Life is a UK promotion of design excellence and best practice in house-building

Zero energy and zero carbon

The UK government and the European Union have expressed some environmental targets for future buildings in terms of 'zero energy', 'zero carbon' or 'near zero energy/carbon'. There are various definitions of a zero-energy building (ZEB), but the common central idea is that such a building can meet all its energy needs from local renewable sources. A ZEB can import off-site energy, such as grid electricity or gas, if it is necessary; but when the on-site energy generation is in excess this local electricity can be exported to the public electricity grid.

- At the most rigorous level, a ZEB is a building that balances its annual energy use with energy generated *on site* from renewable sources such as photovoltaic panels. This may be termed net-zero site energy.
- At a less rigorous level, a ZEB is a building that balances its annual energy use with energy generated *externally* by renewable sources such as wind turbines. This may be termed net-zero source energy.

In this context *net* means the balance between energy used and energy supplied

A ZEB has zero net carbon dioxide emissions associated with it and can also be known as a *zero carbon building*. Other systems and definitions of 'zero' can include the emissions generated in the construction of the building and the embodied energy of its materials.

Design interactions

Although the various topics that make up environmental technology are usually defined and described separately, it is important to appreciate their dependence upon one another when they are combined together in a building. All of the environmental factors have ideal standards to be separately satisfied, but they may also need to be balanced against one another. As a simple example, larger windows provide better daylighting but they also cause greater heat losses in winter and larger heat gains in summer. Table 14.4 indicates some of the major interactions between different design decisions. This type of table can be extended to include sustainability factors and whole-life performance.

Table 14.4 Interactions of environmental decisions

Design options	Possible environmental effects			
	Heating	Ventilation	Lighting	Sound
Sheltered site	Less heat loss and gain	–	Less daylight	Less noise intrusion
Deep building plan	Less heat loss and gain	Less natural ventilation	Less daylight	Less noise intrusion
Narrow building plan	More heat loss and gain	More natural ventilation	More daylight	More noise intrusion
Heavy building materials	Slower heating and cooling	–	–	Better sound insulation
Increased window area	More heat loss and gain	–	More daylight	More noise intrusion
Smaller, sealed windows	Less heat loss and gain	Reduced natural ventilation	Less daylight	Less noise intrusion

The interactions between the environmental features of construction indicate that they should be considered together at an early stage in the design. The need for this integrated and early involvement of environmental studies has often been neglected in the past. For example, engineers have historically specialised in one particular aspect of controlling the environment of a building and have not considered the effects of their decision on other areas of the environment. In order to provide optimum environment conditions for future buildings, all available knowledge and skill needs to be integrated at an early stage.

Useful internet search terms: green building, environmental building, sustainable building, future building

The future

The issues connected with the design, the construction and the management of high-performance buildings and cities are challenging. Green and sustainable buildings also need to be beautiful and exciting. Fortunately such building projects do exist, and you should seek examples in your locality to inspire you. An equal challenge for the future is to upgrade existing buildings to modern standards of environmental performance.

When studying the different parts of the environment it is useful to keep in mind a larger picture of how the many systems on Planet Earth are interlinked. The Gaia hypothesis, for example, is the theory that the living and non-living systems of the Earth form an inseparable whole and that all living things interact to create the environmental conditions that they need. The ways that we create and use our built environment are therefore part of a much wider environment; connected by the processes discussed in this book and by many other systems of the planet and the universe.

Answers to exercises

Chapter 3

3. 0.26 W/m^2 K
4. 1.47 W/m^2 K 0.41 W/m^2 K
6. 5.43 W/m^2 K 3.36 W/m^2 K
7. 1.27 W/m^2 K
8. 63 mm
9. 139 mm
10. 12.5°C
11. 19°C, 427 W
12. 1 495 W, 1 550 pence
13. 8 030 W, 18.25 m^2
14. 5.868 GJ, 3.144 GJ
15. 21.398 GJ, 892 kg

Chapter 4

1. 44 per cent
2. 72.6 per cent
3. (a) 58 per cent, 12°C; (b) 0.008 kg/kg; (c) 30 per cent
4. (a) 27 per cent; (b) 68 per cent; (c) 9.6 mb
5. (a) 81 per cent; (b) 0.003 kg/kg

Chapter 5

1. 100 cd
2. 2 576 lm, 182 lx
3. 125 lx, 7.81 lx
4. 24.5 lx, 16.03 lx
5. 310.9 lx, 214.3 lx
6. 118.3 cd

Chapter 6

1. 25 lamps, 3.6 kWh
3. 48 lamps
4. 530 lx
5. 0.56, 20 lamps

Chapter 7

1. 1235 cd/m^2
2. 8 per cent

Chapter 8

1. 85 dB
2. 0.0796 Pa
3. 89.1 dB; 94.7 dB
4. 3.06×10^{-7} W/m^2; 10.5 dB
5. 8 dB

Chapter 9

1. Energy is 32 times greater
4. 19 dB; 5.012×10^{-4}
5. 27.4 dB

Chapter 10

2. 60 m^2 sabins
3. 2.4 seconds
4. 198.3 m^2 sabins
5. 2319 people
6. 71.1 m^2 carpet

Chapter 11

1. 3833 turns; 8.33 A
2. 3400 W; 17.40 A; 24.61 A

Part II

Resources

Science Information

UNITS

Measurement is a major activity of science and technology. The result of measuring a physical quantity such as length is expressed as a number, followed by a unit. For example: length AB = 15 metres. In general:

> Physical quantity = Number / Unit

SI units

The number expresses the ratio of the measured quantity to some agreed standard or unit. Different systems of units have arisen over the years, including imperial and metric units. A rational and coherent version of the metric system has been developed, called the Système Internationale d'Unités, or SI.

SI units are intended for worldwide scientific, technical and legal use. The units in this book are given in SI and reference to older units is made where such units still linger in technical practice.

There are seven base units in the SI system, two supplementary units, and numerous derived units, some of which are listed in the table of units. Derived units can be formed by combinations of base units; for example, the square metre. Some derived units are given new names; for example, the newton is a combination of the kilogram, the metre and the second; the pascal is a combination of the newton and the square metre.

The symbols for SI units do not have plural form and are not followed by a full stop, except at the end of a sentence. The symbols for derived units may be written in index form or with a solidus (/). For example:

ms^{-2} or m/s^2

Table R1.1 SI units

Quantity	Symbol	SI uni	Symbol
Base units			
length	*l*	metre	m
mass	*m*	kilogram	kg
time	*t*	second	s
electric current	*I*	ampere	A
thermodynamic temperature	*T*	kelvin	K
luminous intensity	*I*	candela	cd
amount of substance		mole	mol
Supplementary units			
plane angle	θ	radian	rad
solid angle	ω	steradian	sr
Some derived units			
area	*A*	square metre	m²
volume	*V*	cubic metre	m³
density		kilogram per	kg/m³
		cubic metre	
velocity	*v*	metre per second	m/s
force	*F*	newton	N (kg m/s²)
energy	*E*	joule	J (Nm)
power	*P*	watt	W (J/s)
pressure	*p*	pascal	Pa (N/m²)

SI prefixes

Multiplication factors are used to express large or small values of a unit. These multiples or sub-multiples are shown by a standard set of prefix names and symbols which can be placed before any SI unit, with the exception of the kilogram. Multiples should be chosen so that the numerical value is expressed as a number between 0.1 and 1000. See Table R1.2

The Greek alphabet

The symbols for some quantities and units are taken from the Greek alphabet. See Table R1.3

Symbols and formulas

Some common symbols and formulas used in technical and mathematical expressions are given in Table R1.4.

Table R1.2 SI prefixes

Prefix	Symbol	Multiplication factor	
tera	T	10^{12}	= 1 000 000 000 000
giga	G	10^{9}	= 1 000 000 000
mega	M	10^{6}	= 1 000 000
kilo	k	10^{3}	= 1 000
hecto*	h	10^{2}	= 100
deca*	da	10^{1}	= 10
–	–	–	= 1
deci*	d	10^{-1}	= 0.1
centi*	c	10^{-2}	= 0.01
milli	m	10^{-3}	= 0.001
micro	μ	10^{-6}	= 0.000 001
nano	n	10^{-9}	= 0.000 000 001
pico	p	10^{-12}	= 0.000 000 000 000 1
femto	f	10^{-15}	= 0.000 000 000 000 001
atto	a	10^{-18}	= 0.000 000 000 000 000 001

*The standard increment between prefixes is 10^{3} or 10^{-3} but for some units intermediate prefixes are in common use.

Table R1.3 The Green alphabet

A	α	Alpha	N	ν		Nu
B	β	Beta	Ξ	ξ		Xi
Γ	γ	Gamma	O	o		Omicron
Δ	δ	Delta	Π	π		Pi
E	ε	Epsilon	Π	ρ		Rho
Z	ξ	Zeta	Σ	σ		Sigma
H	η	Eta	T	τ		Tau
Θ	θ	Theta	Ψ	ψ		Upsilon
I	ι	Iota	Φ	φ		Phi
K	κ	Kappa	Ξ	X		Chi
Λ	λ	Lambda	Ψ	ψ		Psi
M	μ	Mu	Ω	ω		Omega

Table R1.4 Common symbols

Symbols	Meaning
Σ	sum of
>	greater than
<	less than
a^{n}	a raised to the power n
$\sqrt{a}$ or $a^{0.5}$	square root of a
log x	common logarithm (base 10) of x
π	'pi' = 3.141 593 approx

CHEMICAL PROCESSES

Elements and compounds

Some common elements	
Element	*Symbol*
Aluminium	Al
Calcium	Ca
Carbon	C
Chlorine	Cl
Copper	Cu
Hydrogen	H
Iron	Fe
Lead	Pb
Magnesium	Mg
Nitrogen	N
Oxygen	O
Silicon	Si
Sodium	Na
Sulfur	S
Zinc	Zn

There are millions of different materials in the world but all materials are made up from a few basic elements. There are only 92 elements which occur naturally and about a dozen more which have been produced artificially. All materials in the universe are made from these same elements.

In their natural state on our planet, elements do not usually exist by themselves but are found combined with other elements in the form of compounds. Sodium chloride, for example, is a compound made only from the elements sodium and chlorine. Sodium is a reactive silver-coloured metal and chlorine is a poisonous green-coloured gas. Yet the result of their chemical combination is ordinary table salt, which obviously has different properties to its ingredients!

Table R1.5 Common compounds

Compound	Formula	Atoms in one molecule
Sodium chloride	NaCl	1 sodium
		1 chlorine
Water	H_2O	2 hydrogen
		1 oxygen
Carbon dioxide	CO_2	1 carbon
		2 oxygen
Calcium carbonate	$CaCO_3$	1 calcium
		1 carbon
		3 oxygen

We sometimes need to consider the smallest 'bit' of a material. For an element, the smallest part is an atom of that element. We can't say an 'atom' of water or an atom of concrete because they are compounds of several elements. So we also use the idea of a molecule, which is a group of atoms bonded together.

Chemical terms

Element: a substance which cannot be separated into anything simpler by chemical means.

Compound: a substance containing two or more different elements which are chemically joined together to form a new material with new properties.

Atom: the smallest part of an element which can take part in an chemical reaction.

Molecule: the smallest part of a compound which can take part in a chemical reaction.

Mixtures

Different substances can be found existing together in a non-chemical manner, and if they are not chemically bonded together in a compound then they may be a *mixture*. The proportions of the components in a mixture can vary and the components can be separated relatively easily. Examples of mixtures which show these properties include:

- sand and iron filings
- oil and water.
- sand and cement powder

Sand and cement powder, for example, starts as a mixture and forms chemical compounds when water is added to start the chemical reactions.

Solutions and suspensions

Some mixtures are formed by dissolving a substance in a liquid to form a solution. For instance, sugar dissolves in tea and air dissolves in water. The solute is the dissolved substance and may be a solid, liquid or gas. The solvent is the dissolving liquid, and is often water.

- Solution = Solvent + Solute

A solution has a clear, even appearance and remains unchanged with time. A suspension, on the other hand, is a liquid in which other substances are distributed but *not* dissolved. A suspension has a cloudy appearance and separates if left standing. Clay in water, for example, is a suspension and the clay settles to the bottom when left standing. An emulsion paint is a form of suspension from which, eventually, the ingredients separate. This is why paint needs to be stirred thoroughly.

Chemical reactions

Chemistry involves many thousands of combinations among different elements and compounds. The study and the use of chemistry is made easier by grouping substances and reactions into categories which have similar properties. The processes outlined here have a significant effect on the environment around us and the buildings we live in.

An important feature of chemistry is the making of new substances by chemical reactions.

- A **chemical reaction** is an interaction between substances in which atoms are rearranged.

Chemical reactions are shown in shorthand by chemical equations using the symbols for each type of molecule acting in the equation. The atoms are rearranged during a reaction but the total number of atoms, of each element, must remain the same on each side of the equation. For example:

$$CaCO_3 \longrightarrow CaO + CO_2$$

The above equation means that, when heated, one molecule of calcium carbonate (limestone) produces one molecule of calcium oxide (burnt lime) plus one molecule of carbon dioxide.

Acids and alkalis

Acids, bases and alkalis are important classes of chemicals with many uses. Some occur naturally, such as those in our bodies, or they may be manufactured. Both acids and alkalis are corrosive when they are strong, but many substances are weakly acidic or alkaline when they are dissolved in water.

Some acids:
sulfuric acid, H_2SO_4
hydrochloric acid, HCl
citric acid (in lemons)

- An **acid** is a substance that contains hydrogen which can be chemically replaced by other elements.
- An **alkali** or **base** is a substance which neutralises an acid by accepting hydrogen ions from the acid.

The pH scale

Some alkalis:
sodium hydroxide, NaOH
calcium hydroxide, $Ca(OH)_2$
cement powder

The pH value indicates the acidity or alkalinity of a solution, measured on a scale from 0 to 14 that is related to the concentration of hydrogen ions present.

Table R1.6 The pH scale

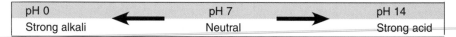

pH 0	pH 7	pH 14
Strong alkali	Neutral	Strong acid

Minerals and salts

Salts are a large group of compounds found naturally in the environment or manufactured as useful materials and products. A salt is the product of the neutralisation between an acid and a base. Some examples of salts are given below.

Table R1.7 Examples of salts

Chemical name	Formula	Common name
Sodium chloride	NaCl	common salt
Calcium carbonate	$CaCO_3$	limestone or chalk
Calcium sulfate	$CaSO_4$	gypsum
Copper sulfate	$CuSO_4$	

Oxidation and reduction

Many different chemical reactions are possible in nature, but some reactions share common features and can be studied as a class of reactions. Oxidation and reduction are common classes of reaction.

- Simple **oxidation** is the gain of oxygen in a chemical reaction.
- Simple **reduction** is the loss of oxygen in a chemical reaction.

Oxidation on the surface of metals forms a layer of oxide. This process can be protective as in the case of aluminium, or it can be destructive as with the rust on steel.

Reduction processes are used to extract pure metals from their ores (rocks) when carbon, for example, 'attracts' the oxygen from the metal oxide. The metal oxide is reduced and the carbon is oxidised.

Electrolysis

There are a number of useful interactions between the electrical and chemical properties of materials. Electrolysis is the production of a chemical reaction by passing an electric current through an electrolyte. For example, when an electric current is passed through water, the chemical reaction causes water to give off the gases hydrogen and oxygen from which it is made.

The electric current enters and leaves the liquid electrolyte by two electrodes called the anode (positive) and the cathode (negative).

Organic chemistry

Most of the chemcial compounds in living (organic) materials, such as our own bodies, are based on carbon. Carbon atoms have a special property of being able to bond to one another and form long chains or circular rings. Each structure and combination of elements makes a new compound with new properties.

- An organic compound is one that is composed of carbon and hydrogen.

Basic organic compounds contain only carbon and hydrogen, and they are called hydrocarbons. An organic compound may also contain atoms of other elements, such as oxygen, chlorine and nitrogen. Although organic chemistry has a huge number of compounds, they are made manageable by grouping organic compounds into 'families'. All members of a particular family have similar chemistry.

The alkanes are a family of compounds which form the basis of petroleum gas and liquid fuels. They have the simplest possible combinations of carbon and hydrogen, and they are named by combining a prefix that describes the number of carbon atoms in the molecule with the root ending 'ane'. The names and prefixes for the first ten alkanes are given in Table R1.8.

Polymers

Modern plastics, paints and adhesives are based on a versatile group of compounds called polymers.

- A **polymer** is a substance with large chain-like molecules consisting of repeated groups.

Table R1.8 Alkane family of organic compounds

Carbon atoms	Prefix	Alkane name	Chemical formula
1	Meth-	Methane	CH_4
2	Eth-	Ethane	C_2H_6
3	Prop-	Propane	C_3H_8
4	But-	Butane	C_4H_{10}
5	Pent-	Pentane	C_5H_{12}
6	Hex-	Hexane	C_6H_{14}
7	Hept-	Heptane	C_7H_{16}
8	Oct-	Octane	C_8H_{18}
9	Non-	Nonane	C_9H_{20}
10	Dec-	Decane	$C_{10}H_{22}$

Most polymers are made by the artificial process of polymerisation which combines simple compounds or monomers into larger units, as in the following examples:

Monomer	polymerises to:	Polymer
ethene		polyethylene (polythene)
styrene		polystyrene

Principles of Heat

NATURE OF ENERGY AND HEAT

Energy is not actually a 'thing' but it is a condition or property of things. A formal definition of energy is:

- **Energy** is the capability of an object, or a system, to do work on another system.

The idea of work being done is connected with wide range of different activities such as our body growing, or moving our body, or using a machine such as a car to move about. The Earth constantly receives energy from the Sun and this energy is used, for example, to grow plants and to drive the weather systems of the world. There are many different forms of energy but they can be divided into two broad types:

- **Potential energy** is stored energy, such as being compressed in a spring, being at a height like water in a dam, or being stored in a chemical like food.
- **Kinetic energy** is energy of motion, such as the movement of objects, waves, electrons, and atoms.

Table R2.1 lists different forms of energy and gives examples. All the forms of energy can also be classified as either potential energy or kinetic energy. For example, chemical energy can also be considered as a form of potential energy. It is a useful exercise for you to consider the other forms of energy listed in Table R2.1 and consider whether they are potential or kinetic energy.

Nature of heat

From earliest times people have understood the importance of the 'hotness' experienced when material is burned in a fire, but it is not so easy to understand and agree what heat is. Theories have included heat being defined as one of the four 'elements' of nature, and heat being considered as an invisible 'caloric' fluid.

Careful observation, good thinking and practical experiments in the 1800s finally established that heat is a form of energy and that there are constant

Table R2.1 Forms of energy and examples

Form of energy	Examples
Motion energy associated with moving objects or substances	• running water • wind • moving vehicles • any moving object
Gravitational energy associated with position and being able to fall	• water behind a dam • apple hanging from a tree • car at top of a hill
Thermal energy or heat energy, associated with the internal vibrations of molecules in a material	• any warm object • fire, steam • the Earth's internal heat – giving geothermal heat
Chemical energy associated with the chemical bonds of molecules	• fuels – such as oil and gas • food – which is also a fuel • electric batteries – which are collections of chemicals
Electrical energy associated with the movement of electrons in some materials	• electric current in causing effects in motors, heaters, lights
Electromagnetic energy associated with the wave movement of electrical and magnetic fields and/or the movement of light particles (photons)	• visible light • infra-red heat, • radiation from the Sun • radio waves, microwaves, x-rays
Nuclear energy or mass energy, associated with changing the strong bonding of particles within the nucleus of atoms	• the Sun, and other stars • nuclear reactors
Sound energy caused when vibrations set up wave motions set up in solids, liquids or gases	• speech and music • machinery vibrations

relationships between heat and other forms of energy. We can now describe heat as follows.

See also:

section on *Energy units* in Chapter 3

• **Heat** (H or Q) is a form of energy also called thermal energy

    Unit: joule (J)

The joule is the standard SI unit of energy, the same as is used for measuring any other form of energy. You will come across some alternative units for measuring heat and these include the following:

• calorie, where 1 cal = 4.187 J
• kilowatt hour, where 1 kWh = 3.6 MJ
• British Thermal Unit, where 1 BTU = 1.055 kJ.

Heat or thermal energy is an internal molecular property of a material. Other forms of energy include mechanical energy, electrical energy and chemical energy. These other forms of energy can all be converted to thermal energy. For example, the mechanical energy of moving surfaces is converted to heat by friction, electric currents flowing in conductors produce heat, and combustion (burning) converts the chemical energy contained in materials to heat.

Thermal energy often forms an intermediate stage in the production of other forms of energy. Most electrical energy, for example, is produced by means of the thermal energy released in the combustion of fuels. The thermal energy radiated from the Sun is also the origin of most energy used on Earth, including the fossil fuels such as coal and oil, which were originally forests grown in sunlight.

> *Forms of energy:*
> thermal
> mechanical
> electrical
> chemical
> nuclear

Power

Power is a measure of the rate at which work is done, or at which energy is converted from one form to another. It can be expressed in the following form.

$$\textbf{Power}\ (P) = \frac{\text{Heat energy }(H)}{\text{time }(t)}$$

Unit: watt (W)

By definition, 1 watt = 1 joule/second. The watt is often used in the measurement of thermal properties, and it is useful to remember that it already contains information about time and there is no need to divide by seconds.

Temperature

Temperature is not the same thing as heat. A red-hot spark, for example, is at a much higher temperature than a pot of boiling water; yet the water has a much higher heat 'content' than the spark and is more damaging.

> **See also:**
> section on
> *Temperature
> scales* below

- **Temperature** is the condition of a body that determines whether heat shall flow from it.

    Unit: degree Kelvin (K)

Heat flows from objects at high temperature to objects at low temperature. When there is no net heat transfer between two objects they are at the same temperature.

Thermometers

The human body is sensitive to temperature but is unreliable for measuring it. The brain tends to judge temperature by the rate of heat flow in or out of the skin. So, for example, a metal surface always 'feels' colder than a plastic surface even though a thermometer may show them to be at the same temperature.

A thermometer is an instrument that measures temperature by making use of some property of a material that changes in a regular manner with changes in temperature. Properties available for such use include changes in size, changes in electrical properties such as resistance, and changes in light emissions. Some of the more common types of thermometer are described below.

Mercury-in-glass thermometers

Mercury-in-glass thermometers use the expansion of the liquid metal mercury inside a narrow glass tube. Mercury responds quickly to changes in temperature and can be used between –39°C and 357°C, which is the range between its freezing point and boiling point.

Alcohol-in-glass thermometers

Alcohol-in-glass thermometers use coloured alcohol as the liquid in the glass tube. Alcohol expands more than mercury and can be used between –112°C and 78°C, which is the range between its freezing and boiling points.

Thermoelectric thermometers

Thermoelectric thermometers use the electric current generated in a thermo-couple, which is made by joining two different metals such as iron and constantan alloy. The current quickly varies with temperature and can be incorporated in remote or automatic control systems.

Resistance thermometers

Resistance thermometers use the change in electrical resistance which occurs when a metal changes temperature. Pure platinum is commonly used, and the changes in its resistance can be measured very accurately by including the thermometer in an electrical circuit.

Optical pyrometers

Optical thermometers measure high temperature by examining the brightness and colour of the light emitted from objects at high temperatures. The light varies with temperature and is compared with a light from a filament at a known temperature.

Temperature scales

In order to provide a thermometer with a scale of numbers, two easily obtainable temperatures are chosen as upper and lower *fixed points*. The interval between these two points on the thermometer is then divided into equal parts, called degrees. The properties of water are used to define two common fixed points – the temperature at which ice just melts and the temperature of steam from boiling water – where both are measured at normal atmospheric pressure.

Celsius scale

The Celsius temperature scale numbers the temperature of the melting point of ice as 0, and the boiling point of water as 100.

- **Celsius temperature** (θ) is a point on a temperature scale defined by reference to the melting point of ice and the boiling point of water.

    Unit: degree Celsius (°C)

Degrees Celsius are also used to indicate the magnitude of a particular change in temperature, such as an increase of 20°C. The less correct term 'centigrade' is also found in use.

Thermodynamic scale

Considerations of energy content and measurement of the expansion of gases lead to the concept of an absolute zero of temperature. This is a temperature at which no more internal energy can be extracted from a body and it occurs at –273.16°C. The absolute (or thermodynamic) temperature scale therefore numbers this temperature as zero.

The other fixed point for the thermodynamic scale is the triple point of water; the temperature at which ice, water, and water vapour are in equilibrium (0.01°C at 611.73 Pa pressure).

- **Thermodynamic temperature** (T) is a point on a temperature scale defined by reference to absolute zero and to the triple point of water.

    Unit: degree Kelvin (K)

degrees Kelvin = degrees Celsius + 273

The degree Kelvin is the formal SI unit of temperature but the degree Celsius is also used in common practice. The interval of a degree Kelvin is the same size as a degree Celsius, therefore a change in temperature of 1 K is the same as a change in temperature of 1°C. The general relationship between the two temperature scales is given by the following formula.

$$T = \theta + 273$$

where

    T = Thermodynamic temperature (K)
    θ = Celsius temperature (°C)

Heat capacity

The same mass of different materials can 'hold' different quantities of heat. Hence water must be supplied with more heat than oil in order to produce the same rise in temperature. Water has a greater heat capacity than oil; a property that is not to be confused with other thermal properties such as conductivity.

The heat capacity of a particular material is measured by a value of specific heat capacity, and Table R2.2 gives values for a variety of materials.

- The **specific heat capacity** (c) of a material is the quantity of heat energy

Table R2.2 Specific heat capacities

Material	Specific heat capacity J/kg K
Water	4190
Concrete and brickwork	3300
Ice	2100
Paraffin oil	2100
Wood	1700
Aluminium	910
Marble	880
Glass	700
Steel	450
Copper	390

Note: The values for particular building materials vary.

required to raise the temperature of 1 kg of that material by 1 degree Kelvin (or 1 degree Celsius).

Unit: J/kg K (or J/kg°C)

The heat capacity of water is higher than the heat capacities of most other substances, so water is a good medium for storing heat. The temperatures on the planet Earth are stabilised by the huge quantities of heat energy stored in the oceans, and the presence of this water around islands, such as the British Isles, prevents seasonal extremes of temperature. In summer the water absorbs heat and helps to prevent air temperatures rising; in winter the heat stored in the water is available to help prevent temperatures falling. Heat exchange devices, such as boilers and heating pipes, also make use of the high heat capacity of water for transferring heat from one place to another.

Density

The heat capacities of different materials are compared on the basis of equal masses. However, the same mass of different materials may occupy different volumes of space, depending upon their densities.

$$\text{Density } \rho = \frac{\text{Mass } (m)}{\text{Volume } (v)}$$

Unit: kilogram per cubic metre (kg/m³)

Heavyweight masonry materials, such as brick, concrete and stone, have high densities. This means that relatively small volumes of these materials have a large mass and they therefore provide a relatively high heat capacity within a small volume. An electric storage heater, for example, contains bricks which are heated by cheap-rate electricity and then hold this heat for use later in the day.

The heat storage provided by the brick, concrete and stone used in construction is particularly relevant to the thermal behaviour of buildings, as discussed in Chapter 2.

Change of state

All matter is made from small particles called *atoms* and for most materials the smallest particle that exists independently is a group of atoms which are combined to form a *molecule*. The spacing of the molecules in a substance and the forces between them determine the phase, or state of matter, of that substance.

In the normal ranges of temperature and pressure there are three possible states of matter and they have the basic characteristics described below.

- **Solid state**: the molecules are held together in fixed positions; the volume and shape are fixed.
- **Liquid state**: the molecules are held together but have freedom of movement; the volume is fixed but the shape is not fixed.
- **Gas state**: the molecules move rapidly and have complete freedom; the volume and shape are not fixed.

The state of a substance depends upon the conditions of temperature and pressure which act on it. Consider, for example, the common forms of iron, water and oxygen. At certain temperatures a material will undergo a change of state and in this change its energy content is increased or decreased.

The *absorption* of heat by a solid or a liquid can produce the following changes of state.

$$\text{SOLID} \xrightarrow[\text{(melting)}]{\text{Liquefaction}} \text{LIQUID} \xrightarrow[\text{(boiling, evaporation)}]{\text{Vaporisation}} \text{GAS}$$

The *release* of heat from a gas or a liquid can produce the following changes of state.

$$\text{GAS} \xrightarrow{\text{Condensation}} \text{LIQUID} \xrightarrow[\text{(fusion)}]{\text{Solidification}} \text{SOLID}$$

Sensible and latent heat

In order to understand how most substances behave, it is useful to consider the changes of state for water. Figure R2.1 shows the effects of supplying heat energy at a constant rate to a fixed mass of ice.

Sensible heat

When the sample exists entirely in a single state of ice, water or steam, the temperature rises uniformly as heat is supplied. This heat is termed 'sensible' because it is apparent to the senses.

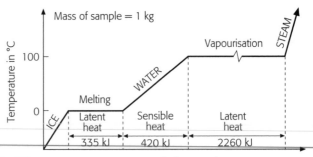

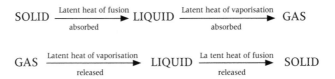

Figure R2.1 Changes of state for water

- **Sensible heat** is the heat energy absorbed by or released from a substance during a change in temperature.

Latent heat

When the sample is changing from one state to another the temperature remains constant, even though heat is being supplied. This heat is termed 'latent' because it seems to be hidden.

- **Latent heat** is the heat energy absorbed by or released from a substance during a change of state, with no change in temperature.

The latent heat absorbed by melting ice or by boiling water is energy which does work in overcoming the bonds between molecules. It is a less obvious, but very important, fact that this same latent heat is given back when the steam changes to water, or the water changes to ice. The latent heat changes occur for any substance and have the following general names:

$$\text{SOLID} \xrightarrow[\text{absorbed}]{\text{Latent heat of fusion}} \text{LIQUID} \xrightarrow[\text{absorbed}]{\text{Latent heat of vaporisation}} \text{GAS}$$

$$\text{GAS} \xrightarrow[\text{released}]{\text{Latent heat of vaporisation}} \text{LIQUID} \xrightarrow[\text{released}]{\text{Latent heat of fusion}} \text{SOLID}$$

A liquid may change to a gas without heat being supplied, by evaporation for example. The latent heat required for this change is then taken from the surroundings and produces an important cooling effect.

Enthalpy

Enthalpy can be described as the total heat content of a sample, with reference to 0°C. For the particular example of water shown in Figure R2.1, the steam at 100°C has a much higher total heat content than liquid water at 100°C. Steam at high temperature and pressure has a very high enthalpy, which makes it useful for transferring large amounts of energy such as from a boiler to a turbine. This steam is also very dangerous if it escapes.

Calculation of heat quantities

Both sensible and latent heat are forms of heat energy that are measured in joules, although they are calculated in different ways.

Sensible heat

When a substance changes temperature, the amount of sensible heat absorbed or released is given by the following formula.

$$H = m\,c\,\Delta\theta$$

where

H = quantity of sensible heat (J)
m = mass of substance (kg)
c = specific heat capacity of that substance (J/kg K)
$\Delta\theta$ = $\theta_2 - \theta_1$ = temperature change (°C)

Worked Example R2.1

A storage heater contains concrete blocks with total dimensions of 800 mm by 500 mm by 220 mm. The concrete has a density of 2400 kg/m³ and a specific heat capacity of 3300 J/kg K. Ignoring heat losses, calculate the quantity of heat required to raise the temperature of the blocks from 15°C to 35°C.

Know

Volume = $0.8 \times 0.5 \times 0.22 = 0.088$ m³

$$\text{Density} = \frac{\text{Mass}}{\text{Volume}}$$

Therefore

mass = density × volume
m = $2400 \times 0.088 = 211.2$ kg

Using formula for sensible heat and substituting values into formula

$$\begin{aligned}
H &= mc\,(\theta_2 - \theta_1) \\
&= 211.2 \times 3300 \times (35 - 15) \\
&= 13\,939\,200 \text{ J}
\end{aligned}$$

So quantity = **13.94 MJ**

Latent heat

During a change of state in a substance the amount of latent heat absorbed or released is given by the following formula.

$$H = ml$$

where

H = quantity of latent heat (J)
m = mass of substance (kg)
l = specific latent heat for that change of state (J/kg)

- **Specific latent heat** (l) is a measure of the latent heat absorbed by or released from a particular material for a given change of state.

    Unit: J/kg.

Specific latent heat is sometimes termed *specific enthalpy change* and some common values are as follows.

Specific latent heat of ice = 335 000 J/kg = 335 kJ/kg

Specific latent heat of steam = 2 260 000 J/kg = 2 260 kJ/kg

Worked Example R2.2

Calculate the total heat energy required to completely convert 2 kg of ice at 0°C to steam at 100°C. The specific heat capacity of water is 4190 J/kg°C. The specific latent heats are 335 kJ/kg for ice and 2260 kJ/kg for steam.

Step 1

Divide the process into sensible and latent heat changes:

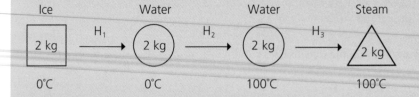

Step 2

Use formulas for latent or sensible heat to calculate each change separately.

Changing ice at 0°C to water at 0°C requires a latent heat calculation:

Using $H = ml$

$\quad H_1 = 2 \times 335\,000 = 670\,000$ J

$\quad H_1 = 670$ kJ

Changing water at 0°C to water at 100°C requires a sensible heat calculation:

Using $H = m\,c\,\Delta\theta$

$\quad H_2 = 2 \times 4190 \times 100 = 838\,000$ J

$\quad H_2 = 838$ kJ

Changing water at 100°C to steam at 100°C requires a latent heat calculation:

Using $H = ml$

$\quad H_3 = 2 \times 2\,260\,000 = 4\,520\,000$ J

$\quad H_3 = 4520$ kJ

Step 3

Combine the separate stages and present the answer in suitable units:

Total heat $= H_1 + H_2 + H_3$

$\quad\quad\quad\quad = 670 + 838 + 4520$

So total heat = **6 028 kJ**

Expansion

Most substances expand on heating and contract on cooling. If the natural expansion and contraction of a body is restricted, then very large forces may occur. Different substances expand by different amounts, and the coefficient of linear thermal expansion is a measure of the relative change of length. Superficial (area) expansion and cubical (volumetric) expansion can be predicted from the linear expansion.

Solids

The coefficient of linear expansion for steel is about $12 \times 10^{-6}/°C$, which means that a steel bar increases its length by 12/1,000,000 for each degree of temperature rise. Concrete expands at a similar rate to steel. The expansion of

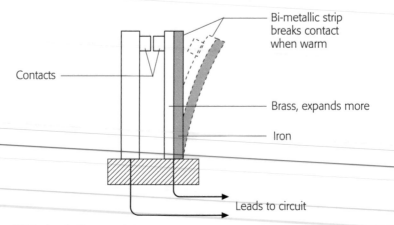

Figure R2.2 simple thermostat

aluminium is about twice that of steel, and the expansion of plastics is up to ten times that of steel.

Allowance must be made in constructions for the effects of expansion, particularly for concrete, metals and plastics. The result of destructive expansion can be seen in the twisted girders of a building after a fire. Expansion effects can also be useful. Heated rivets, for example, contract on cooling and tighten the joint between metal plates. The unequal expansion of two metals deforms a bi-metallic strip, which can then be used as a temperature switch like the thermostat shown in Figure R2.2.

Liquids

Liquids tend to expand more than solids, for the same temperature rise. The expansion rates of different liquids vary, and the expansion of alcohol is about five times that of water. Most liquids contract upon cooling, but water is unusual in that its volume increases as it cools from 4°C to 0°C. At 0°C the volume of water expands by a larger amount as it changes to the solid state of ice.

Thermometers make use of the expansion of liquids, but in the hot water systems of buildings and car engine blocks the expansion is troublesome if it is not allowed for. Rainwater that freezes and expands in the pores and crevices of concrete or stonework will disrupt and split the material. This process has, over long periods of time, destroyed whole mountain ranges.

Gases

The expansion of gases is hundreds of times greater than the expansion of liquids. This will not be apparent if the gas is confined in a container because the pressure will then increase instead of the volume; this behaviour will be described in the later section on gases and vapours. If a gas is allowed to expand under conditions of constant pressure, then the coefficient of volumetric expansion is found to be 1/273 per degree, starting at 0°C.

The concept of an absolute zero of temperature at -273°C was originally a result of imagining the effect of cooling an ideal gas. Starting at 0°C, this ideal gas would shrink in size by 1/273 for each drop in temperature of 1°C. At -273°C the volume of the gas would therefore be zero, and matter would have disappeared. It is not possible to achieve this condition, although it is possible to approach close to it. Real gases do not actually stay in the gaseous state at very low temperatures, but the concept of an absolute zero of temperature remains valid.

HEAT TRANSFER

Heat energy always tends to transfer from high temperature to low temperature regions. There is no real thing as 'cold' that flows into warm places, even though the human senses may interpret the loss of heat energy as a 'cold flow'. If several bodies at different temperatures are close together, then heat will be exchanged between them until they are at the same temperature.

This equalising of temperature can occur by the three basic processes of heat transfer given below:

- conduction
- convection
- radiation.

Heat may also be transferred by the process of evaporation when latent heat is absorbed by a vapour in one place and released elsewhere.

Conduction

If one end of a metal bar is placed in a fire then, although no part of the bar moves, the other end will become warm. Heat energy travels through the metal by the process of conduction.

- **Conduction** is the transfer of heat energy through a material without the molecules of the material changing their basic positions.

Conduction can occur in solids, liquids and gases although the speed at which it occurs will vary. At the place where a material is heated, the molecules gain energy and this energy is transferred to neighbouring molecules which then become hotter. The transfer of energy may be achieved by the drift of *free electrons*, which can move from one atom to another, especially in metals. Conduction also occurs by vibrational waves called *phonons*, especially in non-metals.

Different materials conduct heat at different rates and the measurement of thermal conductivity is described in Chapter 3. Metals are the best conductors of heat, because of the free electrons that they possess. Good conductors have many applications for the efficient transfer of heat, for example in boilers and heating panels.

Poor conductors are called insulators and include most liquids and gases. Porous materials that trap a lot of air tend to be good insulators and are of particular interest in controlling heat conduction through the fabric of buildings.

Measurement of thermal conductivity

Thermal conductivity is a measure of the rate at which heat is conducted through a particular material under specified conditions. The symbols used for coefficient of thermal conductivity are k or λ (lamda), leading to the term 'k-value', which is measured in units of W/m K.

The coefficient of thermal conductivity is measured as the heat flow in watts across a thickness of 1 m of material for a temperature difference of 1 degree K and a surface area of 1 m². Different techniques of practical testing are needed for different types of material, but the measurements required are shown schematically in Figure R2.3. The following general formula is then used to calculate the value of thermal conductivity for the material tested:

$$\frac{H}{t} = \frac{\lambda A (\theta_1 - \theta_2)}{d}$$

where

λ	= coefficient of thermal conductivity for that material (W/m K)
H/t	= rate of heat flow between the faces (J/s = W)
A	= cross-sectional area of the sample (m²)
$\theta_1 - \theta_2$	= temperature difference between the faces (°C or K)
d	= distance between the faces (m)

Resistivity (r) is an alternative index of conduction in materials and is the reciprocal of thermal conductivity, so that r = 1/λ. Similarly, the unit of resistivity is the reciprocal of the unit k-value unit: m K/W.

$$\text{Resistivity } r = 1/k$$

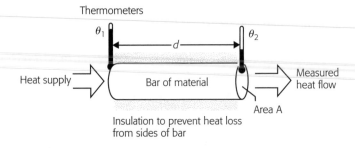

Figure R2.3 Thermal conductivity measurements

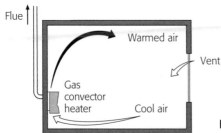

Figure R2.4 Convection currents in room

Convection

Air is a poor conductor of heat yet it is still possible to heat all the air in a room from a single heating panel by the process of convection.

- **Convection** is the transfer of heat energy through a material by the bodily movement of particles.

Convection can occur in fluids (liquids and gases), but never in solids. Natural convection occurs when a sample of fluid, such as air, is heated and so expands. The expanded air is less dense than the surrounding air and the cooler air displaces the warmer air causing it to rise. The new air is then heated and the process is repeated, giving rise to a 'convection current'.

Natural convection occurs in hot-water storage tanks, which are heated by an electric element or heat exchanger coils near the bottom of the tank; convection currents ensure that all the water in the tank is heated. The term *stack effect* describes the natural convection that occurs in buildings, causing warm air to flow from the lower to the upper stories. Forced convection uses a mechanical pump to achieve a faster flow of fluid, as in the water cooling of a car engine or in a small-bore central heating system.

> **See also:**
> section on *Natural ventilation* in Chapter 4

Radiation

Heat is transferred from the Sun to the Earth through space, where conduction and convection are not possible. The process of radiation is responsible for the heat transfer through space and for many important effects on Earth.

- **Radiation** is the transfer of heat energy by electromagnetic waves.

Heat radiation occurs when the thermal energy of surface atoms in a material generates electromagnetic waves in the infra-red range of wavelengths. These waves belong to the large family of electromagnetic radiations, including light and radio waves, whose general properties are described in Chapter 6.

The rate at which a body emits or absorbs radiant heat depends upon the nature and temperature of its surface. Rough surfaces present a larger total area and absorb or emit more heat than polished ones. Surfaces which appear dark, because they absorb most light, also absorb most heat. Good absorbers are also good emitters. Poor absorbers are also poor emitters. In general, the following rules are true.

- Dull black surfaces have the **highest** absorption and emission of radiant heat.
- Shiny silver surfaces have the **lowest** absorption and emission of radiant heat.

These surface properties are put to use for encouraging heat radiation as, for example, in a blackened solar energy collector. Or they can be used for discouraging heat radiation as, for example, by the use of aluminium foil insulation.

The rate at which a body emits heat increases with the temperature of the body. Every object is continuously emitting and absorbing heat to and from its surroundings. Prévost's Theory of Exchanges explains that the balance of these two processes determines whether or not the temperature of the object rises, falls, or stays the same.

The wavelengths of the radiation emitted by a body also depend upon the temperature of the body. High temperature bodies emit a larger proportion of short wavelengths, which have a better penetration than longer wavelengths. The short wavelengths emitted by hot bodies also become visible at about 500°C, when they first appear as dull red.

The greenhouse effect

The 'greenhouse effect', as illustrated in Figure R2.5, is one result of the differing properties of heat radiation when it is generated by bodies at different temperatures.

See also:

References to global warming in Chapters 1 and 14

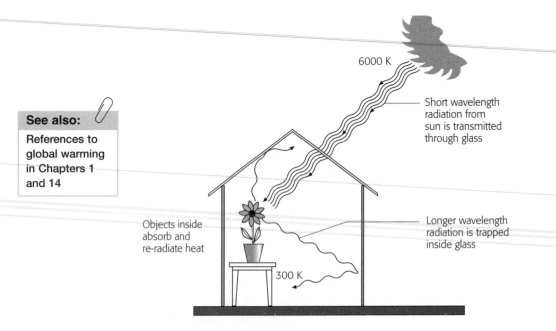

Figure R2.5 Greenhouse effect

The high-temperature sun emits radiation of short wavelength which can pass through the atmosphere and through glass. Inside a greenhouse or other building this is absorbed by objects such as plants, which then re-radiate the heat. Because the objects inside the greenhouse are at a lower temperature than the sun, the radiated heat is of longer wavelengths which do not easily penetrate glass. This re-radiated heat is therefore trapped and causes the temperature inside the greenhouse to rise.

The atmosphere surrounding the earth also behaves as a large 'greenhouse' around our world. There is a balance between the heat absorbed and given off by the planet Earth. Increasing certain 'greenhouse' gases in the atmosphere, such as carbon dioxide from the burning of fossil fuels, reduces the quantity of heat that would otherwise radiate back into space. The planet retains more heat and so this particular greenhouse effect therefore contributes to *global warming*.

GASES AND VAPOURS

Gases

The gas state is one of the three principal states in which all matter exists. According to the *kinetic theory*, the molecules of a gas are always in motion and their velocity increases with temperature. When the molecules are deflected by the walls of a container there will be a change in their momentum and a force imparted to the wall. The collisions of many molecules acting on a particular area will be detected as pressure.

$$\textbf{Pressure } (p) = \frac{\text{Force } (F)}{\text{Area } (A)}$$

Unit: pascal (Pa)

By definition: 1 pascal = 1 newton/metre^2 (1 N/m^2)

Other units which may be found in use include the following:
    millibars (mb), where 1 mb = 100 Pa
    mm of mercury, where 1 mm = 133 Pa

Gas laws

Heating a gas increases the velocity and the kinetic energy of the molecules. If the gas is free to expand then the heated molecules will move further apart and increase the volume of gas. If the volume is fixed, the heated molecules will exert a greater force at each collision with the container and so the pressure of the gas increases. The gas laws are an expression of the relationships between the temperature, volume and pressure of a constant mass of gas.

Boyle's law

For a fixed mass of gas at constant temperature, the volume (V) is inversely proportional to the pressure (p).

$$pV = \text{constant} \qquad \text{or } p_1 V_1 = p_2 V_2$$

Charles' law

For a fixed mass of gas at constant pressure, the volume (V) is directly proportional to the thermodynamic temperature (T).

$$V = \text{constant} \times T$$

Pressure law

For a fixed mass of gas at constant volume, the pressure (p) is directly proportional to the thermodynamic temperature (T).

$$p = \text{constant} \times T$$

General gas law

The relationships between the pressure, volume, and temperature of a gas can be combined into one expression.

$$\frac{pV}{T} = \text{constant} \qquad \text{or} \frac{p_1 V_1}{T_1} = \frac{p_2 V_2}{T_2}$$

Note: temperature must always be in degrees kelvin.

Dalton's law of partial pressures

Where there is a mixture of different gases each gas exerts an individual partial pressure and has the following features.

- The partial pressure exerted by each gas is independent of the pressure of the other gases.
- The total pressure of the mixture equals the sum of the partial pressures.

Standard temperature and pressure (STP)

In order to compare gases measured under different conditions of temperature and pressure it is convenient to convert them to the following standard temperature and pressure (STP):

$$
\left.
\begin{array}{ll}
\text{Standard temperature} & = 0°C = 273K \\
\text{Stardard pressure} & = 101.3 \text{ kPa} \\
& = 760 \text{ mm of mercury}
\end{array}
\right\} 1 \text{ atmosphere}
$$

Worked Example R2.3

At 20°C temperature and 200 kPa pressure a certain sample of gas occupies a volume of 3 litres. What volume will this sample occupy at standard temperature and pressure?

Initial conditions: $p_1 = 200$ kPa, $V_1 = 3$ litres,

$$T_1 = 273 + 20 = 293 \text{ K}$$

Final conditions: $p_2 = 101.3$ kPa, $V_2 = ?$, T2 $= 273$ K

Using general gas law and substituting values

$$\frac{p_1 V_1}{T_1} = \frac{p_2 V_2}{T_2}$$
$$\frac{200 \times 3}{293} = \frac{101.3 \times V_2}{273}$$

Therefore

$$V_2 = \frac{200 \times 3 \times 273}{293 \times 101.3} = 5.519$$

So final volume = **5.519 litres**

Vapours

A vapour is a material in a special form of the gas state and has some different properties from those of a gas. For example, when a vapour is compressed the pressure increases until, at a certain point, the vapour condenses to a liquid.

- A **vapour** is a material in the gas state which can be liquefied by compression, without change in temperature.

The *critical temperature* of a substance is the temperature above which a vapour is not able to exist. Table R2.3 gives the critical temperatures of some substances relevant to heating and refrigeration. So it is seen, for example, that steam is a vapour at 100°C but a gas at 500°C.

The atmosphere

Air is a mixture of gases and has the percentage composition shown in Table R2.4 when it is clean and dry.

Table R2.3 Critical temperatures

Substance	Critical temperature (°C)
Oxygen (O_2)	-119
Air	-141
Carbon dioxide (CO_2)	31
Ammonia (NH_3)	132
Water (H_2O)	374

The atmosphere is the collection of gases that surrounds the surface of the Earth. In addition to air the atmosphere contains up to several per cent of water vapour, and may contain local pollution products.

At the surface of the Earth the atmosphere produces a fluid pressure that depends upon the average density and the height of the atmosphere above that point. This pressure acts in all directions and varies with altitude and with local weather conditions. At sea level atmospheric pressure has a standard value of 101.3 kPa and is measured by barometers.

> 101.3 kPa is also expressed as 1013 hPA (hecto pascals)

The simple *mercury barometer* consists of a glass tube sealed at one end which is filled with mercury and inserted upside down into a dish of mercury. A certain height of mercury remains in the tube, its weight balanced by the force resulting from atmospheric pressure. A standard atmosphere supports a 760 mm column of mercury and this height changes with changes in atmospheric pressure. The absolute unit of pressure is the pascal but it is also convenient to refer to pressure in terms of height, such as mm of mercury.

The *aneroid barometer* uses a partially evacuated box of thin metal whose sides move slightly with changes in atmospheric pressure. The movements are magnified by levers and shown by a pointer on a scale. Aneroid barometers are commonly used as altimeters for aircraft and as household weather gauges.

Table R2.4 Composition of air

Gas	Percentage of air
Nitrogen (N_2)	78
Oxygen (O_2)	21
Carbon dioxide (CO_2) and other gases	1

Principles of Light and Sound

NATURE OF LIGHT

Light is energy in the form of electromagnetic radiation. This energy is radiated by processes in the atomic structure of different materials and causes a wide range of effects. The different forms of electromagnetic radiation all share the same properties of transmission although they behave quite differently when they interact with matter.

Light is that particular electromagnetic radiation which can be detected by the human sense of sight. The range of electromagnetic radiation to which the eye is sensitive is just a very narrow band in the total spectrum of electromagnetic emissions, as is indicated in Figure R3.1.

Electromagnetic waves

The transmission of light energy can be described as a wave motion or as 'packets' of energy called photons. The two theories co-exist in modern physics and

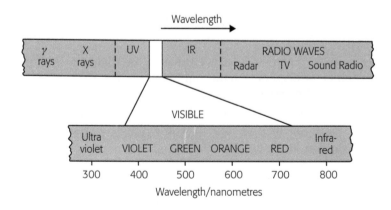

Figure R3.1 Electromagnetic spectrum

are used to explain different effects. The most convenient theory for everyday effects is that of electromagnetic wave motion. This can be considered as having the general properties listed below.

Transverse waves: vibrations are at right angles to direction of travel

- The energy resides in fluctuations of electric and magnetic fields, which travel as a transverse wave motion.
- The waves require no medium and can therefore travel through a vacuum.
- Different types of electromagnetic radiation have different wavelengths or frequencies.
- All electromagnetic waves have the same velocity, which is approximately 3×10^8 m/s in vacuum.
- The waves travel in straight lines unless affected by reflection, refraction and diffraction.

Although not visible to us, the wave motion of light and other electromagnetic waves is the same mechanism found in sea waves, sound waves and earthquakes. All waves can change their direction when subject to the following effects.

- **Reflection** is reversal of direction which occurs at a surface. Examples include mirrors and coloured surfaces.
- **Refraction** is deflection that occurs at the boundaries of different materials. For example: prism effects occur at the edge of air and glass, red sunsets are caused by differing layers of the atmosphere.
- **Diffraction** is deflection that occurs at apertures, at edges and in thin layers. Examples include coloured patterns in thin layers of oil, and coloured spectrums caused by narrow slits.

Visible radiation

The wavelengths of electromagnetic radiation that are visible to the eye range from approximately 380 nm to 760 nm. If all the wavelengths of light are seen at the same time, the eye cannot distinguish the individual wavelengths and the brain has the sensation of white light.

1 nanometre (nm) is 10^{-9} metre

- **White light** is the effect on sight of combining all the visible wavelengths of light.

White light can be separated into its component wavelengths. One method is to use the different degrees of refraction of light that occur in a glass prism, as shown in Figure R3.2. The result is a spectrum of light, which is traditionally described in the seven colours of the rainbow although, in fact, there is a continuous range of hues (colours) whose different wavelengths cause different sensations in the brain.

- **Monochromatic light** is light of one particular wavelength and colour.

If the colours of the spectrum are recombined then white light is again produced. Varying the proportions of the individual colours can produce different qualities of 'white' light.

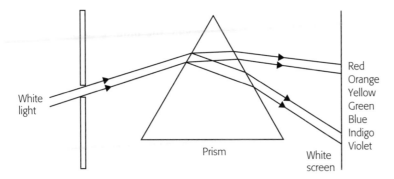

Figure R3.2 Dispersion of white light

Non-visible radiation

Electromagnetic radiation with wavelengths outside the range of visible wavelengths cannot, by definition, be detected by the human eye. However, there are wavelengths emitted by the sun which are adjacent to the visible range of wavelengths, and although they cannot be seen they are relevant to lighting processes.

Infrared

Infrared (IR) radiation has wavelengths slightly greater than those of red light and can be felt as heat radiation from the sun and from other heated bodies. Infrared radiation is made use of in radiant heating devices, for detecting patterns of heat emissions, for 'seeing' in the dark, and for communication links.

Ultraviolet

Ultraviolet (UV) radiation has wavelengths slightly less than those of violet light. It is emitted by the sun and also by other objects at high temperature. Ultraviolet radiation helps keep the body healthy but excessive amounts can damage the skin and the eyes. The composition of the earth's atmosphere normally protects the planet from excessive UV radiation emitted by the Sun.

Ultraviolet radiation can be used to kill harmful bacteria in kitchens and in hospitals. Certain chemicals can convert UV energy to visible light, and the effect is made use of in fluorescent lamps.

NATURE OF VISION

The portion of the electromagnetic spectrum known as light is of environmental interest to human beings because it activates our sense of sight, or vision. Vision is a sensation caused in the brain when light reaches the eye. The eye initially treats light in an optical manner, producing a physical image in the same way as a camera. This image is then interpreted by the brain in a manner which is psychological as well as physical.

The eye

Figure R3.3 shows the main features of the human eye with regard to its optical properties. The convex lens focuses the light from a scene to produce an inverted image of the scene on the retina. When in the relaxed position, the lens is focused on distant objects. To bring closer objects into focus the ciliary muscles increase the curvature of the lens, a process called *accommodation*. The closest distance at which objects can be focused, called the *near point*, tends to retreat with age as the lens become less elastic.

> *Eye vs camera:*
> lens – lens
> iris – aperture
> retina – film or
>   CCD

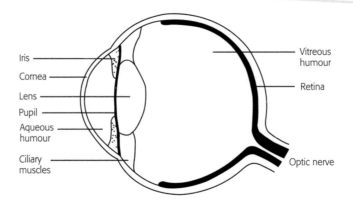

Figure R3.3 Structure of the eye

Labels: Iris, Cornea, Lens, Pupil, Aqueous humour, Ciliary muscles, Vitreous humour, Retina, Optic nerve

The amount of light entering the lens is controlled by the iris, a coloured ring of tissue, which automatically expands and contracts with the amount of light present. The retina, on which the image is focused, contains light receptors which are concentrated in a central area called the *fovea*, and are deficient in another area called the *blind spot*.

Operation of vision

> *Photometry:*
> the science of
> measuring visible
> light in units that
> are linked to the
> sensitivity of the
> human eye

The light energy falling on the retina causes chemical changes in the receptors, which then send electrical signals to the brain via the optic nerve. A large portion of the brain is dedicated to the processing of the information received from the eyes, and the eyes are useless if this sight centre in the brain is damaged.

The initial information interpreted by the brain includes the brightness and colour of the image. The stereoscopic effect of two eyes gives further information about the size and position of objects. The brain controls selection of the many items in the field of view and the sense of vision greatly depends on interpretations of images learned from previous experience.

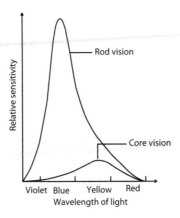

Figure R3.4 Sensitivity of the eye

Sensitivity of vision

The light-sensitive receptors on the retina are of two types. These receptors respond to different wavelengths of light in the manner shown in Figure R3.4, and they give rise to two types of vision.

Cone vision

The cones are the light receptors that operate when the eye is adapted to normal levels of light. The spectrum appears coloured. There is a concentration of cones on the fovea at the centre of the retina and these are used for seeing details.

Rod vision

The rods are the light receptors that operate when the eye is adapted to very low levels of light. The rods are much more sensitive than the cones but the spectrum appears uncoloured. The colourless appearance of objects in moonlight or starlight is an example of this vision. There is a concentration of rods at the edges of the retina, which cause the eyes to be sensitive to movements at the boundary of the field of view.

Terminology of vision

Visual field

Visual field is the total extent in space that can be seen when looking in a given direction.

Visual acuity

Acuity is the ability to distinguish between details that are very close together. This ability increases as the amount of available light increases.

Adaption

Adaption is the process occurring as the eyes adjust to the relative brightness or colour of objects in the visual field. The cones and the rods on the retina take a significant amount of time to reach full sensitivity.

Contrast

Contrast is the difference in brightness or colours between two parts of the visual field.

NATURE OF SOUND

Origin of sound

Sound is a variation in the pressure of the air of a type which has an effect on our ears and brain. These pressure variations transfer energy from a source of vibration that can be naturally occurring, such as the wind, or produced by humans, such as speech. Sound in the air can be caused by a variety of vibrations, including:

- **Moving objects**: examples include loudspeakers, guitar strings, vibrating walls and human vocal chords.
- **Moving air**: examples include horns, organ pipes, mechanical fans and jet engines.

A vibrating object compresses adjacent particles of air as it moves in one direction and leaves the particles of air 'spread out' as it moves in the other direction. This effect is illustrated in Figure R3.5. The displaced particles pass on their extra energy and a pattern of compressions and rarefactions travels out from the source, while the individual particles return to their original positions.

In addition to its link with human hearing, the term *sound* is also used for other phenomena in air governed by similar physical principles. Disturbances in the air with frequencies of vibration which are too low (*infrasound*) or too high (*ultrasound*) to be heard by human hearing are also regarded as sound. Other sound terms in common usage include: underwater sound, sound in solids or structure-borne sound.

Infrasound: frequency too low for human hearing
Ultrasound: frequency too high for human hearing

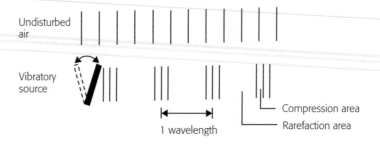

Figure R3.5 Pressure variations in a sound wave

Wave motion

The mechanical vibrations of sound move forward using wave motion. This means that, although the individual particles of material such as air molecules return to their original position, the sound energy obviously travels forward. The front of the wave spreads out equally in all directions unless it is affected by an object or by another material in its path.

The waves are *longitudinal* in type because the particles of the medium carrying the wave vibrate in the same direction as the travel of the wave, as shown in Figure R3.5. The sound waves can travel through solids, liquids and gases, but not through a vacuum.

It is difficult to depict a longitudinal wave in a diagram so it is convenient to represent the wave as shown in Figure R3.6, which is a plot of the vibrations against time. For a pure sound of one frequency, as in Figure R3.6, the plot takes the form of a sine wave.

Sound waves are like any other wave motion and therefore can be specified in terms of wavelength, frequency and velocity.

Wavelength

- **Wavelength** (λ) is the distance between any two repeating points on a wave.

    Unit: metre (m).

In Figure R3.5 a wavelength is shown measured between two compressions, but the length between any two repeating points would be the same.

Frequency

- **Frequency** (f) is the number of cycles of vibration per second.

    Unit: hertz (Hz)

Figure R3.6 shows two complete vibrations, or cycles.

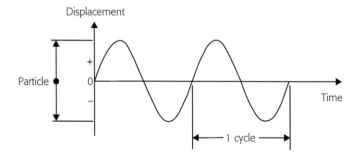

Figure R3.6 Vibrations of a sound wave

Velocity

- **Velocity** (*v*) is the distance moved per second in a fixed direction.

    Unit: metres per second (m/s)

For every vibration of the sound source the wave moves forward by one wavelength. The number of vibrations per second therefore indicates the total length moved in 1 second; which is the same as velocity. This relationship is true for all wave motions and can be written as the following formula.

$$v = f \times \lambda$$

where
  v = velocity in m/s
  f = frequency in Hz
  λ = wavelength in m

Worked Example R3.1

A particular sound wave has a frequency of 440 Hz and a velocity of 340 m/s. Calculate the wavelength of this sound.

Know

  $v = 340$ m/s, $f = 440$ Hz, $\lambda = ?$

Using formula for velocity and substituting

  $v = f \times \lambda$

  $340 = 440 \times \lambda$

  $\lambda = \dfrac{340}{440} = 0.7727$

So wavelength = **0.7727 m**

Velocity of sound

A sound wave travels away from its source at a speed of 344 m/s (770 miles per hour) when measured in dry air at 20°C. This is a respectable speed within a room but slow enough over the ground for us to notice the delay between seeing a source of sound, such as a distant firework, and later hearing the explosion.

The velocity of sound is independent of the rate at which the sound vibrations occur, which means that the frequency of a sound does not affect its speed. The velocity is also unaffected by variations in atmospheric pressure such as those caused by the weather.

But the velocity of sound is affected by the properties of the material through which it is travelling, and Table R3.1 gives an indication of the velocities of sound in different materials. The velocity of sound in gases decreases with increasing density as, when the molecules are heavier, then they move less readily. Moist air contains a greater number of light molecules and therefore sound travels slightly faster in moist humid air.

> *Velocity of sound*: varies directly with the square root of elasticity; varies inversely with the square root of density

Table R3.1 Velocity of sound

Material	Typical velocity (m/s)
Air (0°C)	331
Air (20°C)	344
Water (25°C)	1498
Pine	3300
Glass	5000
Steel	5000
Granite	6000

Sound travels faster in liquids and solids than it does in air because of the effect of density and elasticity of those materials. The particles of such materials respond to vibrations more quickly and so convey the pressure vibrations at a faster rate. For example, steel is very elastic and sound travels through steel about 14 times faster than it does through air.

Frequency of sound

If an object that produces sound waves vibrates 100 times a second, for example, then the frequency of that sound wave will be 100 Hz. The human ear hears this as sound of a certain pitch.

- **Pitch** is the frequency of a sound as perceived by human hearing.

Low-pitched notes are caused by low-frequency sound waves and high-pitched notes are caused by high-frequency waves. The pitch of a note determines its position in the musical scale. The frequency range to which the human ear responds is approximately 20 to 20,000 Hz and frequencies of some typical sounds are shown in Figure R3.7.

Most sounds contain a combination of many different frequencies, and it is usually convenient to measure and analyse them in ranges of frequencies, such as the octave.

- An **octave band** is the range of frequencies between any one frequency and double that frequency.

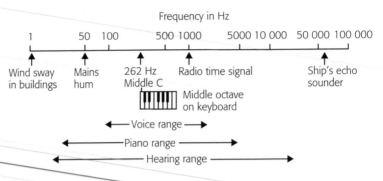

Figure R3.7 Frequency ranges of sound waves

> bass = low
> frequency
> treble = high
> frequency

For example, 880 Hz is one octave above 440 Hz. Octave bands commonly used in frequency analysis have the following centre frequencies:

31.5   63   125   250   500   1 000   2 000   4 000   8 000

The upper frequency of an octave band can be found by multiplying the centre frequency by a factor of $\sqrt{2}$.

Quality of sound

A pure tone is sound of only one frequency, such as that given by a tuning fork or electronic signal generator. Most sounds heard in everyday life are a mixture of more than one frequency, although a lowest *fundamental* frequency predominates when a particular 'note' is recognisable. This fundamental frequency is accompanied by *overtones* or *harmonics*.

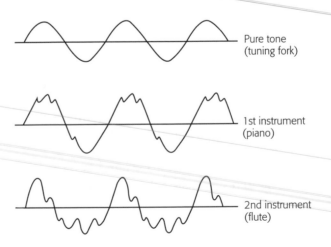

Figure R3.8 Different waveforms with different overtones

- **Overtones** and **harmonics** are frequencies equal to whole-number multiples of the fundamental frequency.

For example, the initial overtones of the note with a fundamental of 440 Hertz are as follows:

440 Hz = fundamental or 1st harmonic
880 Hz = 1st overtone or 2nd harmonic
1 320 Hz = 2nd overtone or 3rd harmonic etc.

Different voices and instruments are recognised as having a different quality when making the same note. This individual *timbre* results because different instruments produce different mixtures of overtones that accompany the fundamental, as shown in Figure R3.8. The frequencies of these overtones may well rise to 10,000 Hz or more and their presence is often an important factor in the overall effect of a sound. A telephone, for example, transmits few frequencies above 3000 Hz and the exclusion of the higher overtones noticeably affects reproduction of the voice and of music.

Cancellation of sound

The nature of a sound wave, as shown in Figure R3.6, means that the vibration of the wave has alternate changes in amplitude called *phases*. If a wave vibration in one direction meets an equal and opposite vibration, then they will cancel. The effect of this *phase inversion* in sound waves is to produce little or no sound and gives the possibility of 'cancelling' noise.

> **See also:**
> section on
> *Active Noise
> Reduction (ANR)*
> in Chapter 9

Resonance

Every object has a *natural frequency* which is the characteristic frequency at which it tends to vibrate when disturbed. For example, the sound of a metal bar dropped on the floor can be distinguished from a block of wood dropped in the same way. The natural frequency depends upon factors such as the shape, density and stiffness of the object.

Resonance occurs when the natural frequency of an object coincides with the frequency of any vibrations applied to the object. The result of resonance is extra large vibrations at this frequency.

Resonance may occur in many mechanical systems. For instance, it can cause loose parts of a car to rattle at certain speeds when they resonate with the engine vibrations. The swaying of a suspension bridge can resonate with footsteps from walkers. A wineglass can be shattered by the resonance from a singer's top note! Less dramatic, but of practical application in buildings, is that resonance affects the transmission and absorption of sound within partitions and cavities.

Principles of Electricity

CURRENT ELECTRICITY

Structure of matter

Electricity involves changes at the atomic level of materials. An atom is the smallest part of matter that has a separate chemical existence. Atoms contain many smaller particles and among the forces that bind the sub-atomic particles together is an electric property called charge. There are two kinds of charges: positive (+) and negative (−). The forces between charges obey the following rules.

- Like charges repel one another.
- Unlike charges attract one another.

There are three fundamental sub-atomic particles which help determine the nature of matter and give rise to electrical effects. The important properties of the fundamental sub-atomic particles are shown in Figure R4.1 and summarised below.

Protons

Protons have a positive electric charge, equal and opposite to that of the electron.

- Protons have a mass of 1 atomic mass unit.
- Protons are found in the central nucleus.

Neutrons

- Neutrons have no electric charge so are 'neutral'.
- Neutrons have the same mass as the proton.
- Neutrons are found in the nucleus with the proton.

Electrons

- Electrons have a negative electric charge, equal and opposite to that of the proton.

- Electrons have very small mass, approximately 1/1840 atomic mass unit.
- Electrons are found surrounding the nucleus.

The nucleus occupies a very small volume at the centre of the atom but contains all the protons and neutrons. Therefore, despite its small size, the nucleus contains nearly all the mass of an atom. The electrons can be considered as circulating in orbits around the nucleus, held in position by the opposing charge of the protons in the nucleus. An atom contains the same number of electrons as protons, so the positive and negative charges are balanced and the overall charge of an atom is zero.

The electrons in the outer orbits are held by relatively weak forces, so outer orbits can sometimes lose or gain electrons. *Free electrons* are electrons in outer orbits that are able to move from one atom to another. *Ionisation* is a process in which an atom permanently gains or loses electrons and so acquires an overall charge. This charged atom is then called an *ion*.

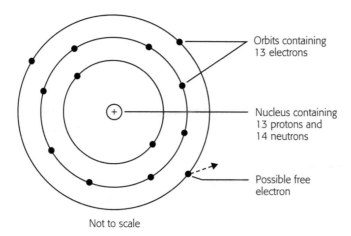

Figure R4.1 Aluminium atom, simple model

Electric charge

When electrons move from one place to another, by whatever mechanism, they transfer their electric charge. It is this charge that electricity is composed of.

- **Charge** (Q) is the basic quantity of electricity.

      Unit: coulomb (C)

> Electrical energy is stored as charge

The idea of charge or static electricity is more fundamental than current electricity. Although current electricity is more usual in everyday applications, electricity does not have to move to exist. A thundercloud, for example, contains a huge quantity of electricity, which does not flow except during a strike of lightning.

Electric current

If electric charge transfers through a material then an electric current is said to 'flow'. The movement of free electrons is the usual mechanism for the transfer of charge. The amount of electric current is described by the quantity of charge which passes a fixed point in a given time.

- **Electric current** (I) is the rate of flow of charge in a material.

    Unit: ampere (A)

This relationship is written as the following formula.

> **Current direction:** positive to negative

$$I = \frac{Q}{t}$$

where
I = electric current flowing (A)
Q = electric charge (C)
t = time taken (s)

Direction of current

Electrons have a negative charge and, by the rules of charge, they are attracted to a positive charge. Therefore, when electrons move through a material such as a cable, they are attracted to the positive side of the electrical supply, as shown in Figure R4.2. However, by convention, it is usual to say that direct electric current flows from positive to negative, even though electrons actually flow in the opposite direction. This convention works for practical problem solving as long as it is maintained consistently.

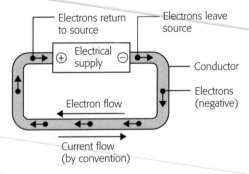

Figure R4.2 Electron and current flow

Effects of current

Electrons are too small to be detected themselves, but the presence of an electric current can be known by its effects. Three important effects are listed below.

- **Heating effect**: current flowing in a conductor generates heat. The amount of heat depends upon the amount of current but does not depend upon its direction. Applications of the heating effect include electric heaters and fused switches.
- **Magnetic effect**: current flowing in a conductor produces a magnetic field. The size and direction of the field is affected by the size and direction of the current. An electric motor is one example of this effect.
- **Chemical effect**: current flowing through some substances causes a chemical change and produces new substances. The type of change depends upon the amount and direction of the current. This type of effect is used in electroplating and refining processes.

Conductors and insulators

If a material allows a significant flow of electric current then the material is termed a conductor of electricity. A material that passes relatively little current is termed an insulator.

Conductors

Solid conductors are materials whose free electrons readily produce a flow of charge. If the conductor is a liquid or a gas then the charge is usually transferred by the movements of ions. Common types of conductor are listed below.

- **Metals**: examples include copper and aluminium cable conductors.
- **Carbon**: examples include sliding contacts in electric motors.
- **Liquids and gases**: current can flow when ions are present, for example in salty water or in a gas discharge lamp.

Insulators

Insulators are materials that have few free electrons available to produce a flow of charge. Common types of insulator are listed below.

- **Rubber and plastic polymers**: examples include PVC cable insulation.
- **Mineral powder**: examples include magnesium oxide cable insulation (MICC).
- **Oil**: examples include underground cable insulation.
- **Dry air**: examples include overhead power-line insulation.
- **Porcelain and glass**: examples include overhead power-line insulation.

Electrical potential

Potential difference

For an electric current to flow in a conductor there must be a difference in charge between two points. This potential difference is similar to the pressure difference that must exist for water to flow in a pipe.

- **Potential difference** (*pd* or *V*) is a measure of the difference in charge between two points in a conductor.

    Unit: volt (V)

The volt is defined in terms of the energy needed to move an electric charge. Because potential difference is measured in volts it is sometimes termed 'voltage drop'. Current flows from the point of higher potential to the point of lower potential, as shown in Figure R4.3.

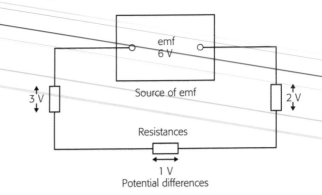

Figure R4.3 Potential difference and electromotive force

Electromotive force

In order to produce a potential difference and the resulting current, there must be a source of electrical 'pressure' acting on the charge. This source, shown in Figure R4.3, is called an electromotive force (emf).

- An **electromotive force** (*E*) is a supply of energy capable of causing an electric current to flow.

    Unit: volt (V)

Because the unit of emf is the volt, the same unit as for potential difference, emf of a circuit is sometimes called the 'voltage'. Common sources of emf are batteries and generators.

Resistance

Some materials oppose the flow of electric current more than others. Resistance is a measure of opposition to the flow of electric current and is related to the potential difference and the current associated with that opposition.

> Symbol for the ohm is the Greek letter omega Ω

Resistance $(R) = \dfrac{\text{Potential difference } (V)}{\text{Current } (I)}$

Unit: ohm (Ω) where $1\ \Omega = 1\ \text{V/A}$.

A *resistor* is a component that is used to provide resistance, and its resistance depends upon the following factors:

- **Length**: doubling the length doubles the resistance.
- **Cross-sectional area**: Doubling the area decreases the resistance by half.
- **Temperature**: the resistance of most metals increases with temperature.
- **Material**: the resistance provided by a particular material is given by a value of *electrical resistivity*. The resistivity of copper, for example, is about 1/2 that of aluminium, about 1/6 that of iron, and about 10^{-20} that of a typical plastic.

Ohm's law

The potential difference and the current associated with a resistance may be measured and the values compared. For most metal conductors it is found that if, for example, the potential difference (voltage) is doubled then the current also doubles. This relationship is expressed as Ohm's law.

- **Ohm's law**: For a metal conductor, at constant temperature, the current flowing is directly proportional to the potential difference across the conductor.

The constant of proportionality is, by definition, the resistance of the conductor, so that the following useful expression is a result from Ohm's law.

$$V = IR$$

where
V = potential difference across a component (V)
I = current flowing in the component (A)
R = resistance of the component (Ω)

Note that this expression can also be expressed in two other forms as follows.

$$I = \frac{V}{R} \quad \text{and } R = \frac{V}{I}$$

Worked Example R4.1

An electronic calculator has an overall resistance of 6 kΩ and is connected to a supply which provides a potential difference of 9 V. Calculate the current flowing to the calculator.

We know $V = 9$ $VI = ?$ $R = 6\,000\ \Omega$

Using Ohm's law formula and substituting values

$$V = IR$$
$$9 = I \times 6\,000$$
$$I = \frac{9}{6\,000} = 1.5 \times 10^{-3}\,A$$

So current = **1.5 mA**

Circuits

For a continuous electric current to be able to flow there must be a complete circuit path from, and back to, the source of electromotive force. In most circuits this complete path is supplied by two obvious conductors connecting the electrical device to the supply. Some systems, however, use less obvious means, such as the conduction of a metal structure or of the earth itself as one part of the circuit. For example, the metal chassis of a car or a television set is part of the circuit, and the earth itself forms part of the circuit for the distribution of electrical energy to buildings.

For simple problems the resistance of the connecting conductors is assumed to be negligible. The voltage drop produced by the resistance of practical lengths of cable can be predicted from published tables and, if it is excessive, a cable of lower resistance may be used. The theoretical layout of circuits is shown in a geometrical manner using standard symbols, as used for the diagrams in this chapter. In an actual wiring system the connections are made on the components, rather than on the conductors as shown in the circuit diagrams.

Circuit rules

There are basic principles regarding the behaviour of current, voltage and resistance that apply to any circuit or part of a circuit. If some of the electrical values are known for a circuit then others can be predicted by the use of these rules.

There are two basic layouts for connecting components in a circuit: parallel and series connection. Figures R4.4 and R4.5 show the two types of circuit using three components, but the circuit rules apply to any number of components.

Series connection

- **Current**: the current flowing in each resistor has the same value.
- **Voltage**: the sum of the voltage drops across all the resistors is equal to the applied voltage

$$V = V_1 + V_2 + V_3.$$

- **Resistance**: the total resistance of the circuit is equal to the sum of the individual resistances

$$R = R_1 + R_2 + R_3.$$

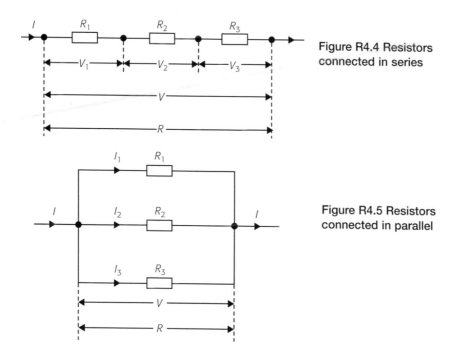

Figure R4.4 Resistors connected in series

Figure R4.5 Resistors connected in parallel

Parallel connection

- **Current**: the sum of the currents in all the resistors is equal to the total current flowing in the circuit

$$I = I_1 + I_2 + I_3$$

- **Voltage**: the voltage drop (potential difference) across each resistor has the same value.
- **Resistance**: the total resistance of the circuit is obtained from the reciprocals of the individual resistances.

$$\frac{1}{R} = \frac{1}{R_1} + \frac{1}{R_2} + \frac{1}{R_3}$$

Worked Example R4.2

A $3\,\Omega$ and a $6\,\Omega$ resistor are connected together in parallel and then connected in series with a $4\,\Omega$ resistor. If an emf of $24\,V$ is applied to this circuit then calculate the following items.

(a)   The total resistance of the circuit.
(b)   The total current flowing in the circuit.
(c)   The potential difference across the $3\,\Omega$ resistor.
(d)   The current flowing in the $3\,\Omega$ resistor.

Draw the circuit and label the known and unknown values, as in Figure R4.6.

(a) Let total resistance be R_T

$$\frac{1}{R_1} = \frac{1}{3} + \frac{1}{6} = \frac{2+1}{6} = \frac{3}{6} \text{ so } R_1 = 2\,\Omega$$

$R_2 = 4\,\Omega$

So $R_T = R_1 + R_2$

$\quad = 2 + 4 = \mathbf{6\,\Omega}$

(b) Let total current be I_T

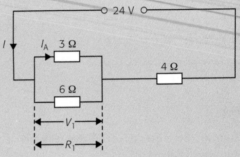

Figure R4.6 Current in circuit

Total $R = 6\,\Omega$, Total $V = 24$ V

$$I_T = \frac{V}{R} = \frac{24}{6} = \mathbf{4A}$$

(c) Potential difference V_1

Resistance $R_1 = 2\,\Omega$, Current in $R_1 = 4\,$A

$V_1 = I \times R_1$

$\quad = 4 \times 2 = \mathbf{8\ V}$

(d) Current I_A

$$I_A = \frac{V_1}{R} = \frac{8}{3}$$

$\quad = \mathbf{2.67\ A}$

Cells

A cell is a device that converts chemical energy to electrical energy and supplies an electromotive force (emf) capable of causing a direct electric current to flow. There are two classes of cell: primary and secondary cells.

Primary cells

- **Primary cells** are cells that *cannot* be recharged.

In a primary cell, the conversion of energy is not reversible and the cell must be replaced. Examples of these cells include simple cells, zinc-carbon cells, dry mercury cells, alkaline manganese cells and silver oxide cells.

Secondary cells

- **Secondary cells** are cells that *can* be recharged.

In a secondary cell, the conversion process can be reversed, the energy content replaced, and the cell used again. Examples of secondary cells include lead-acid accumulators used as car batteries, nickel-cadmium (NiCad) and Nickel-Metal Hydride (Ni-MH) cells found in mobile phones, digital cameras and other portable electronic items.

Types of cell

Some common forms of cell are described below; all of them have the same basic components of *electrodes* and *electrolyte*.

- **Electrodes**: electrodes are conductors which form the terminals of the cell. The *anode* is the positive electrode and the *cathode* is the negative electrode.
- **Electrolyte**: the electrolyte is a compound that undergoes chemical change and releases energy.

Simple cell

A simple or 'voltaic' cell is a primary cell which is constructed as shown in Figure R4.7. This cell is not a practical source of supply as it has a limited life, mainly because of polarisation. *Polarisation* is a reverse emf set up in the cell because hydrogen is liberated and deposited on the copper electrode. The general nature of a simple cell is an important mechanism in *electrolytic corrosion* when simple cells can occur between neighbouring areas of different metals.

Dry cell

The dry cell or zinc-carbon cell is the commonest type of primary cell and its construction is shown in Figure R4.8. This is a form of *Leclanché* cell with the electrolyte made as a paste rather than a liquid. The nominal emf of any single dry cell is 1.5 V and the cell contains a depolarising agent. It is suitable for intermittent use in torches, radios and similar devices.

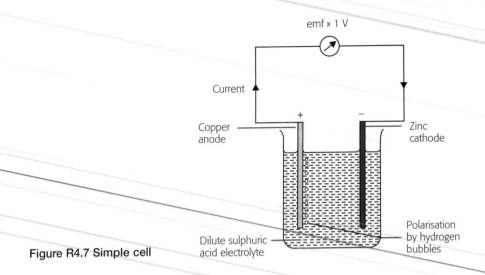

Figure R4.7 **Simple cell**

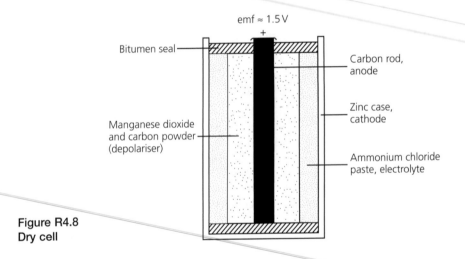

Figure R4.8
Dry cell

Lead-acid cells

The lead-acid cell is the commonest type of secondary cell and is used in most motor vehicles. The electrodes are made of lead and lead oxide set in lead alloy grids; the electrolyte is dilute sulphuric acid. Each electrode changes from one form of lead to the other as the cell is charged or discharged; the concentration of the acid electrolyte indicates the state of charge. The emf of any single lead-acid cell is 2 V.

Connections between cells

A *battery* is a combination of more than one cell connected together to give an increased emf (voltage) or increased capacity (amp-hours).

- **Series connection**: this type of connection is shown in Figure R4.9. The total emf of the battery is the sum of the individual emfs.
- **Parallel connection**: this type of connection is shown in Figure R4.10. The total emf of the battery is equal to the emf of a single cell. The maximum current or the life of the battery is three times that of a single cell.

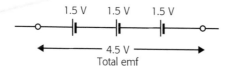

Figure R4.9 Cells connected in series

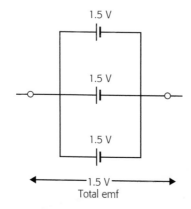

Figure R4.10 Cells connected in parallel

Power and energy

Electrical energy and power have the same meaning and the same units as other forms of power and energy. Power is the rate of using energy and, because the volt is defined in terms of electric charge and energy, it is possible to express power in terms of electric current (flow of charge) and voltage.

Electrical power

Power (P) = Current (I) × Potential difference (V)

$$P = IV$$

Unit: watt (W)

Where, by definition, 1 watt = 1 joule/second.

Two other useful expression are obtained by combining the expressions $P = IV$ and $V = IR$.

$$P = I^2R \text{ and } P = \frac{V^2}{R}$$

The *power rating* of an electrical appliance is often quoted in specifications and it is an indication of the relative energy consumption of the device. Some typical power ratings are given below.

Table R4.1 Power ratings

electric kettle element	2 500 W
electric fire (1 bar)	1 000 W
colour television	100 W
reading lamp	60 W
calculator charger	5 W

Electrical energy

Power is defined by the energy used in a certain time, so it is also possible to express energy in terms of power and time.

Energy (E) = Power (P) × Time (t)

$$E = Pt$$

Unit: joule (J)

where 1 joule = 1 watt × 1 second.

The *kilowatt hour* (kWh) is an alternative unit of energy in common use for electrical purposes.

1 kWh = 1 kilowatt × 1 hour = 3.6 MJ.

Worked Example R4.3

A 3 kWh electric heater is connected to a 230 V supply and is run continuously for 8 hours. Calculate:

(a) The current flowing in the heater.
(b) The total energy used by the heater.

(a) Know $P = 3000$ W, $I = ?$ $V = 230$ V.

Using suitable formula for power
$P = IV$

$3000 = I \times 230$

$I = \dfrac{3000}{230} = 13.04$ A

So current = **13 A**

(b) Know $E = ?$ $P = 3$ kW = 3000 W, $t = 8\,h = 28{,}800$ s.

Using relationship for energy and power

$E = Pt$

$E = 3000 \times 28{,}800$

$= 86.4 \times 10^6$ J

Alternatively, using kilowatt hours

$E = 3 \times 8$
$= 24$ kWh

So energy = **86.4 MJ** or **24 kWh**

MAGNETISM

There are certain pieces of iron, and other materials, that can push or pull one another when there is no physical contact between them. This force of magnetic attraction or repulsion is caused by the movement of charged particles inside the material. This motion may be due to the natural spin of sub-atomic particles, like the electron, or it may be due to the flow of electrons in an electric current. Magnetism is therefore linked with electric currents. This section describes those effects where electricity and magnetism combine to produce forces that can be used in devices such as motors.

Magnetic fields

The individual atoms of all materials have magnetic properties but they are not usually detected outside the atoms. In certain materials the magnetic effects of the individual atoms can be aligned to produce an overall effect. Iron, nickel and cobalt, for example, can retain this magnetic alignment and are used to make permanent magnets.

A *magnetic field* is a region where a magnetic force can be detected. The common sources of magnetic fields include the following:

- permanent magnetic materials, such as iron
- a conductor carrying an electric current
- the interior of some planets, such as the Earth.

A simple *compass* is made from a magnet balanced on a string or a pivot. In the presence of a magnetic field, the compass will turn and align itself in the direction of the magnetic field. The direction of the magnetic field changes with position, and it is useful to consider this pattern of the magnetic field as being made up of 'lines of force' or *magnetic flux*.

Magnetic direction: **north pole to south pole**

The magnetic field pattern for a simple bar magnet, in one plane only, is shown in Figure R4.11. The lines of magnetic flux begin and terminate at the two *poles* of the magnet: a 'north-seeking pole' and a 'south-seeking pole'. By convention, the magnetic field is said to flow from north pole to south pole.

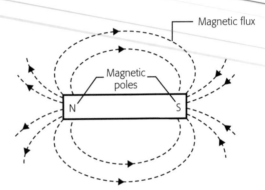

Figure R4.11 Magnetic field of bar magnet

When two magnets are placed close together the forces between them act towards or away from the poles of the magnets. The direction of the force follows the general rules:

Forces between magnetic poles:
N + S = attraction
N + N = repulsion
S + S = repulsion

- Unlike poles attract one another.
- Like poles repel one another.

Two north poles, for example, will push away from one another; while a north and a south pole will pull towards one another.

Electromagnetism

A conductor carrying an electric current produces a magnetic field around itself. A compass placed near a wire shows that the field is in the form of concentric circles, in a clockwise direction for current travelling away from view. A single turn of a loop of wire sets up two magnetic fields which combine, in the same direction, between the loops.

A *solenoid* is a coil made up of many loops of wire. When a current flows in a solenoid, the magnetic field produced around the solenoid, shown in Figure

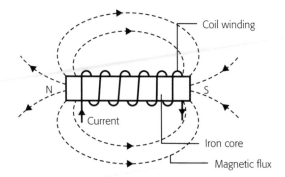

Figure R4.12 Magnetic field of solenoid

R4.12, is found to be similar to that of a bar magnet, with north and south poles produced at the ends of the solenoid.

An *electromagnet* is the general name given to a solenoid that is connected to an electric supply. If the solenoid is wound on an *armature* core of iron, the lines of magnetic flux are concentrated and change the shape of the magnetic field.

Applications of electromagnets

The main property of an electromagnet is its dependence on the supply of current: it can be turned on and off. Electromagnets are used instead of permanent magnets as a source of magnetic fields. Some important applications of electromagnets are:

- lifting iron and steel, for example in scrap-metal yards
- solenoid valves for turning supplies of gas and water on
- electric bells and buzzers
- relay switches for controlling circuits at a distance
- circuit breakers for preventing excessive current flows.

Electric bell or buzzer

A classic type of electric bell or buzzer uses an electromagnet as shown in Figure R4.13. Pressing the bell-push connects the bell to the supply and magnetises the electromagnet which then attracts the armature. But the movement of the armature breaks the circuit at the contact points and the magnetism dies away so that the armature springs back to its original position, reconnects the circuit and starts the cycle again.

The continuous trembling of the armature can be used to ring a bell, or to produce a buzzing sound, for as long as the switch is pushed. This principle of using a device to switch its own electrical supply on and off also finds uses beyond simple bells and buzzers.

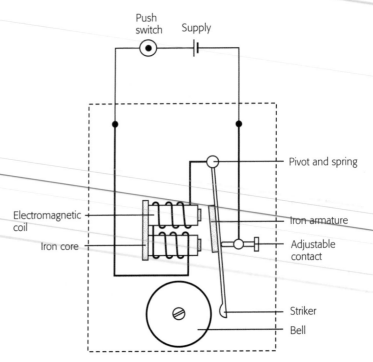

Figure R4.13 Electric bell

Relay

A relay switch is a device used for controlling one circuit by means of another. A relay uses the force produced by an electromagnet, and a simple example is shown in Figure R4.14. Switching on the control circuit magnetises the electromagnet,

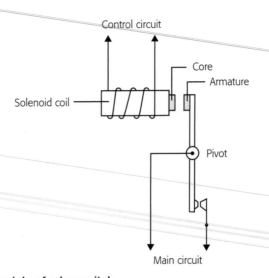

Figure R4.14 Principle of relay switch

which then attracts the armature. The movement of the armature closes the contacts of the relay switch and causes the main circuit to be turned on.

The advantage of this type of switch is that the control circuit can use low voltage and current to control a main circuit which carries a larger load and is situated some distance away.

Force on a conductor

When a conductor in a magnetic field carries a current, a force is found to act on that conductor. This *motor effect* is caused by the interaction of two magnetic fields: the field due to the magnet and the field due to the current. The strength of the force on the conductor is increased by the following measures:

- increase in current
- increase in magnetic field strength
- increase in length of the conductor sited in the magnetic field.

If the conductor in the magnetic field is in the form of a loop or a coil, then a force acts on each side of the coil. These forces act in opposite directions and they combine to produce a turning effect, called a torque, on the coil. The turning action of a current-carrying coil in a magnetic field is the basis of devices such as direct-current motors and electric meters.

Electric motors

An electric motor is a device for converting electrical energy to mechanical energy by using the force produced when two magnetic fields interact.

Simple DC motor

The simple motor illustrated in Figure R4.15 is a design that works but would not be used for a practical machine. However, the essential features and the operation of a simple motor are convenient to describe and they also apply to practical motors. The coil is set in a magnetic field and is free to turn on an axle. The DC supply is connected to the coil via conducting 'brushes' which slide against the rotating *commutator*.

When current flows in the coil the motor effect produces forces that turn the coil. When the coil is vertical the commutator disconnects the supply and the coil continues turning by momentum. The commutator then reconnects the coil to the opposite supply terminals. This reversal of connections ensures that the same direction of force always acts on the same side of the motor.

The repeating cycle of reversed connections keeps the axle turning in one direction and it can be used to do mechanical work.

Practical DC motors

A practical motor works on the same general principle as the simple motor but is made more efficient by additional features.

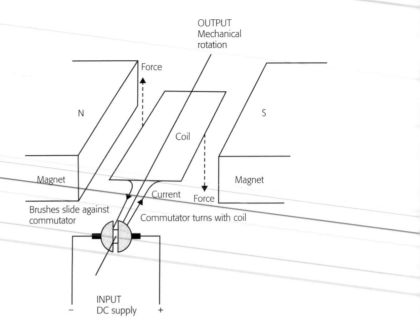

Figure R4.15 Simple DC electric motor

Field windings supply the magnetic field. These electromagnets are arranged in a circular form in order to keep the air gap to a minimum.

The *armature* turns on an axle inside the magnetic field and has the main coils wound upon it. More than one set of coils is wound on the same armature, and a commutator, with many splits, connects the electrical supply to one coil at a time.

The field coils, the armature windings and the electrical supply can be interconnected in different ways, which gives the motor different characteristics when starting and when under load. The speed of the motor can be varied by the use of a variable resistor in series with the armature.

AC motors

Most household electric motors, and many industrial motors, operate from an alternating current supply. The magnetic interactions of such motors involve the properties of induction, discussed in the next section. In general, the effect of alternating current is to set up rotating magnetic fields in the stator (field windings). The magnetic field of the rotor (armature) tends to follow the rotating fields and so causes the motor to turn. The construction of AC motors can be simple, but speed control of AC motors is more difficult than for DC motors.

INDUCTION

One general meaning of the word 'induction' is the production of an effect without physical contact. It is possible for an electrical effect to be induced in one circuit by the action of another circuit, even though there is no apparent contact between the two. In fact there is always a very real magnetic link of magnetic flux. Induction is a special form of electromagnetic effect which explains the operation of important features of public electricity supplies such as generators, transformers and alternating current properties.

Electromagnetic induction

If a conductor is moved across a magnetic field then an emf is induced in that conductor. This induced electricity also occurs in the types of circuit shown in Figure R4.16.

Moving magnet

In Figure R4.16a, a magnet is moved towards the coil and an emf is induced in the coil. A current therefore flows in the circuit and is detected by the meter. The direction of the current reverses when the direction of the magnet is reversed or the poles of the magnet are reversed. The induction occurs when *either* the coil or the magnet moves.

Switching circuit

If the coil and the meter are connected directly to a DC supply, as in Figure R4.16b, there is no induced emf. If, however, a switch is inserted in the circuit, then the emf changes every time that the switch is put either on or off. This

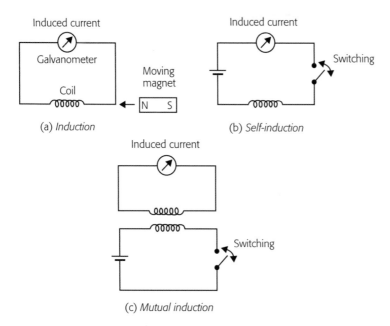

Figure R4.16 Electromagnetic induction

self-induction is produced when the electromagnetic field of the coil cuts across adjoining turns of the coil.

Adjacent coils

If two coils are placed adjacent to one another, as in Figure R4.16c, then an emf is induced in the meter whenever the switch in the battery circuit is put either on or off. This *mutual inductance* is produced when the electromagnetic field of the first coil cuts the turns of the second coil. The 'flux linkage' between the two coils can be improved by using a soft-iron core between them.

Principles of induction

The different demonstrations of induction described above share a common mechanism – in each case, the induction coil experiences a magnetic field which is changing. The change in magnetic field may be caused by a movement or by switching a circuit on or off.

General principle of induction

- An electric current will be induced in a conductor which is subjected to a *changing* magnetic field.

Magnitude of induction

The size of the induced current depends upon the following factors:

- relative speed of movement of the magnetic field
- strength of the magnetic field
- length of conductor in the magnetic field
- angle between the conductor and the field.

Lenz's law

The direction of all induced currents can be predicted by Lenz's law, given below:

- **Lenz's law**: the direction of the induced current is such that it will always oppose the change that produced it.

This rule applies to all methods of induction. In the simple example shown in Figure R4.16a, a magnet is moved towards the coil with the north pole leading. The current induced in the coil sets up a magnetic field and, by Lenz's law, this field will oppose the approaching magnet by having its own north pole outwards. If the movement of the magnet is reversed then the induced field will also be reversed, so as to oppose the movement.

Principles of Water Technology

FLUIDS AT REST

A fluid is a material whose particles are free to move their positions. Liquids and gases are both fluids and share common properties as fluids, although liquids and gases are also classed as different 'states' of matter. The sections on fluid properties that follow draw most examples from the flow of water in pipes, but it is useful to remember that the general principles described also explain other effects, such as the flow of air in ventilation ducts.

> Fluid = liquid or gas

Pressure

The pressure on any surface is defined as the force acting at right angles on that surface divided by the area of the surface. An area which is submerged in a fluid, such as the base of a tank of water, experiences a pressure caused by the weight (which is a force) of water acting on the area of the base.

For example, if an area A is submerged at depth h in a fluid, the force on that area is equal to the weight of the column of liquid or gas above the area. The pressure is then this weight divided by the area.

We know that:
    Force   = weight
             = mass × gravitational acceleration
             = $m\,g$
    We also know that:
    Mass   = density × volume
             = ρV

We also know that:
    Volume = $A\,h$
So using the relationship:

$$\text{Pressure} = \frac{\text{Force}}{\text{Area}}$$

We can write:

$$\text{Pressure} = \frac{mg}{A} = \frac{pVg}{A} = \frac{pAhg}{A}$$

The area A in the expression cancels so that, in general, the pressure at any point in a fluid is given by the following formula:

$$p = \rho\,gh$$

where
  pressure at a point in a fluid (Pa)
  ρ = density of the fluid (kg/m³)
  g = gravitational acceleration (m/s²) [9.81 m/s² is a working value]
  h = vertical depth from surface of fluid to the point (m)

Units of pressure

- The **pascal** (Pa) is the SI unit of pressure.

By definition, 1 pascal = 1 newton/square metre (N/m²).

Because the values of density and gravitational acceleration in the pressure equation are usually constant, it is a common working practice to quote only the height associated with the pressure. This leads to the following expressions.

- Pressure 'head' in metres is a practical unit of pressure.
- Specific weight (w) of a fluid is the product of density and gravitational acceleration, where $w = \rho g$.

Principles of fluid pressure

- Pressure at any given depth is equal in all directions.
- Pressure always acts at right angles to the containing surfaces.

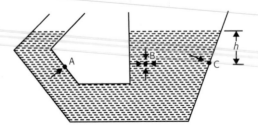

Figure R5.1 Principles of pressure

• Pressure is the same at points of equal depth, irrespective of volume or shape.

In Figure R5.1, for example, the pressure is the same at all three points A, B and C, and is not affected by the irregular shapes involved.

Pressure measurement

In addition to the various units of pressure, described above, it is sometimes necessary to specify the base pressure being used as a reference.

• **Absolute pressure** is the value of a particular pressure measured in units above an absolute *zero* of pressure.
• **Gauge pressure** is the value of a particular pressure measured in units above or below the *atmospheric* pressure.

For example, the pressure of the air in a car tyre is quoted as a gauge pressure and needs be added to the value of atmospheric pressure to give absolute pressure.

Manometer

The manometer is an instrument that measures pressure by comparing levels in a U-tube containing liquid, as shown in Figure R5.2. Water is a convenient liquid; oils may be used to measure some gases, and mercury is used for measuring high pressures.

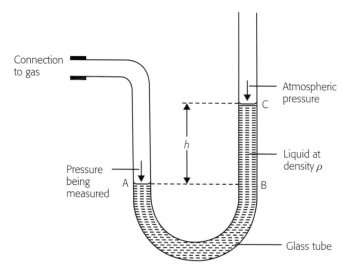

Figure R5.2 Manometer

In Figure R5.2 the pressure at point A must be the same as the pressure at point B. The pressure at point B is produced by a column of liquid, height *h*, plus atmospheric pressure. This relationship can be written as follows.

Measured pressure = Atmospheric pressure + $\rho g h$

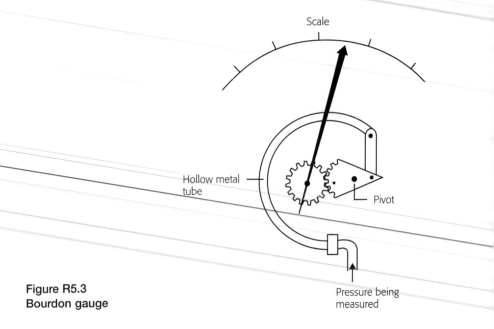

Scale

Hollow metal tube

Pivot

Pressure being measured

Figure R5.3
Bourdon gauge

Bourdon gauge

The Bourdon gauge is an instrument that measures pressure by using the changing curvature of a bronze tube, as shown in Figure R5.3. When the pressure increases the tube tends to straighten and, by means of a mechanical linkage, this force moves a needle on a scale. The gauge needs calibrating against some absolute form of measurement and it is common to measure the *gauge* pressure (compared to atmospheric pressure).

Force on immersed surfaces

When a fluid is contained by a structure such as a dam or a water tank, there is a significant pressure on the side of the structure, and this pressure acts at right-angles to the surfaces. For efficient structural design, it is necessary to know the size of the force due to the fluid pressure and any turning effect (moment) produced by the force. These considerations also apply to the design of earth-retaining walls as most soils behind a wall tend to act like a fluid acting on a dam.

The force on such a surface can be calculated from the pressure on the surface and the area of the surface. A submerged plate is subject to different pressures and forces acting at many different points. The total force or thrust on a submerged plate is given by the following expression.

$$F = p_c A$$

where
   F = total resultant force or thrust (newtons N)

p_c = pressure at the centre of area (centroid) of the immersed area (Pa)
A = total immersed area of the plate (m²)

For a rectangular area, as shown in Figure R5.4, the centre of area is positioned at half the immersed depth. Although the centre of area (*centroid*) is used to calculate the value of the resultant force, this force is exerted at a different position called the centre of pressure.

- The **centre of pressure** is the point where the line of action of the resultant force passes.

The centre of pressure is always located below the centre of area of the figure. For a rectangular surface, the centre of pressure is at 2/3 of the immersed depth, measured from the top.

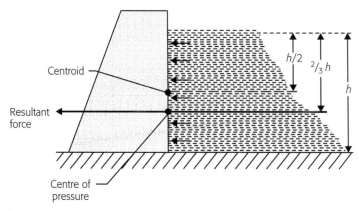

Figure R5.4 Force on an immersed surface

Summary

- The *size* of the resultant force is calculated using the centroid – such as the midpoint of a rectangular surface.
- The *position* of the resultant force is below the centroid – for example at 2/3 of the immersed depth.

Worked Example R5.1

A rectangular sluice gate is 1.6 m wide and is retaining water to a depth of 800 mm. Calculate the sideways thrust on the gate produced by the water. Given: density of water is 1000 kg/m³, and gravitational acceleration is 9.81 m/s².

Know

$p = 1000$ kg/m³, $g = 9.81$ m/s², $A = 1.6 \times 0.8 = 1.28$ m²
$h =$ depth to centre of area $= 0.4$ m.

Using a suitable formula for pressure and substituting values

$p_c = \rho g h$

$= 1000 \times 9.81 \times 0.4 = 3924$ Pa

Using the formula for force and substituting

$F = p_c \times A$

$= 3924 \times 1.28 = 5023$

So force on gate = **5023 N**

FLUID FLOW

The behaviour of moving fluids is complex and difficult to analyse, but a knowledge of fluid flow is needed for the correct design and construction of water supplies, and drainage and ventilation systems. The theoretical study of fluids (hydrodynamics) is usually combined with experimental studies (hydraulics) to produce designs for practical situations. One method of classifying fluid movement is to divide it into two major types: laminar flow and turbulent flow, described in the following sections.

Laminar flow

Laminar or *streamline* flow produces orderly flow paths in the same direction. All particles move in fixed layers at a constant distance from the wall and they do not cross one another's path. For the full pipe shown in Figure R5.5 the particles at the centre have the highest velocity and some microscopic particles at the boundary have zero velocity. In general, laminar flow occurs only at low velocities.

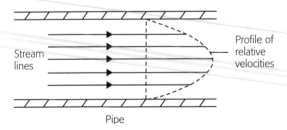

Figure R5.5 Laminar flow

Turbulent flow

Turbulent flow produces irregular flow paths, as shown in Figure R5.6. The particles move at random, colliding with one another and exchanging momentum. Turbulent flow occurs at higher velocities and most flow in practical situations is turbulent in nature.

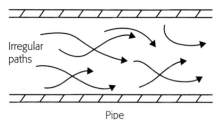

Pipe

Figure R5.6 Turbulent flow

Transitional flow

Between the states of laminar flow and turbulent flow there is a transitional zone where the nature of the flow is complex. The change from laminar flow to change to turbulent flow depends upon the following factors.

- **Velocity**: increase in velocity above the *critical velocity* causes turbulent flow.
- **Pipe size**: increase in pipe diameter causes turbulent flow.
- **Viscosity**: decrease in viscosity (internal friction or 'stickiness') causes turbulent flow.

Reynolds' number

Reynolds' number (R_e) is used to predict whether a particular flow of fluid will be laminar or turbulent in nature.

- R_e less than 2000 indicates laminar flow.
- R_e greater than 2000 indicates transitional or turbulent flow.

Reynolds' number can be calculated from the following formula.

$$R_e = \frac{\rho \, v \, D}{\mu}$$

where
  ρ = density of the flowing fluid (kg/m³)
  v = mean velocity of fluid past a cross-section (m/s)
  D = diameter of the pipe (m)
  μ = coefficient of absolute viscosity (Ns/m² or Pa s)

Flow rate

The amount of liquid flowing in a pipe or in a channel depends upon the dimensions of the pipe and upon the velocity of the liquid flow. This flow capacity can be described as the *discharge* or flow rate.

- **Flow rate** (Q) is the volume of water flowing per second.

    Unit: cubic metres per second (m³/s)

An alternative practical unit of flow rate is litres per second (l/s), where 1000 litres/s = 1 m³/s.
    The definition of flow rate can also be expressed as the following formula.

$$Q = \frac{V}{t}$$

where

   Q = flow rate (m³/s)
   V = volume of liquid passing a point (m³)
   t = time taken for the volume to pass the point (s)

Consider some liquid flowing with a mean (average) velocity v in a pipe of uniform cross-sectional area A. If the pipe discharges a length l of liquid each second, then the volume of water flowing per second is equivalent to a tube or cylinder whose volume is length l multiplied by area A. These relationships can be rearranged, as shown below, to give a formula for flow rate.

We know $Q = \dfrac{V}{t} = \dfrac{lA}{t}$ We also know $v = \dfrac{l}{t}$

So $Q = vA$

where

   Q = flow rate (m³/s)
   v = mean velocity of flow (m/s)
   A = cross-sectional area of pipe (m²)

Continuity equation

Figure R5.7 shows a pipe which smoothly decreases in cross-sectional area. If the liquid flowing in the pipe is incompressible and does not vary in density then the mass of liquid flowing past each point must remain constant. The flow rate at each point in the pipe must be the same.

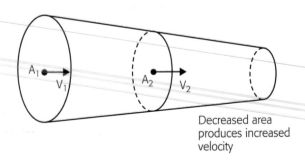

Decreased area produces increased velocity

Figure R5.7
Continuity of flow

For continuity the quantity must remain unchanged, so

$$Q_1 = Q_2$$

By definition, $Q = vA$ so we can also write

$$v_1A_1 = v_2A_2$$

The practical result of this continuity equation is that when the diameter of a pipe decreases, then the velocity of the flow in the pipe must increase.

Worked Example R5.2

A storage tank measuring 2 m by 2 m is filled to a depth of 1.5 m in 5 minutes by a supply pipe with a diameter of 100 mm which runs full bore.

(a) Calculate the flow rate in the pipe.
(b) Calculate the mean velocity of flow in the pipe.

(a)   We know $V = 2 \times 2 \times 1.5 = 6$ m³, $t = 5 \times 60 = 300$ s, $Q = ?$

Using volume formula and substituting

$$Q = V/t$$
$$Q = 6/300 = 0.02$$

So flow rate = **0.02 m³/s**

(b)   We know $Q = 0.02$ m³/s, $v = ?$, $A = \pi \times r^2 = \pi \times (0.05)^2$

Using volume formula and substituting

$$Q = vA$$
$$0.02 = v \times \pi(0.05)^2$$
$$v = \frac{0.002}{\pi(0.05)^2} = 2.546$$

So velocity = **2.546 m/s**

FLUID ENERGY

The energy possessed by a moving liquid is made up of the three components listed below.

- **Potential energy**: the energy associated with a mass of liquid at a height above a reference level (datum).
- **Pressure energy**: the energy or the work associated with moving a mass of liquid by a force.
- **Kinetic energy**: the energy associated with a mass of liquid having a velocity.

The energy components of a moving liquid are usually quantified in terms of equivalent pressure 'heads' of liquid, measured in metres. The three types of head are listed below and shown in Figure R5.8.

A: The *datum head* (z) is a measure of potential energy
B: The *pressure head* (H) is a measure of pressure energy
C: The *velocity head* ($v^2/2g$) is a measure of kinetic energy

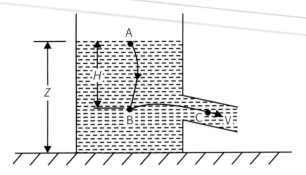

Figure R5.8 Energy components of a liquid

The Bernoulli principle

A particle of fluid, as shown in Figure R5.8, loses potential energy but gains pressure energy when it is lowered from point A to point B in Figure R5.8. If the particle gains kinetic energy in moving from point B to point C, then it must lose some of its pressure energy or potential energy. This is a particular example of the general law of conservation of energy, which states that the total energy of a closed system must remain constant. When this principle is applied to moving fluids it is stated as the *Bernoulli theorem*:

- **Bernoulli theorem**: the total energy possessed by the particles of a moving fluid is constant.

This statement assumes that there is no loss of energy from the liquid by effects such as friction. The Bernoulli theorem can also be expressed by the following equation.

$$\text{Potential energy} + \text{Pressure energy} + \text{Kinetic energy} = \text{Constant}$$

The Bernoulli equation

By referring to Figure R5.9, and by writing effective pressure heads, the constancy of energy can also be written as the *Bernoulli equation*:

$$z_1 + H_1 + \frac{v_1^2}{2g} = z_2 + H_2 + \frac{v_2^2}{2g}$$

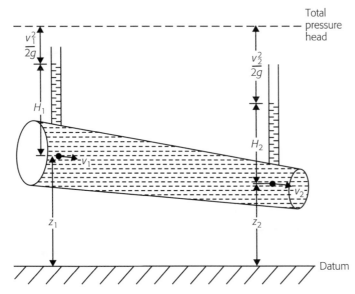

Figure R5.9 Total pressure of moving fluid

Pressure and velocity

A horizontal pipe has a constant datum head and constant potential energy, so that the Bernoulli theorem simplifies to the following expression.

- Pressure energy + Kinetic energy = Constant.

If the velocity of flowing water or air increases, the kinetic energy must also increase. To keep the total energy constant, the pressure energy must therefore decrease. As a general rule for a moving fluid:

- Increase in velocity gives a decrease in pressure.

This principle, derived from the Bernoulli theorem, may initially seem surprising but it explains a number of important effects associated with moving liquids and gases.

For example, the shape of an aerofoil such as an aircraft wing causes the airflow to have a higher velocity at the top of the wing than at the bottom. The increase in velocity produces a lower pressure on top of the wing, so that there is an upwards force on the wing. This Bernoulli effect also causes a lifting force when strong

Applications of Bernoulli:
aircraft flying
ventilation of
  buildings

winds blow across a pitched roof. Similarly, strong winds blowing around a building can lower pressures enough to cause windows to be 'sucked' outwards. When wind blows across the top of a chimney the pressure in the chimney decreases and causes the chimney to 'draw' smoke outwards more effectively.

Flow measurement

The Bernoulli principle is also the basis of many devices that are used to measure the flow of liquid in rivers, channels, water mains and sewers.

Venturimeter

The venturimeter is a device which measures the flow rate in a pipe by applying the Bernoulli principle to the pressures measured in the pipe. A constriction is constructed in the pipe and pressure gauges are fitted in the pipe and in the throat of the constriction, as shown in Figure R5.10. The liquid flowing in the reduced cross-section increases in velocity and therefore, by the Bernoulli principle, it has a lower pressure. The pressure heads in the pipe and the throat are measured and used to find the flow rate.

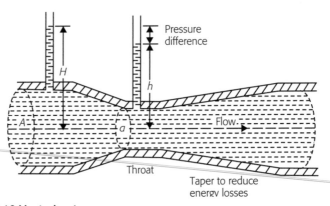

Figure R5.10 Venturimeter

The flow rate or discharge varies in direct proportion to the difference in pressures measured by the meter. A direct reading of flow rate can be made from a suitably calibrated gauge, or else the flow rate can be calculated from the following formula.

$$Q = C_d Aa \sqrt{\frac{2g\,(H-h)}{A^2 - a^2}}$$

where
  Q = flow rate in the pipe (m³/s)
  A = cross-sectional area of pipe (m²)
  a = cross-sectional area of throat (m²)
  H = pressure head in pipe (m)

h = pressure head in throat (m)

g = gravitational acceleration (m/s^2) (9.81 m/s^2 is a working value)

C_d = the discharge coefficient for a particular meter (usually 0.98)

Venturimeters are commonly used in water mains, and the loss of energy that occurs in the meter is minimised by using a long taper downstream from the throat of the meter.

Other measurement devices

The venturimeter is one form of device for measuring the flow of liquid in a pipe or open channel. The following devices can also be installed in pipes and channels to measure flow rate.

- notches
- orifices
- Pitot tubes
- weirs.

Like the venturimeter, these devices make use of the Bernoulli relationship between velocity and pressure. If we make a liquid or gas change velocity and also measure the changes in pressure or height caused by the change in velocity, we can then use the Bernoulli equation to work out the flow rate. The Bernoulli equation will need to contain a practical coefficient which makes allowance for friction, viscosity and turbulence associated with the particular device.

The *velocity meter* or current meter is another form of device that measures fluid flow by using a spinning propeller whose speed has been calibrated with velocity.

Energy losses

According to Bernoulli's theorem, the total energy of the liquid flowing in a pipe is constant. In practice there is a continuous loss of energy, even in a horizontal pipe. Friction between the liquid and the pipe wall is the main cause of energy loss, and the size of the resulting pressure drop depends upon the following factors:

- roughness of the pipe wall
- velocity of flow
- area of the pipe wall
- turbulence of flow
- length of the pipe
- viscosity and temperature of the liquid.

The combination of these factors makes the theory of the pressure loss complicated. In order to predict the pressure losses that can be expected in a given pipeline or channel, engineers use a number of practical formulas which have been found to give reasonable results for different situations. A general relationship that arises from theory and practice is that the pressure losses in a pipe are proportional to the square of the velocity of the liquid.

The sections below describe formulas used to predict pressure changes in pipelines and open channels. Despite their variety and apparent complexity, you will see the following square or square root relationship at the heart of the formulas:

Pressure change varies with the square of velocity

$$H \propto v^2 \quad \text{or rearranged:} \quad v = \sqrt{H}$$

For a given pipe it is the velocity that determines the flow rate out of a tap or shower for example. A practical example of the effect of the velocity–pressure relationship is seen when the pressure on a system is doubled, for instance by raising the height of a header tank. Although the pressure H is doubled, the velocity v does not double but only increases by $\sqrt{2}$ or 1.14. Therefore to get a worthwhile increase in flow rate from a given tap, the pressure needs to be increased by a significant factor.

Flow rate change varies with the square root of pressure change

Flow in pipes

A water main is an example of liquid flowing in a pipe that is completely full and under pressure. Continuity of flow is maintained within the pipeline even if slopes uphill for some of its length. Friction and other losses in the pipeline cause the pressure head that drives the flow to be gradually lost, as shown in Figure R5.11. To maintain the flow it may be necessary to replace the losses with energy from an outside system such as a pump.

Energy loss factors: pipe roughness, pipe length, velocity, turbulence, viscosity, temperature

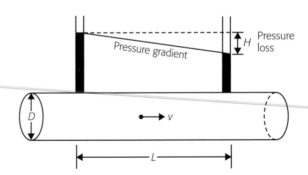

Figure R5.11 Flow in full pipe

The Bernoulli equation and the continuity equation are the basis for analysing and predicting pipeline flow. Numerical values need to be found or estimated for the various factors causing energy losses. Darcy's formula, given below, is one of the theoretical relationships that can be used for predicting the pressure head lost from a liquid flowing in a full pipe because of friction between the liquid and the pipe surfaces.

The Darcy formula

$$H = \frac{4fLr^2}{2gD}$$

where

H = loss of pressure head (m)

L = length of the pipe (m)

v = mean velocity of flow in pipe (m/s)

D = internal diameter of pipe (m)

g = gravitational acceleration (m/s²) [9.81 m/s² is a working value]

f = Darcy frictional coefficient (dimensionless with no units)

The value of the Darcy frictional coefficient f is found in tables, and ranges from about 0.005 for smooth pipes to 0.01 for rough pipes. The value of f is also different for laminar and turbulent flow.

Other formulas

In practice, and depending upon the accuracy required, engineers use various practical formulas derived from the Darcy formula or other theoretical analysis of pipeline flow. The following equations are found in use.

- **Blasius equation**: for smooth turbulent flow
- **Hazen-Williams equation**: for transitional turbulent flow
- **Manning equation**: for rough turbulent flow.

These formulas are used together with tables of practical coefficients for factors such as roughness. The relationships between variables can also be converted to graphs and tables of pressure-head losses for various types of pipe and fittings.

Pipeline formulas

Blasius equation:	$v = 75\ D^{5/7}\ S^{4/7}$
Hazen-Williams equation:	$v = 0.355\ C_{HW}\ D^{0.63}\ S^{0.54}$
Manning equation:	$v = (0.397/n)\ D^{2/3}\ S^{1/2}$

Where

v = mean velocity of flow in pipe (m/s)

D = internal diameter of pipe (m)

S = pressure-head loss / length of pipe = H/L

C_{HW} = Hazen-Williams coefficient

n = Manning roughness coefficient

Flow in open channels

Water flows in rivers, canals, gutters, drainpipes and sewers are examples of liquid flowing in open channels or partly filled pipes. Even though it is underground, a drain still acts like an 'open' channel because there is constant pressure, usually atmospheric pressure, on the surface of the liquid. The liquid must always flow downhill, and the force producing the motion of the liquid is balanced by the friction forces resisting the motion.

To maintain the flow of liquid in the channel, the energy lost by friction is balanced by the potential energy of the liquid under gravity, and its total pressure head decreases as it flows. For the analysis of open-channel flow the following variables, shown in Figure R5.12, are used.

- *Wetted perimeter* (P) of a cross section is the length of contact between the liquid and the channel surfaces.
- *Hydraulic radius* (R) is the ratio of the cross-sectional area of flow A and the wetted perimeter P, so that $R = A/P$.
- S = slope or hydraulic gradient is the ratio of the loss of pressure head H and the horizonal length of flow L, so that $S = H/L$.

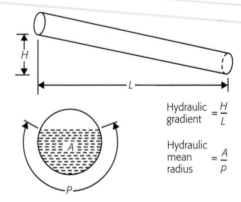

Figure R5.12 Flow in open channel

Chézy formula

The Chézy formula expresses the relationship between the fall in height of a channel, such as a gutter, and the velocity of flow that this produces. The formula assumes that the channel has a constant gradient and is of uniform cross section and roughness:

$$v = c \sqrt{RS}$$

where
- v = mean velocity of flow (m/s)
- R = hydraulic radius
- S = slope or hydraulic gradient
- c = Chézy coefficient for a particular pipe (for example: $c = 50 \text{ m}^{1/2}/\text{s}$).

Manning equation for open channel

The Manning equation for an open channel is a widely-used adaptation of the Chézy equation based on practical measurements.

$$v = (1/n)R^{2/3}S^{1/2}$$

where

v = mean velocity of flow (m/s)
R = hydraulic radius
S = slope or hydraulic gradient
n = Manning number for different types of surface ($s/m^{1/3}$)

Some typical values of the Manning number (in units of $s/m^{1/3}$):

- concrete pipe 0.012
- glass 0.010
- earth canal 0.020
- river stones 0.040.

References

The following signposts are useful starting points when looking for information. Most commercial providers of materials and equipment also have websites with technical information. Many sites, such as the professional body sites, also have links or 'portals' to other sites. Some of the Internet references will evolve and change so you should also try and build up your own set of reference sources. Newspapers, magazines and journals list up-to-date links.

Some of the references are links to standards or arrangements for a particular country, but they can also be useful in giving examples of good practice, for highlighting trends and suggesting sources of information for your region. Note that government departments and agencies may change their name following reorganisations and their publications may continue to be listed under a previous name. Entering key words in an Internet search engine such as Google should quickly give you links.

All these links are also listed on the companion website for this book at www.palgrave.com/engineering/builtenvironment/mcmullan.

General built environment

Sources (UK-based except where indicated)	Contacts
Barbour Expert: commercial portal	www.barbourexpert.com
BBA, British Board of Agrément: approval body	www.bbacerts.co.uk
BRE: testing, certification research and publications	www.bre.co.uk
BSI, British Standards Institution	www.bsi-global.com
Building design and construction services directory: portal to range of commercial sites	www.buildingdesign.co.uk
Building Centre and Bookshop: commercial site	www.buildingcentre.co.uk
Building Regulations: England and Wales	www.planningportal.gov.uk
Building Regulations: Scotland	www.scotland.gov.uk/Topics/Built-Environment/Building/Building-standards

Sources	Contacts
Building Regulations: Northern Ireland	www.buildingregulationsni.gov.uk
Building Standards: Ireland	www.environ.ie
Building Code: Australia	www.greenhouse.gov.au
Building Regulations: New Zealand	www.dbh.govt.nz
Design Council: includes former Commission for Architecture and the Built Environment (CABE)	http://www.designcouncil.org.uk
Health and Safety Executive: publications	www.hse.gov.uk
NBS – Architecture, Engineering, Construction: portal to variety of sites	www.thenbs.com/resources
NPL, National Physical Laboratory	www.npl.co.uk
WSSN, World Standards Services Network: links to standards organisations around the world	www.wssn.net

Energy and carbon use, green and sustainable buildings

Sources	
AECB, the sustainable building association	www.aecb.net
British Standards Institution – various standards and publications including: U-value calculations; Thermal conductivity; Assessing thermal comfort	www.bsi-global.com
BRE: various publications and web information including: BRE Digests; SAP, The Government's Standard Assessment Procedure for Energy Rating of Dwellings	www.bre.co.uk
Carbon Trust: government funded organisation with aim to help UK move to lower carbon emissions	www.carbontrust.co.uk
Centre for Sustainable Energy	www.cse.org.uk
Energy Saving Trust	www.est.org.uk
Greenspec: resources and portals	www.greenspec.co.uk
iiSBE International Initiative for a Sustainable Built Environment	www.iisbe.org
Micropower Council: small-scale energy generation	www.micropower.co.uk
SBIC Sustainable Buildings Industry Council, USA	www.sbicouncil.org
SEAI Sustainable Energy Authority of Ireland	www.seai.ie
Sustainable Building Support Centre, Netherlands/ International	www.sustainablebuilding.info
The Environmental Organisation Web Directory: environmental search engine	www.webdirectory.com
UK Green Building Council	http://www.ukgbc.org
World Energy Council (WEC)	www.worldenergy.org

Building services

Sources	Contacts
ACRIB Air Conditioning and Refrigeration Industry Board: information	www.acrib.org.uk
ASHRAE American Society of Heating, Refrigerating and Air-Conditioning Engineers: standards and publications.	www.ashrae.org
CIOB Chartered Institute of Building Services Engineers: standards and publications	www.cibse.org
SAP: includes Boiler efficiency database	www.seduk.com
Lighting Academy: international portal to lighting information	www.lightingacademy.org
Society of Light and Lighting: publications and information	www.cibse.org
The Lighting Association: trade portal with information	www.lightingassociation.com
LIF Lighting Industry Federation: lighting information	www.lif.co.uk
OFTEC, Oil Firing Technical Association: technical information about heating	www.oftec.co.uk

Noise and acoustics

Sources	Contacts
British Standards Institution, BSI: standards for Sound Insulation and Noise Reduction for Buildings	www.bsi-global.com
BRE: publication on measuring sound insulation in buildings	www.bre.co.uk
Building Regulations: England and Wales	www.planningportal.gov.uk
IOA Institute of Acoustics: UK organisation concerned with acoustics	www.ioa.org.uk
Control the Noise: information on sound and noise control	www.controlthenoise.com
DFT Department for Transport: information on transport and airport noise	www.dft.gov.uk
Living 4 Quiet: information on hearing loss and protection, quiet buildings	www.living4quiet.com
NoiseNet: information and commercial directory	www.noisenet.org

Water and drainage

Sources	Contacts
International Water Association: global information	www.iwahq.org
UK Water Industry Research	www.ukwir.org.uk

Sources	Contacts
Water UK: water industry information	www.water.org.uk
World Water Council: information on global water issues	www.worldwatercouncil.org

Professional bodies: UK and Ireland

Sources	Contacts
AAI, Architectural Association of Ireland	architecturalassociation.ie
ABE, Association of Building Engineers	www.abe.org.uk
CIAT, Chartered Institute of Architectural Technologists	www.ciat.org.uk
CIBSE, Chartered Institute of Building Services Engineers	www.cibse.org
CIOB, Chartered Institute of Building	www.ciob.org.uk
CIWEM, The Chartered Institution of Water and Environmental Management	www.ciwem.org
ICE, Institution of Civil Engineers	www.ice.org.uk
RIAS, Royal Incorporation of Architects in Scotland	www.rias.org.uk
RIBA, Royal Institute of British Architects	www.architecture.com
RICS, Royal Institute of Chartered Surveyors	www.rics.org
RTPI, Royal Town Planning Institute	www.rtpi.org.uk

Standards organisations

Standards are written agreements containing technical specifications or other criteria that are used to ensure that materials, products and services are fit for their purpose, safe and reliable. Examples of areas where standards are important are machinery, transport and medical equipment.

Standards are generated by obtaining agreements among stakeholders, such as suppliers and users. A government-approved body usually encourages the standards for a particular country. An international standard, such as those produced by ISO (International Organisation for Standardisation), is a wider agreement among the various national standards bodies.

Some national standards organisations

Country	Standards organisation
Australia	SAI: Standards Australia
Canada	SCC: Standards Council of Canada

Country	Standards organisation
Europe	CEN: European Committee for Standardization
Hong Kong	ITCHKSAR: Innovation and Technology Commission
Ireland	NSAI: National Standards Authority of Ireland
New Zealand	SNZ: Standards New Zealand
Singapore	PSB: Singapore Productivity and Standards Board
South Africa	SABS: South African Bureau of Standards
United Kingdom	BSI: British Standards Institution
USA	ANSI: American National Standards Institute

International standards organisations

IEC: International Electrotechnical Commission
ISO: International Organisation for Standardisation
ITU: International Telecommunication Union

World Standards Services Network (WSSN) has links to standards organisations around the world via their website: www.wssn.net

Index

THE THREE BEARS

Once upon a time there were three bears who lived together in a house of their own in a wood. One of them was a little wee bear, and one was a middle-sized bear, and the third was a great big bear. They each had a bowl for their porridge—a little bowl for the little wee bear, and a middle-sized bowl for the middle-sized bear, and a great big bowl for the great big bear. And they each had a chair to sit on— a little chair for the little wee bear, and a middle-sized chair for the middle-sized bear, and a great big chair for the great big bear. And they each had

a bed to sleep in—a little bed for the little wee bear.
and a middle-sized bed for the middle-sized bear.
and a great big bed for the great big bear.

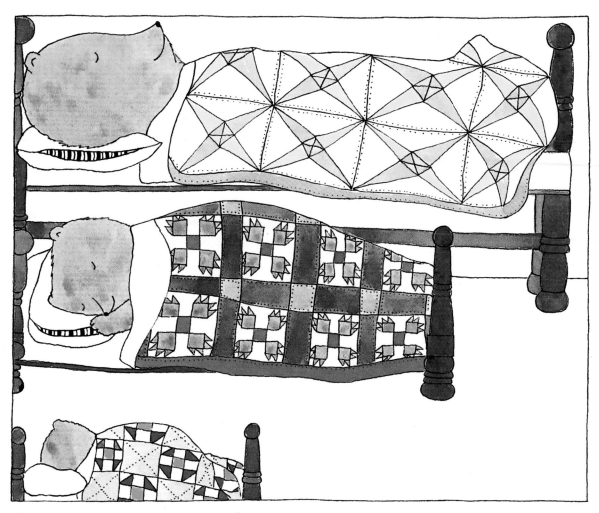

One day, after they had made the porridge for their breakfast and poured it into their bowls, they walked out in the woods while the porridge was cooling. A little girl named Goldilocks passed by the house and looked in at the window. And then she looked in at the keyhole, and when she saw that there was no one home, she lifted the latch on the door.

The door was not locked because the bears were good bears who never did anyone any harm and never thought that anyone would harm them. So Goldilocks opened the door and walked in. She was very glad to see the porridge on the table, as she was hungry from walking in the woods, and so she set about helping herself.

First she tasted the porridge of the great big bear, but that was too hot for her. Next she tasted the porridge of the middle-sized bear, but that was too cold for her. And then she tasted the porridge of the little wee bear, and that was neither too hot nor too cold but just right, and she liked it so much that she ate it all up.

Then Goldilocks sat down on the chair of the great big bear, but that was too hard for her. And

then she sat down on the chair of the middle-sized bear, and that was too soft for her. And then she

sat down on the chair of the little wee bear, and that was neither too hard nor too soft, but just right. So she seated herself in it, and there she sat until she sat the bottom out of the chair and down she came upon the floor.

Then Goldilocks went upstairs to the bedroom where the three bears slept. And first she lay down upon the bed of the great big bear, but that was too high for her. And next she lay down upon the bed of the middle-sized bear, but that was too low for her. But when she lay down upon the bed of the little wee bear, it was neither too high nor too low, but just right. So she covered herself up comfortably and fell fast asleep.

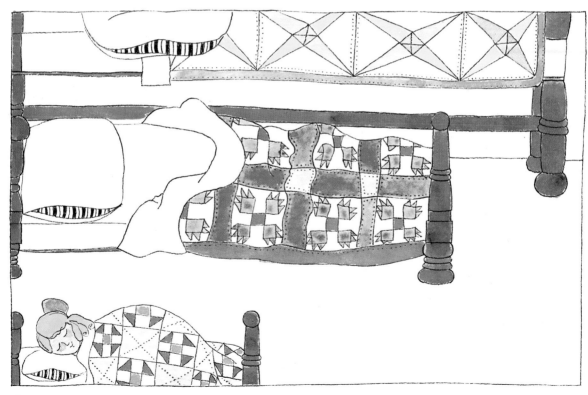

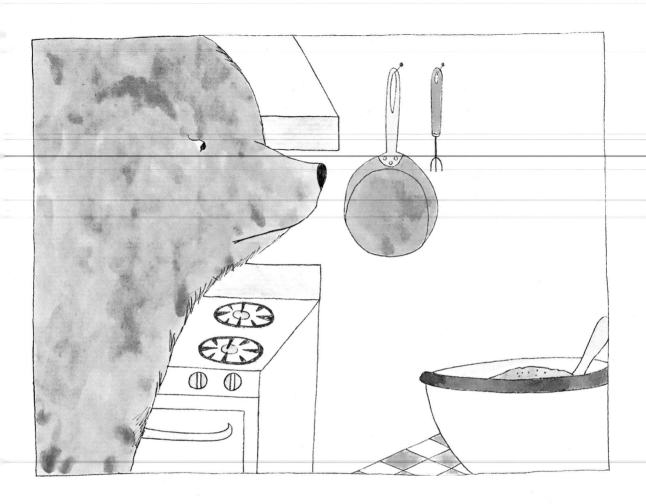

When the three bears thought their porridge would
be cool enough for them to eat, they came home for
breakfast. Now Goldilocks had left the spoon of the
great big bear standing in the porridge.

"Somebody has been eating my porridge!" said the great big bear in a great, rough gruff voice.

Then the middle-sized bear looked at its porridge and saw the spoon was standing in it, too.

"Somebody has been eating my porridge!" said the middle-sized bear in a middle-sized voice.

Then the little wee bear looked at its bowl, and there was the spoon standing in the bowl, but the porridge was all gone.

"Somebody has been eating my porridge and has eaten it all up!" said the little wee bear in a little wee voice.

Upon this, the three bears, seeing that someone had come into their house and eaten up all the little wee bear's breakfast, began to look around them. Now Goldilocks had not put the cushion straight when she rose from the chair of the great big bear.

"Somebody has been sitting in my chair!" said the great big bear in a great, rough gruff voice.

And Goldilocks had squashed down the soft cushion of the middle-sized bear.

"Somebody has been sitting in my chair!" said the middle-sized bear in a middle-sized voice.

"Somebody has been sitting in my chair, and has sat the bottom through!" said the little wee bear in a little wee voice.

Then the three bears thought that they had better look farther in case it was a burglar, so they went upstairs into their bedroom. Now Goldilocks had pulled the pillow of the great big bear out of its place.

"Somebody has been lying in my bed!" said the great big bear in a great, rough gruff voice.

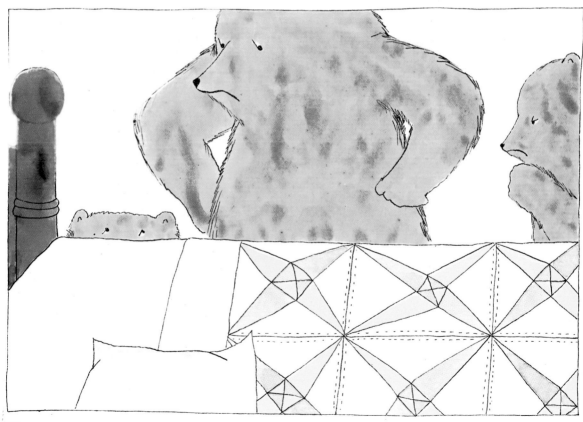

Contents

for Hannah,
Elizabeth & Oliver

And Goldilocks had pulled the cover of the middle-sized bear out of its place.

"Somebody has been lying in my bed!" said the middle-sized bear in a middle-sized voice.

But when the little wee bear came to look at its bed, there was the pillow in its place. But upon the pillow? There was Goldilocks' head, which was not in its place, for she had no business there.

"Somebody has been lying in my bed, and here she is still," said the little wee bear in a little wee voice.

Now Goldilocks had heard in her sleep the great, rough gruff voice of the great big bear, but she was so fast asleep that it was no more to her than the rumbling of distant thunder. And she had heard the middle-sized voice of the middle-sized bear, but it was only as if she had heard someone speaking in a dream. But when she heard the little wee voice of the little wee bear, it was so sharp and so shrill that it woke her up at once.

Up she sat, and when she saw the three bears on one side of the bed, she tumbled out at the other and ran to the window. Now the window was open, for the bears were good, tidy bears who always opened their bedroom window in the morning to let in the fresh air and sunshine. So Goldilocks jumped out through the window and ran away, and the three bears never saw anything more of her.

A lion lay sleeping. A little mouse ran across his paw. That tickled the lion and woke him up. He roared and grabbed the little mouse.

"Please," said the little mouse, "do not hurt me. I am sorry I woke you up, but if you do not hurt me, I promise I will do something good for you some day."

The lion laughed. "How silly," he thought. "What could a tiny little mouse do for a big, strong lion like me?" But he let him go.

Soon after, the lion was walking through the forest when suddenly he was caught in a net some hunters had made to trap him. He roared and roared, and his roars made the leaves on the trees tremble, but still he could not free himself from the net.

Far away the little mouse heard him roar. He hurried to the place where the big lion lay trapped.

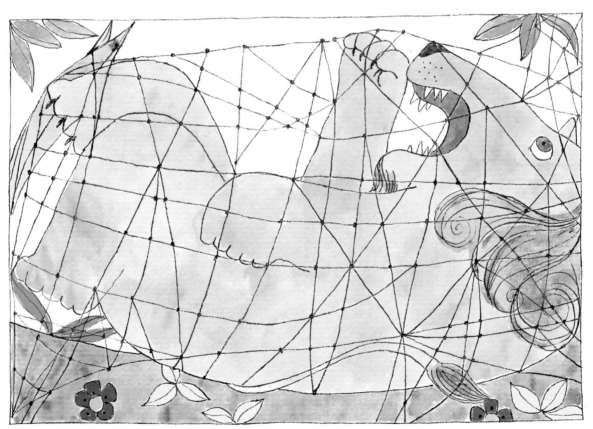

"Remember, I promised you I would some day do
something good for you," said the little mouse, and
he began to nibble on the ropes of the net with his
sharp little teeth. He nibbled and nibbled and nibbled
some more until there was a big hole in the net.

Then the lion was free, and the lion and the mouse
walked away together.

16

THE COCK AND THE MOUSE AND THE LITTLE RED HEN

Once upon a time there was a hill, and on the hill there was a pretty little house. It had one little green door and four little windows, and in this house there lived a cock and a mouse and a little red hen. On another hill, across the valley, there was another little house. It was very ugly. It had a door that wouldn't shut and two broken windows, and all the paint was peeling off. And in this house there lived a big bad fox and four bad little foxes.

One morning these four bad little foxes came to the big bad fox and said, "Oh, Father, we are so hungry!"

"We had nothing to eat yesterday," said one.

"And not much the day before," said another.

The big bad fox shook his head, for he was thinking. At last he said in a big gruff voice, "On the hill over there I see a house. And in that house there lives a cock."

"And a mouse," screamed two of the little foxes.

"And a little red hen!" screamed the other two.

"And they are nice and fat," said the big bad fox. "This very day I'll take my sack, and I will go down this hill, across the stream, up that hill, and in at that door, and into my sack I will put the cock and the mouse and the little red hen."

So the four little foxes jumped for joy, and the big bad fox took his sack and went on his way.

Now that morning the cock and the mouse had both got out of the wrong side of the bed. The cock said the day was too hot, and the mouse said the day was too cold. But when they came grumbling down to the kitchen, the little red hen was bright and busy and cheerful.

"Who will get some sticks to light the fire?" she asked.

"I won't," said the cock.

"I won't," said the mouse.

"Then I will do it myself," said the little red hen. So off she went and got the sticks.

"And now who will fill the kettle?" she asked.

"I won't," said the cock.

"I won't," said the mouse.

"Then I will do it myself," said the little red hen. And she filled the kettle.

"And who will get the breakfast ready?" she asked as she put the kettle on to boil.

"I won't," said the cock.

"I won't," said the mouse.

"Then I will do it myself," said the little red hen.

At breakfast the cock upset the jug of milk, and the mouse scattered crumbs all over the floor, and they both quarrelled and grumbled.

"Who will clear the table?" asked the little red hen.

"I won't," said the cock.

"I won't," said the mouse.

"Then I will do it myself," said the little red hen.

She cleared the table and mopped up the milk and swept away all the crumbs.

"And now who will help me make the beds?"
she asked.

"I won't," said the cock.

"I won't," said the mouse.

"Then I will do it myself," said the little red hen.

And she went upstairs while the lazy cock and
mouse sat down each in a comfortable chair by the
fire, where they soon fell fast asleep.

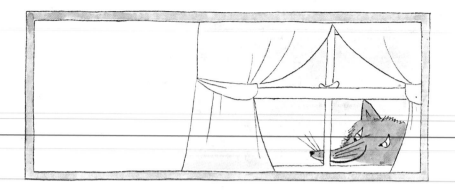

Now the big bad fox had crept up the hill and into the garden. If the cock and the mouse had not been sleeping, they would have seen his sharp eyes peering in at the window.

Knock, knock! went the fox at the door.

"Who could that be?" said the mouse, yawning.

"Go and see for yourself, if you want to know," said the cock.

"Perhaps it is the postman," said the mouse, "and perhaps he has a letter for me!" So without looking to see who it was, he turned the knob and opened the door.

As soon as he opened the door, in jumped the big bad fox.

"Eeek, eeek, eeek!" said the mouse as he tried to run up the chimney.

"Cock-a-doodle-doo!" crowed the cock as he jumped on the back of the biggest armchair.

But the fox only laughed and caught the little mouse by the tail and popped him into the sack, and he took the cock by the neck and popped him in, too.

Then the little red hen came running down the stairs to see what all the noise was about. So the fox caught her, too, and put her into the sack with the others.

Then he took a piece of string out of his pocket and tied up the mouth of the sack very tight indeed. After that, he threw the sack over his back, and off he set down the hill.

"Oh, I wish I hadn't been so cross," said the cock. "Oh, I wish I hadn't been so lazy," said the mouse. And the cock and the mouse began to cry.

"Don't be too sad," whispered the little red hen. "See, here is my little workbag, and in it there is a pair of scissors, and a thimble, and a needle and thread. Now, you wait and see what I am going to do. But be quiet!"

By then the sun was very hot, and the sack was heavy, so the big bad fox thought he would lie down under a tree and go to sleep for a while. As soon as the fox began to snore, the little red hen took out her scissors and snipped a hole in the sack just big enough for the mouse to creep through.

"Quick!" she said to the mouse "Run as fast as you can and bring back a stone as big as yourself."

And the mouse did.

"Push it in here," said the little red hen, and the mouse pushed it into the sack.

Then the little red hen snipped away at the hole until it was large enough for the cock to go through.

"Quick!" she said "Run as fast as you can and bring back a stone as big as yourself."

Out flew the cock, and he soon came back with a big stone which he too pushed into the sack.

Now the little red hen popped out and got a stone as big as herself and pushed it into the sack. Then she put on her thimble, took out her needle and thread, and quickly sewed up the hole.

When the hole was sewed up, the cock and the mouse and the little red hen ran home very fast. They slammed the door and locked it tight. Not long after, the big bad fox woke up. He went grumbling and groaning down the hill, for the sack was very heavy.

When he came to the stream, *splash,* in went one foot, *splash,* in went the other, but the stones in the sack were so heavy that at the very next step down tumbled the big bad fox into a deep pool. And the four greedy little foxes had to go to bed without any supper.

But the cock and the mouse never grumbled again. They lit the fire and filled the kettle, set the table, cooked the breakfast and made the beds, while the good little red hen had a holiday and sat resting in the big armchair in front of the fire.

No foxes ever troubled them again and they are still living happily in the pretty little house that stands on the hill.

The Gingerbread Man

Once upon a time there was a little old woman and a little old man, and they lived all alone. They were very happy together, but they wanted a child and since they had none, they decided to make one out of gingerbread. So one day the little old woman and the little old man made themselves a little gingerbread man, and they put him in the oven to bake.

When the gingerbread man was done, the little old woman opened the oven door and pulled out the pan. Out jumped the little gingerbread man—and away he ran. The little old woman and the little old man ran after him as fast as they could, but he just laughed and said, "Run, run, as fast as you can. You can't catch me! I'm the Gingerbread Man!"

And they couldn't catch him.

The gingerbread man ran on and on until he came to a cow.

"Stop, little gingerbread man," said the cow. "I want to eat you."

But the gingerbread man said, "I have run away from a little old woman and a little old man, and I can run away from you, too. I can, I can!"

And the cow began to chase the gingerbread man,
but the gingerbread man ran faster, and said, "Run,
run, as fast as you can. You can't catch me! I'm the
Gingerbread Man!"

And the cow couldn't catch him.

The gingerbread man ran on until he came to a horse.

"Please, stop, little gingerbread man," said the horse. "I want to eat you."

And the gingerbread man said, "I have run away from a little old woman, a little old man, and a cow, and I can run away from you, too. I can, I can!"

And the horse began to chase the gingerbread man, but the gingerbread man ran faster and called to the horse, "Run, run, as fast as you can. You can't catch me! I'm the Gingerbread Man!"

And the horse couldn't catch him.

By and by the gingerbread man came to a field full of farmers.

"Stop," said the farmers. "Don't run so fast. We want to eat you."

But the gingerbread man said, "I have run away from a little old woman, a little old man, a cow, and a horse, and I can run away from you, too. I can, I can!"

And the farmers began to chase him, but the gingerbread man ran faster than ever and said, "Run, run, as fast as you can. You can't catch me! I'm the Gingerbread Man!"

And the farmers couldn't catch him.

The gingerbread man ran faster and faster. He ran past a school full of children.

"Stop, little gingerbread man," said the children. "We want to eat you."

But the gingerbread man said, "I have run away from a little old woman, a little old man, a cow, and a horse, a field full of farmers, and I can run away from you, too. I can, I can!"

And the children began to chase him, but the gingerbread man ran faster as he said, "Run, run, as fast as you can. You can't catch me! I'm the Gingerbread Man!"

And the children couldn't catch him.

By this time the gingerbread man was so proud of himself he didn't think anyone could catch him. Pretty soon he saw a fox. The fox looked at him and began to run after him. But the gingerbread man said, "You can't catch me! I have run away from a little old woman, a little old man, a cow, a horse, a field full of farmers, a school full of children, and I can run away from you, too. I can, I can! Run, run, as fast as you can. You can't catch me! I'm the Gingerbread Man!"

"Oh," said the fox, "I do not want to catch you. I only want to help you run away."

Just then the gingerbread man came to a river. He could not swim across, and he had to keep running.

"Jump on my tail," said the fox. "I will take you across."

So the gingerbread man jumped on the fox's tail,

and the fox began to swim across the river. When he had gone a little way, he said to the gingerbread man, "You are too heavy on my tail. Jump on my back."

And the gingerbread man did.

The fox swam a little farther, and then he said, "I am afraid you will get wet on my back. Jump on my shoulder."

And the gingerbread man did.

In the middle of the river, the fox said, "Oh, dear, my shoulder is sinking. Jump on my nose, and I can hold you out of the water."

So the little gingerbread man jumped on the fox's nose, and the fox threw back his head and snapped his sharp teeth.

"Oh, dear," said the gingerbread man, "I am a quarter gone!"

Next minute he said, "Now I am half gone!"

And next minute he said, "Oh, my goodness gracious! I am three quarters gone!"

And then the gingerbread man never said anything more at all.

THE HOUSE THAT JACK BUILT

1. This is the house that Jack built.

2. This is the malt
That lay in the house that Jack built.

3. This is the rat,
   That ate the malt
   That lay in the house that Jack built.

4. This is the cat,
   That killed the rat,
   That ate the malt
   That lay in the house that Jack built.

5. This is the dog,
   That worried the cat,
   That killed the rat,
   That ate the malt
   That lay in the house that Jack built.

6. This is the cow with the crumpled horn,
   That tossed the dog,
   That worried the cat,
   That killed the rat,
   That ate the malt
   That lay in the house that Jack built.

7. This is the maiden all forlorn,
   That milked the cow with the crumpled horn,
   That tossed the dog,
   That worried the cat,
   That killed the rat,
   That ate the malt
   That lay in the house that Jack built.

8. This is the man all tattered and torn,
That kissed the maiden all forlorn,
That milked the cow with the crumpled horn,
That tossed the dog,
That worried the cat,
That killed the rat,
That ate the malt
That lay in the house that Jack built.

9. This is the priest all shaven and shorn,
   That married the man all tattered and torn,
   That kissed the maiden all forlorn,
   That milked the cow with the crumpled horn,
   That tossed the dog,
   That worried the cat,
   That killed the rat,
   That ate the malt
   That lay in the house that Jack built.

10. This is the cock that crowed in the morn,
    That waked the priest all shaven and shorn,
    That married the man all tattered and torn,
    That kissed the maiden all forlorn,
    That milked the cow with the crumpled horn,
    That tossed the dog,
    That worried the cat,
    That killed the rat,
    That ate the malt
    That lay in the house that Jack built.

11. This is the farmer sowing his corn,
    That kept the cock that crowed in the morn,
    That waked the priest all shaven and shorn,
    That married the man all tattered and torn,
    That kissed the maiden all forlorn,
    That milked the cow with the crumpled horn,
    That tossed the dog,
    That worried the cat,
    That killed the rat,
    That ate the malt
    That lay in the house that Jack built.

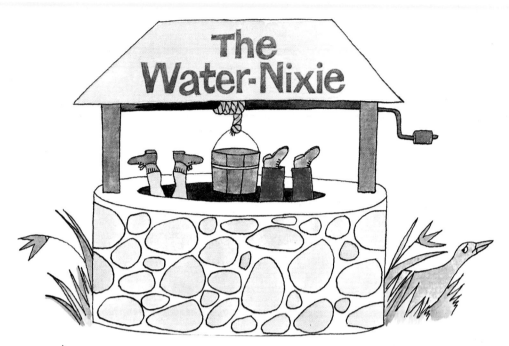

The Water-Nixie

A little brother and sister were once playing by a well, and they both fell in. Down below, there lived a water-nixie who said, "Now I have got you! You shall both work for me."

She gave the little girl a hank of dirty tangled flax to spin, and made her fetch water in a bucket that had a hole in it. The boy had to chop down a tree with a blunt axe, and they both got nothing to eat but dumplings, hard as stones.

One Sunday, when the nixie was at church, the children ran away. But as soon as church was over, the nixie followed them with long strides. The children

saw her coming from far away, and the little girl
quickly threw behind them her hairbrush, which
turned into an immense hill of bristles with thousands
and thousands of spikes. It was very difficult for the
nixie to scramble across the spikes, but at last she got
over, and began to run after them again. Then the
boy threw down his comb which made a great ridge
with a thousand times a thousand teeth, but the nixie
managed to cross over that, too, and ran after them

again. Then the girl threw her mirror behind her. This turned into a great mountain made all of glass mirrors, and it was so slippery that the nixie could not climb it.

Then the nixie thought, "I will go home quickly and fetch my axe and chop that hill of glass mirrors in two!" But before she came back and chopped her way through, the children had run far away, and the nixie had to go home to her well again.

THE THREE BILLY GOATS GRUFF

Once upon a time there were three billy goats who wanted to go up to the hillside to make themselves fat, and the name of all three billy goats was Gruff.

On the way up was a bridge over a stream that they had to cross, and under the bridge there lived a terrible, ugly troll, with eyes as big as saucers and a nose as long as a poker.

First of all came the youngest billy goat Gruff to cross the bridge.

Trip, trap, trip, trap went the sound of his hooves on the bridge.

"Who's that tripping over the bridge?" roared the troll.

58

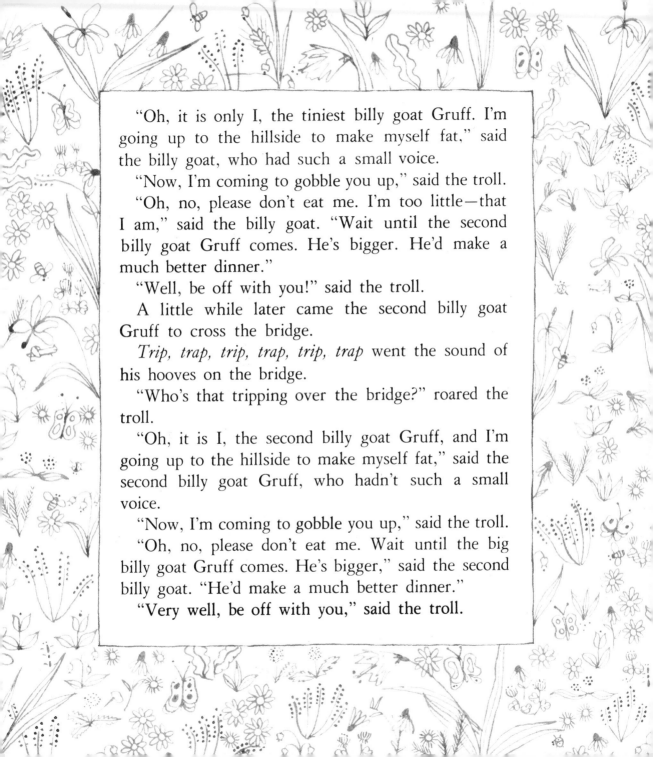

"Oh, it is only I, the tiniest billy goat Gruff. I'm going up to the hillside to make myself fat," said the billy goat, who had such a small voice.

"Now, I'm coming to gobble you up," said the troll.

"Oh, no, please don't eat me. I'm too little—that I am," said the billy goat. "Wait until the second billy goat Gruff comes. He's bigger. He'd make a much better dinner."

"Well, be off with you!" said the troll.

A little while later came the second billy goat Gruff to cross the bridge.

Trip, trap, trip, trap, trip, trap went the sound of his hooves on the bridge.

"Who's that tripping over the bridge?" roared the troll.

"Oh, it is I, the second billy goat Gruff, and I'm going up to the hillside to make myself fat," said the second billy goat Gruff, who hadn't such a small voice.

"Now, I'm coming to gobble you up," said the troll.

"Oh, no, please don't eat me. Wait until the big billy goat Gruff comes. He's bigger," said the second billy goat. "He'd make a much better dinner."

"Very well, be off with you," said the troll.

But just then up came the big billy goat Gruff. *Trip, trap, trip, trap, trip, trap, trip, trap!* went the sound of his hooves on the bridge very loudly, for the big billy goat was so heavy that the bridge creaked and groaned under him.

"Who's that tramping over the bridge?" roared the troll.

"It is I, the big billy goat Gruff," said the big billy goat, who had a big hoarse voice of his own.

"Now, I'm coming to gobble you up!" roared the troll.

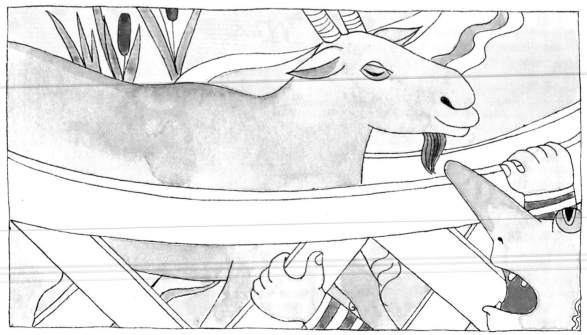

"Well come along! I've got two spears,
And I'll poke your eyeballs out at your ears.
I've got besides two great, big stones,
And I'll crush you to bits, both body and bones!"

That was what the big billy goat said. And that
was just what he did. And that was the last anyone

ever heard of the troll. Then the big billy goat went up to the hillside. There the three billy goats Gruff got so fat they were scarcely able to walk home. And if the fat hasn't fallen off them, why they're still fat, and so,

Snip, snap, snout,
This tale's told out.

Henny-Penny was pecking at grain in the farmyard when suddenly something hit her on the head.

"Oh, my goodness!" said Henny-Penny. "The sky is falling. I must go and tell the king."

So she went along and she went along until she met Cocky-Locky.

"Where are you going?" asked Cocky-Locky.

"Oh, I'm going to tell the king the sky is falling," said Henny-Penny.

"May I come with you?" asked Cocky-Locky.

"Yes, you may," said Henny-Penny. So Henny-Penny and Cocky-Locky went to tell the king the sky was falling.

They went along and they went along until they
met Ducky-Daddles.

"Where are you going?" asked Ducky-Daddles.

"Oh, we're going to tell the king the sky is falling,"
said Henny-Penny and Cocky-Locky.

"May I come with you?" asked Ducky-Daddles.

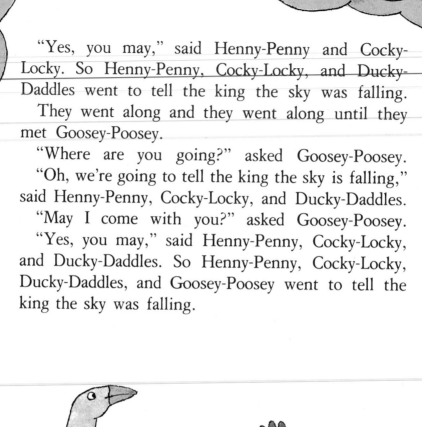

"Yes, you may," said Henny-Penny and Cocky-Locky. So Henny-Penny, Cocky-Locky, and Ducky-Daddles went to tell the king the sky was falling.

They went along and they went along until they met Goosey-Poosey.

"Where are you going?" asked Goosey-Poosey.

"Oh, we're going to tell the king the sky is falling," said Henny-Penny, Cocky-Locky, and Ducky-Daddles.

"May I come with you?" asked Goosey-Poosey.

"Yes, you may," said Henny-Penny, Cocky-Locky, and Ducky-Daddles. So Henny-Penny, Cocky-Locky, Ducky-Daddles, and Goosey-Poosey went to tell the king the sky was falling.

They went along and they went along until they
met Turkey-Lurkey.

"Where are you going?" asked Turkey-Lurkey.

"Oh, we're going to tell the king the sky is falling,"
said Henny-Penny, Cocky-Locky, Ducky-Daddles, and
Goosey-Poosey.

"May I come with you?" asked Turkey-Lurkey.

"Yes, you may," said Henny-Penny, Cocky-Locky,
Ducky-Daddles, and Goosey-Poosey. So Henny-Penny,
Cocky-Locky, Ducky-Daddles, Goosey-Poosey, and
Turkey-Lurkey went to tell the king the sky was falling.

They went along and they went along until they met Foxy-Woxy, and Foxy-Woxy asked Henny-Penny, Cocky-Locky, Ducky-Daddles. Goosey-Poosey, and Turkey-Lurkey, "Where are you going?"

And Henny-Penny, Cocky-Locky, Ducky-Daddles, Goosey-Poosey, and Turkey-Lurkey said, "Oh, we're going to tell the king the sky is falling."

"But this is not the right way to go to the king," said Foxy-Woxy. "I know the right way. Do you want to follow me?"

"Oh, yes, Foxy-Woxy," said Henny-Penny, Cocky-Locky, Ducky-Daddles, Goosey-Poosey, and Turkey-Lurkey. So Foxy-Woxy, Henny-Penny, Cocky-Locky, Ducky-Daddles, Goosey-Poosey, and Turkey-Lurkey all went to tell the king the sky was falling.

They went along and they went along until they came to a dark hole. Now this was the door of Foxy-Woxy's cave. But Foxy-Woxy said to Henny-Penny, Cocky-Locky, Ducky-Daddles, Goosey-Poosey, and Turkey-Lurkey, "This is the right way to go to the king. You will get there quickly if you follow me."

"Oh, yes, indeed!" said Henny-Penny, Cocky-Locky, Ducky-Daddles, Goosey-Poosey, and Turkey-Lurkey.

Foxy-Woxy went into his cave and waited for Henny-Penny, Cocky-Locky, Ducky-Daddles, Goosey-Poosey, and Turkey-Lurkey to follow him.

Turkey-Lurkey went through the dark hole first. He hadn't gone far when Foxy-Woxy ate up Turkey-Lurkey. Then Goosey-Poosey went in, and Foxy-Woxy ate up Goosey-Poosey. Then Ducky-Daddles went in, and Foxy-Woxy ate up Ducky-Daddles. Then Cocky-Locky went in, but before Foxy-Woxy could eat him up, Cocky-Locky crowed, "Cock-a-doodle-do!"

As soon as Henny-Penny heard Cocky-Locky crow, "Cock-a-doodle-do!" she said, "Oh, my goodness! It is time to lay my egg!"

And she ran home to the farmyard and never did get to tell the king the sky was falling.

A shoemaker had become so poor that he had only enough leather for one pair of shoes. In the evening he cut out the leather for the shoes he would make next morning, and then he went to bed. But in the morning, when he was about to go to work, the pair of shoes stood all finished on his cobbler's bench. When he saw them, the shoemaker was very surprised, and did not know what to think. And when he looked closely at the shoes, he saw there was not one bad stitch in them; they were perfect in every way.

Before long, a customer came into his shop and bought the shoes. But because they were so beautiful and so well made the customer paid extra for them, and the shoemaker was able to buy leather for two pairs of shoes.

He cut them out at night, and woke up early to
sew the shoes together, but once again, there was no
need to do so. For when he got up, the shoes were
already made. Customers came, too, and he made
enough money that day to buy leather for four pairs
of shoes. Again next morning he found the four pairs
of shoes made, and so it went on—what leather he
cut in the evening was finished by morning, and the
shoes were always perfect and well made, so the shoemaker became a rich man.

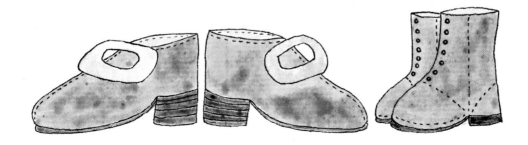

One winter evening, when the man had finished cutting out the leather, he said to his wife, "What if we were to stay up tonight and see who it is that helps us so?"

His wife liked that idea, and so the two of them hid themselves in a corner of the shop, and waited and watched.

When it was exactly midnight, two little naked elves came. They sat down promptly on the cobbler's bench, took the cutout leather, and began to stitch and began to sew and began to hammer. And all this they did so nimbly and so well that the shoemaker could not turn his eyes away for amazement. The two little elves did not stop working until the shoes were all done. Then quickly and quietly they ran away.

75

Next morning the woman said, "Those little elves have made us rich, and we should show them that we are grateful for it. They run about so without any clothes on. They must be cold. I will make them each a little hat and shirt and vest and trousers, and I will knit them each a pair of stockings. And you can make two little pairs of shoes."

"I will be happy to," said the shoemaker.

A few nights later they laid their presents on the cobbler's bench in place of the cut out leather and hid themselves where they could watch the two little elves.

At midnight the little elves came bounding in, ready to go to work at once. They were very surprised to find no cut out leather, only the pretty little clothes. They touched the little vests and hats and shirts and trousers, the striped stockings and leather shoes. Suddenly they smiled with delight when they realized the clothes were theirs. They dressed themselves quickly and sang:

> "We are boys so fine to see,
> Why should we now cobblers be?"

Then they danced and skipped off the cobbler's bench and around the shop and out the door. From then on they came no more, but the shoemaker and his wife lived happily and well.

Teeny-Tiny

There was once a teeny-tiny woman who lived in a teeny-tiny house in a teeny-tiny town. One day this teeny-tiny woman put on her teeny-tiny hat and went out of her teeny-tiny house to take a teeny-tiny walk. When the teeny-tiny woman had gone a teeny-tiny way, she came to a teeny-tiny gate. Then the teeny-tiny woman opened the teeny-tiny gate and went into a teeny-tiny churchyard. In the teeny-tiny churchyard she saw a teeny-tiny bone upon a teeny-tiny grave, and the teeny-tiny woman said, "This teeny-tiny bone will make some teeny-tiny soup for my teeny-tiny supper." And the teeny-tiny woman took the teeny-tiny bone from the teeny-tiny grave and put it in her teeny-tiny pocket and went back to her teeny-tiny house.

When the teeny-tiny woman got inside her teeny-tiny house, she was a teeny-tiny bit sleepy, so she put the teeny-tiny bone into her teeny-tiny cupboard, and then she went up her teeny-tiny stairs and climbed into her teeny-tiny bed.

After the teeny-tiny woman had closed her teeny-tiny eyes for a teeny-tiny nap, she was awakened by a teeny-tiny voice from the teeny-tiny cupboard down the teeny-tiny stairs that said:

"Give me my bone!"

When she heard this the teeny-tiny woman was a teeny-tiny bit frightened, so she hid her teeny-tiny head under her teeny-tiny blanket and went to sleep again. And when she had slept a teeny-tiny time, the teeny-tiny voice cried out again a teeny-tiny bit louder:

"Give me my bone!"

This made the teeny-tiny woman a teeny-tiny bit more frightened, so she hid her teeny-tiny head a teeny-tiny bit farther under the teeny-tiny blanket. And when the teeny-tiny woman had gone to sleep again for a teeny-tiny time, the teeny-tiny voice from the teeny-tiny cupboard down the teeny-tiny stairs said a teeny-tiny bit louder:

"Give me my bone!"

By now the teeny-tiny woman was a teeny-tiny bit more frightened, but she lifted up her teeny-tiny head from under her teeny-tiny blanket and said in her loudest teeny-tiny voice:

"TAKE IT!"

Once upon a time there was an old sow with three little pigs, and she sent them out into the world to seek their fortune.

The first little pig met a man with a bundle of straw and said to him, "Please, man, give me that straw to build a house." And the man did, and the little pig built a house with the straw.

Along came a wolf who knocked at the door and said, "Little pig, little pig, let me come in."

And the little pig answered, "No, no, by the hair of my chinny, chin, chin!"

So the wolf said, "Then I'll huff and I'll puff and I'll blow your house in!"

So the wolf huffed and he puffed and he blew the house in and ate up the little pig.

The second little pig met a man with a bundle of sticks and said to him, "Please, man, give me those sticks to build a house." And the man did, and the little pig built a house with the sticks.

Then along came the wolf who knocked at the door and said, "Little pig, little pig, let me come in."

And the little pig answered, "No, no, by the hair of my chinny, chin, chin!"

So the wolf said, "Then I'll huff and I'll puff and I'll blow your house in!"

So the wolf huffed and he puffed, and he puffed and he huffed, and he blew the house down and ate up the little pig.

The third little pig met a man with a load of bricks
and said to him, "Please, man, give me those bricks
to build a house." And the man gave him the bricks,
and the little pig built a house with them.

Then the wolf came, just as he had to the other little pigs, and knocked on the door and said, "Little pig, little pig, let me come in."

And the little pig answered, "No, no, by the hair of my chinny, chin, chin!"

So the wolf said, "Then I'll huff and I'll puff and I'll blow your house in!"

Well, the wolf huffed and puffed, and he puffed and he huffed, and he huffed and he puffed, but he could not blow that house down. So when he found that he could not, he said, "Little pig, I know where there are some nice turnips."

"Where?" said the little pig.

"Down in Farmer Smith's field," said the wolf. "If you will be ready tomorrow morning at six o'clock I will come and get you, and we can go together to pick some turnips for dinner."

"Very well," said the little pig, and the wolf went away.

The next morning the little pig got up at five o'clock and picked the turnips and got back home before the wolf came to get him at six.

When the wolf arrived, he said, "Little pig, are you ready?"

The little pig said, "I have been there and back, and I have a nice potful of turnips for dinner."

The wolf felt very angry at this, but he said in his nicest voice, "Little pig, I know where there is a fine apple tree."

"Where?" said the little pig.

"Over in Mary's garden," said the wolf. "And if you will not trick me, I will come for you tomorrow morning at five o'clock, and we will pick some apples."

Well, the little pig got up next morning at four o'clock and went off for the apples, hoping to get home before the wolf came for him. But he had farther to go this time, and he had to climb the tree besides, so just as he was climbing down the tree, he saw the wolf coming, and this frightened the little pig very much.

The wolf stood under the tree and said, "Why, little pig! Are you here before me? Are they nice apples?"

"Oh, yes," said the little pig, "they are good and sweet. Here, I will throw one down to you."

And he threw the apple so far that while the wolf was running off to get it, the little pig jumped down from the tree and ran all the way home.

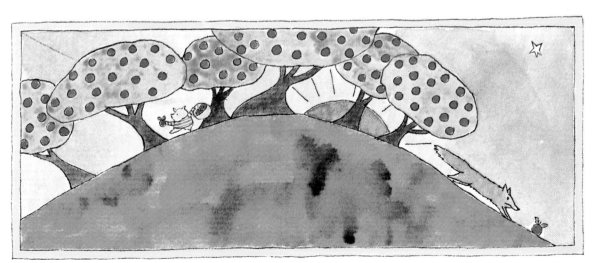

Next day the wolf came again, and said to the little pig, "Little pig, there is a fair in town this afternoon. Will you go with me?"

"Oh, yes," said the little pig, "I will. What time will you come to get me?"

"I will come at three," said the wolf.

So the little pig left early, as usual, and bought a butter churn at the fair. He was walking home with it when he saw the wolf coming. The little pig did not know what to do. So he got inside the butter churn to hide. While he was squeezing himself in, he turned the butter churn around on its side, and it rolled down the hill with the little pig inside. And it rolled right down the hill towards the wolf, and the wolf was so frightened that he ran home without ever going to the fair.

Later he went to the little pig's house and told him about the big scary thing that had come rolling down the hill to chase him.

Then the little pig said, "Ha! So I scared you, did I? That was only the butter churn I bought at the fair, and I was inside."

When he heard this the wolf was very, very angry indeed, and he declared that he would eat up that little pig, and he was coming down the chimney to get him.

As soon as the little pig heard this, he filled up a large pot with water and built up a blazing fire. Just as the wolf was coming down the chimney the little pig took the lid off the pot, and in fell the wolf.

The little pig popped the lid back on the pot, boiled up the wolf, ate him for supper, and lived happily ever after.

LAZY JACK

Once upon a time there was a boy whose name was Jack, and he lived with his mother in a little house. They were very poor. Jack's mother earned her living by spinning, but Jack was so lazy he would do nothing but sit in the sun in the summer and sit by the fire in the winter. And as he always sat and sat and did nothing useful, everyone called him Lazy Jack. At last his mother told him one Monday morning that if he did not begin to work for his food, she would turn him out of the house to feed himself as best he could.

So Jack went out and went to work next day for a neighbour who paid him a penny. But on his way home Jack lost his penny.

"You stupid boy," said his mother, "you should have put it in your pocket."

"I'll do that next time," said Jack.

Well, the next day Jack went out again and went to work for a milkman who gave him a little bucket of milk for his day's work. Jack took the bucket and put it into his pocket. It splashed around as he walked, and it was all gone before he got home.

"Dear me!" said Jack's mother when she saw his wet jacket and the good milk all gone. "You should have carried it on your head."

"I'll do that next time," said Jack.

The next day Jack went to work for a grocer who gave him a soft cheese for his day's work. In the evening Jack took the cheese and went home with it on his head. By the time he got home the cheese had melted. Part of it was lost, and part matted into Jack's hair.

"You silly," said Jack's mother, "you should have carried it carefully in your hands."

"I'll do that next time," said Jack.

Now the next day Jack went out and went to work for a baker who would not give him anything for his day's work but a large tomcat. Jack took the cat and began carrying it very carefully in his hands. But before long the cat had scratched him so badly that he had to let it go.

When he got home, his mother said to him, "You poor fool, you should have tied it with a string and pulled it along after you."

"I'll do that next time," said Jack.

The next day Jack went to work for a butcher who paid him with a piece of beef. Jack took the beef, tied it to a string, and pulled it after him. He pulled it in the dirt, so that by the time he got home the meat was not fit to eat. His mother was especially angry with him this time, for the next day was Sunday, and they would have to have cabbage for their dinner.

"You ninny," Jack's mother said to her son, "you should have carried it on your shoulder."

"I'll do that next time," said Jack.

Well, when Monday came, Lazy Jack went out once more and went to work for a farmer, who gave him a donkey for his trouble. Although Jack was strong, he found it hard to hoist the donkey on to his shoulders, but at last he did it and began walking slowly home.

Along the road he passed a house where a rich man lived with his only daughter, a very beautiful girl. But this beautiful girl had never laughed in her life, not once. So her father had said that any man who made her laugh would be her husband. Now the beautiful girl just happened to be looking out of the

window as Jack came by with the donkey on his shoulders. The poor beast had its legs sticking up in the air, and it was kicking and heehawing with all its might. The sight was so funny she burst out laughing, and she laughed and laughed and laughed some more. Her father was overjoyed, and he kept his promise by marrying her to Lazy Jack. So Jack became a rich gentleman and never had to do another day's work. He and his wife lived in a large house very happily, and Jack's mother came and lived there, too.

THE DOG AND THE BONE

One day a dog stole a big soup bone from a butcher shop and ran off with it. He ran and he ran, out of the town and down the road, and on to a bridge that crossed a bright clear river. Suddenly the dog saw his own reflection in the water. He stopped.

"Who is that dog with that nice big bone?" he thought to himself and growled. "I want that bone, too! It looks bigger and better than mine," he thought and growled a little louder. He snarled, and the dog he saw in the water snarled, too.

Then the dog opened his mouth and made a grab for the other dog's bone. But as soon as he opened his mouth, the bone fell into the river, and the dog had nothing to eat that day. No, nothing at all.

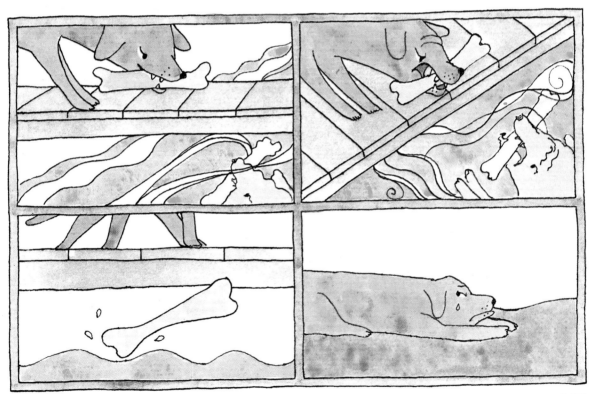

Little Red Riding Hood

Once upon a time there was a dear little girl whose grandmother made her a beautiful cape and hood of red velvet. She wore it wherever she went, and so everyone called her "Little Red Riding Hood."

One day her mother said to her, "Little Red Riding Hood, here is a piece of cake and some good soup. Take them straight to your grandmother and go quickly, for she is sick in bed."

The grandmother lived in the forest, half a mile from the village. Just as Little Red Riding Hood entered the forest, she met a wolf. Little Red Riding Hood did not know how wicked the wolf was, and she was not afraid of him.

"Good morning, Little Red Riding Hood. Where are you going this fine day?" said the wolf.

"I am going to Grandmother's to take her some cake and some good soup," said Little Red Riding Hood.

"Where does your grandmother live?" asked the wolf.

"She lives in a little house under three large oak trees. There is a berry-bush hedge. Surely you must know it," said Little Red Riding Hood.

And the wicked wolf thought, "What a sweet little girl! She would certainly be good to eat. And if I am clever, I will be able to eat her up and the old woman, too."

So he walked pleasantly along the path with Little Red Riding Hood, and he said, "Why are you so serious, Little Red Riding Hood? Look at those pretty flowers growing over there. Listen to those birds singing. Why don't you stop and play? Surely your grandmother can wait until lunchtime for her cake and soup."

Little Red Riding Hood looked at the flowers growing in the sunlight beneath the trees, and she thought, "I will pick Grandmother a bouquet," and she ran off the path and into the forest, picking flowers as she ran. And as soon as she had picked one, she saw another that was prettier still, and so she went deeper and deeper into the forest.

Meanwhile the wolf ran straight to the grandmother's house and gulped the old woman down.
Then he put on her nightgown and cap, and got into
her bed, and pulled up the covers.

At last, when Little Red Riding Hood had picked as many flowers as she could carry, she returned to the path and set out again for her grandmother's house.

When she got there, she saw her grandmother in bed with the blankets pulled up high and thought she looked very strange.

"Oh, Grandmother," she said, "what big ears you have!"

"The better to hear you with, my dear" was the reply.

"But, Grandmother, what big eyes you have!"

"The better to see you with, my dear."

"But, Grandmother, what big teeth you have!"
"The better to eat you with!" said the wolf, and
jumped out of bed and gulped down Little Red Riding
Hood.

Then the wolf climbed back into bed and fell asleep. He soon began to snore very loudly. A friendly woodcutter was passing by the house, and he thought, "Listen to the old woman snore! I will call in and see if she needs anything today."

And the woodcutter went into the house and saw the wolf sleeping and snoring in the grandmother's bed.

Quickly and quietly the woodcutter took out his sharp knife and carefully, he cut a hole in the stomach of the wolf. He saw some red velvet shining there, so he made the hole a bit bigger, and out jumped Little Red Riding Hood, crying, "Ah, how frightened I have been! How dark it was inside the wolf!"

And then the woodcutter made the hole another bit bigger, and out came the grandmother, alive but very weak indeed.

But the grandmother ate the cake and drank the good soup and felt much better. And Little Red Riding Hood went home through the forest, and no harm came to her.

The Little Pot

There was once a poor but good little girl who lived alone with her mother, and they no longer had anything to eat. The little girl went into the forest, and there she met an old woman who felt sorry for her. So the old woman gave her a little pot which when she said, "Cook, little pot, cook," would cook good sweet porridge. And when she said, "Stop, little pot," it would stop cooking. The child took the pot home to her mother, and now they were no longer hungry, for they ate sweet porridge as often as they wished.

One day the little girl had gone out, and her mother said, "Cook, little pot, cook." And the pot did cook, and she ate until she was full. Then she wanted the pot to stop cooking, but she did not know the right words to say to make it stop. Only the little girl knew that, and she was not at home. So the pot went on cooking, and porridge bubbled over the edge,

and still it cooked on until the kitchen and the whole house were full of porridge, and then the next house, and then the whole street. And everyone in town wanted the pot to stop cooking, but no one knew how to stop it.

At last, when only one single house remained that was not covered with porridge, the little girl came home. She said, "Stop, little pot," and it stopped cooking. And whoever wanted to return to town had to eat his way back.

The Star Money

There was once upon a time a good, kind little girl who had no mother and no father and who was so poor she had no house to live in and no bed to sleep in. She was so poor she had nothing, but the clothes she wore and a little piece of bread in her pocket. All alone she went out into the world.

She had not gone far before she met an old man who said, "Ah, give me something to eat. I am so hungry."

And the little girl gave him her piece of bread and went on again. But before long, she met a little child who moaned and said, "My head is cold. Give me something to cover it with."

So she gave him her hat, and when she had walked a little farther, she met another child who had no jacket, and he too was cold. So she gave him her own, and a little farther on, a child begged for a dress, and she gave that away also. Then she came to a forest, and it was dark.

Soon she met another child who cried and shivered
and said, "I am so cold!" And the little girl thought,
"It is dark and no one can see me anyway." And so
she gave away all the clothes she had.

And as she stood, with not one single thing left, suddenly some stars fell down at her feet. And when she looked, she saw they were not stars at all but shiny gold and silver pieces of money. And although she had just given all her clothes away, she found she was dressed in new ones, much better and warmer than the old. So she put the money in her pocket. Never again was she cold or hungry, and she was happy all the days of her life.